Advances in Experimental Medicine and Biology

Volume 1344

Advances in Experimental Medicine and Biology provides a platform for scientific contributions in the main disciplines of the biomedicine and the life sciences. This series publishes thematic volumes on contemporary research in the areas of microbiology, immunology, neurosciences, biochemistry, biomedical engineering, genetics, physiology, and cancer research. Covering emerging topics and techniques in basic and clinical science, it brings together clinicians and researchers from various fields.

Advances in Experimental Medicine and Biology has been publishing exceptional works in the field for over 40 years, and is indexed in SCOPUS, Medline (PubMed), Journal Citation Reports/Science Edition, Science Citation Index Expanded (SciSearch, Web of Science), EMBASE, BIOSIS, Reaxys, EMBiology, the Chemical Abstracts Service (CAS), and Pathway Studio.

2020 Impact Factor: 2.622

More information about this series at http://www.springer.com/series/5584

Olivia Engmann • Marco Brancaccio
Editors

Circadian Clock in Brain Health and Disease

Springer

Editors
Olivia Engmann
Institute of Human Genetics
University Hospital Jena
Friedrich Schiller University
Jena, Germany

Marco Brancaccio
UK Dementia Research Institute
Imperial College London
London, UK

ISSN 0065-2598 ISSN 2214-8019 (electronic)
Advances in Experimental Medicine and Biology
ISBN 978-3-030-81146-4 ISBN 978-3-030-81147-1 (eBook)
https://doi.org/10.1007/978-3-030-81147-1

This Springer imprint is published by the registered company Springer Nature Switzerland AG.
The registered company address is: Gewerbestrasse 11, 6330 Cham, Switzerland

Preface

The brain is a micro-cosmos of evolution. It is fantastically adaptable to environmental circumstances we never encounter in nature, such as zero gravity, mathematics and the Internet. The constant that accompanied all life on earth from day one was the 24-h rhythm of day and night, caused by the rotation of our planet. Until recently, there has been no need to adapt to altered rhythms as environmental changes such as changing day lengths do not alter the 24 h period. It is one of the brain's many formidable achievements to adjust to new time zones that only a century ago we were unable to cross at a significant speed. This plasticity, or room for play and error, is intimately intertwined with other functions of our bodies, which have to be orchestrated in a circadian manner. Hence, disturbed circadian clocks in our bodies are closely associated with illnesses, including those of the brain.

In this book, the world's leading scientists in neuroscience and biological clocks have summarized the latest findings on circadian rhythms and the brain. The first part explores the molecular basis of cellular and molecular clocks and their interaction with metabolism. The second part applies these findings to various aspects of mental health and disease.

Unfortunately, during the production of this volume, we unexpectedly lost our dear co-author Paolo Sassone-Corsi. Paolo was undoubtedly one of the brightest biologists of our time and he will be strongly missed. He once told me, one day, if things should ever take a turn for the worse, he would just disappear. Upon closer inquiry he said, he would move to one of the many small Italian islands such as Capri or Favignana. I like to imagine that this is where he is now.

Jena, Germany　　　　　　　　　　　　　　　　　　　　Olivia Engmann

Preface

As we fall asleep at night and wake up in the morning, it is tempting to feel slave to the natural world and the unceasing progression of the hours, the days, the seasons which marks the time of our lives from the day we are born to the day we die.

But we are not; we are one step ahead of the game. Circadian clocks have developed which anticipate those predictable variations of the environment and let us exploit the opportunities which come with these changes and prepare us from the dangers we may face because of them. The clock mechanisms within us are indeed ancient, but not static: they have evolved to suit increasingly complex needs. Circadian clocks in our bodies do not just control biochemical reactions within our cells; they affect our cognitive performances, our mood, our capability to fight off infections and to clear and regenerate our bodies from the wear and tear of time, and much more.

How is this complexity achieved in the brain? The first chapters of the book deal with the basic clockwork mechanisms underpinning cellular function, from gene expression to metabolic regulation of biochemical reactions and epigenetic mechanisms implicated in regulating brain function.

While a vast literature exists that illustrates the universal nature of core clock mechanisms, less is known about the importance of circadian timekeeping in conferring a diverse palette of temporal properties to different cells within tissues and organs. Our brains contain a dazzling multitude of glial and neuronal cell controlling complex behaviours. How are cell-type specific clock mechanisms contributing to brain computations responsible for daily regulation of these behaviours? In the second part of the book, the role of clock mechanisms in different neuronal cell types involved in sleep and reward systems is discussed. Are circadian clocks within non-neuronal cells also contributing to the daily orchestration of complex behaviours in the brain? The role of circadian timekeeping in astrocytes and their relevance to control daily regulation of behaviour are discussed in this section. These chapters serve to illustrate this emerging, intercellular level of circadian regulation and to address how universal intracellular clock mechanisms may be "plugged in" brain circuits to organise daily patterns of physiology and behaviour.

What are the consequences of chrono-disrupted lifestyles on the function of inner clock mechanisms? How does the degradation of circadian temporal information relate to the emergence of pathology in the brain? The third and final part of the book focuses on the role of clock disruption in pathologies affecting brain and mental health.

While the three parts of the book are respectively focused on these emerging topics, large crossovers are present within chapters: epigenetic regulation and depression; astrocytes and neurodegeneration; metabolism and sleep. This reflects the intertwined nature of clock mechanisms in humans and the ultimate need to tackle this complexity, if we are to understand how circadian timekeeping works in complex organisms and which role its disruption may play in the transition from brain health to vulnerability to disease.

Sadly, during the preparation of this volume our esteemed colleague and co-author Paolo Sassone-Corsi passed away, a big, unexpected loss for the circadian and neuroscience fields. Paolo played a key part in the inception of this book: he introduced me to Olivia and encouraged us to undertake this project. He wanted "fresh voices to be heard, new and exciting ideas to be shared". Many of the chapters are contributed by junior investigators; world leaders generously strived to make novel connections across different fields, stemming from their expertise and knowledge. We would like to thank them all for their generosity.

The best dedication to Paolo's memory is perhaps in these fresh ideas, in the excitement of a young science. Olivia, I like to think that whether in Capri or Favignana, he would enjoy reading this.

London, UK Marco Brancaccio

Contents

About the Editors

Olivia Engmann obtained her PhD in Neuroscience from King's College London. Her main interest lies in the molecular basis of mental illness. To that end she has worked in laboratories of leading experts in this field in Paris and New York. She recently moved back to Germany to study the biological basis of depression. Dr. Engmann is currently teaching neuro-epigenetics, human genetics and biology at the University of Jena.

Marco Brancaccio received his PhD degree in Neuroscience from SISSA in Trieste, Italy. He then joined Michael Hastings group in the MRC Laboratory of Molecular Biology in Cambridge, UK, where he studied the molecular and circuit mechanisms underlying circadian function in the brain master clock in mammals (the suprachiasmatic nucleus—SCN). He has then moved to Imperial College London in 2018, as a Lecturer and a Programme Leader of the UK Dementia Research Institute (UK-DRI), where he established his research group focusing on the mechanisms driving circadian dysregulation in the early stages of dementia and the exploitation of circadian function for the prevention and cure of neurodegenerative conditions.

Part I

Molecular Gears of the Circadian Clock

Introduction to the Clock System

1

Kimberly H. Cox and Joseph S. Takahashi

Abstract

Circadian (24-h) rhythms dictate almost everything we do, setting our clocks for specific times of sleeping and eating, as well as optimal times for many other basic functions. The physiological systems that coordinate circadian rhythms are intricate, but at their core, they all can be distilled down to cell-autonomous rhythms that are then synchronized within and among tissues. At first glance, these cell-autonomous rhythms may seem rather straight-forward, but years of research in the field has shown that they are strikingly complex, responding to many different external signals, often with remarkable tissue-specificity. To understand the cellular clock system, it is important to be familiar with the major players, which consist of pairs of proteins in a triad of transcriptional/translational feedback loops. In this chapter, we will go through each of the core protein pairs one-by-one, summarizing the literature as to their regulation and their broader impacts on circadian gene expression. We will conclude by briefly examining the human genetics literature, as well as providing perspectives on the future of the study of the molecular clock.

Keywords

Circadian rhythms · Molecular clock · Cell-autonomous · Core clock · Feedback loops

1.1 Introduction

Circadian (24-h) rhythms dictate almost everything we do, setting our clocks for specific times of sleeping and eating, as well as optimal times for many other basic functions. The physiological systems that coordinate circadian rhythms are intricate, but at their core, they all can be distilled down to cell-autonomous rhythms that are then synchronized within and among tissues. At first glance, these cell-autonomous rhythms may seem rather straight-forward, but years of research in the field has shown that they are strikingly complex, responding to many different external signals, often with remarkable tissue-specificity. To understand the cellular clock system, it is important to be familiar with the major players, which consist of pairs of proteins in a triad of transcriptional/translational feedback loops. In this chapter, we will go through each of the core

K. H. Cox
Department of Neuroscience and Peter O'Donnell Jr. Brain Institute, University of Texas Southwestern Medical Center, Dallas, TX, USA

J. S. Takahashi (✉)
Department of Neuroscience and Peter O'Donnell Jr. Brain Institute, University of Texas Southwestern Medical Center, Dallas, TX, USA

Howard Hughes Medical Institute, University of Texas Southwestern Medical Center, Dallas, TX, USA
e-mail: Joseph.Takahashi@UTSouthwestern.edu

O. Engmann, M. Brancaccio (eds.), *Circadian Clock in Brain Health and Disease*, Advances in Experimental Medicine and Biology 1344, https://doi.org/10.1007/978-3-030-81147-1_1

protein pairs one-by-one, summarizing the literature as to their regulation and their broader impacts on circadian gene expression. We will conclude by briefly examining the human genetics literature, as well as providing perspectives on the future of the study of the molecular clock.

1.2 The Core of the Clock: CLOCK and BMAL1

The first mammalian core clock gene, *Clock,* was discovered in an *N*-ethyl-*N*-nitrosourea (ENU) mutagenesis screen for mice with abnormal circadian behavior. Mice homozygous for the *Clock* mutation had extremely long period (~28 h) in wheel-running activity and damped circadian rhythms in constant darkness (Vitaterna et al. 1994). Upon cloning of the *Clock* gene (Antoch et al. 1997; King et al. 1997b; Wilsbacher et al. 2000), it became apparent that *Clock* shared sequence homology with another mouse gene, *Npas2*, a member of the class of basic Helix-loop-Helix/Per-Arnt-Sim (bHLH-PAS) gene family. The fact that CLOCK protein had a PAS dimerization domain meant it also shared features with the *Drosophila* circadian protein, PERIOD (Hoffman et al. 1991; Nambu et al. 1991; Burbach et al. 1992). The PAS domain, in addition to a bHLH domain which allows for DNA binding (Murre et al. 1989), made the CLOCK protein a likely transcription factor (King et al. 1997b). Importantly, *Clock* mRNA was expressed in the suprachiasmatic nucleus (SCN) of the hypothalamus, a region of the brain known to be important for coordinating circadian rhythms (Hastings et al. 2019), as well as in other tissues throughout the body (King et al. 1997b; Steeves et al. 1999; Maywood et al. 2003), suggesting that this protein could have wide-spread influence in many different cell types.

CLOCK's binding partner, BMAL1 (encoded by the *Arntl* gene), was discovered in a yeast two-hybrid screen, where it was shown that CLOCK:BMAL1 heterodimers bind directly to enhancer regulatory sites (E-boxes) in the mouse *Per1* gene to promote transcription (Gekakis et al. 1998), Fig. 1.1. It was subsequently shown that loss of *Bmal1* causes a complete loss of circadian rhythmicity in mice (Bunger et al. 2000). A paralog of BMAL1, BMAL2, is also expressed in the SCN and can also dimerize with CLOCK; however, it is expressed at lower levels and cannot compensate for *Bmal1* loss-of-function (Bunger et al. 2000; Hogenesch et al. 2000; Ko et al. 2010). Interestingly, although the original mutant *Clock* mice have a clear behavioral phenotype, deletion of the entire *Clock* gene does not completely disrupt rhythms (Debruyne et al. 2006). This is because the original *Clock* mutation (a point mutation causing loss of exon 19) is a dominant negative (King et al. 1997a), while in the knockout, its homolog, NPAS2, also dimerizes with BMAL1 (Reick et al. 2001) and can compensate somewhat for the loss of CLOCK in the SCN. However, this is not true in peripheral tissues where CLOCK is required (DeBruyne et al. 2007a, b; Bertolucci et al. 2008). Intriguingly, *Clock* mutant mice have reduced *Bmal1* expression in the SCN, but not peripheral tissues (Oishi et al. 2000), where other genes are specifically reduced instead (Noshiro et al. 2005), suggesting tissue-specific mechanisms of gene regulation by these proteins (McDearmon et al. 2006). Of note, mice lacking *Npas2* do have slightly abnormal circadian rhythms (Dudley et al. 2003), suggesting some nonredundant functions of CLOCK and NPAS2 (Landgraf et al. 2016).

Although, as discussed later, expression of both *Clock* and *Bmal1* are regulated by other clock proteins (Shearman et al. 2000; Preitner et al. 2002), Fig. 1.1, little else is known about the control of their expression. However, transcription of *Bmal1* correlates with binding of the nuclear membrane protein, MAN1 (Lin et al. 2014), suggesting that nuclear remodeling plays a role in its regulation. Post-translationally, both CLOCK and BMAL1 proteins are modified by various enzymes whose activities are also rhythmic and these modifications (phosphorylation, SUMOylation, O-GlcNAcylation, and acetylation) directly regulate their turnover (Reddy et al. 2006; Reischl and Kramer 2011; Masri et al. 2013; Hirano et al. 2016a; Robles et al.

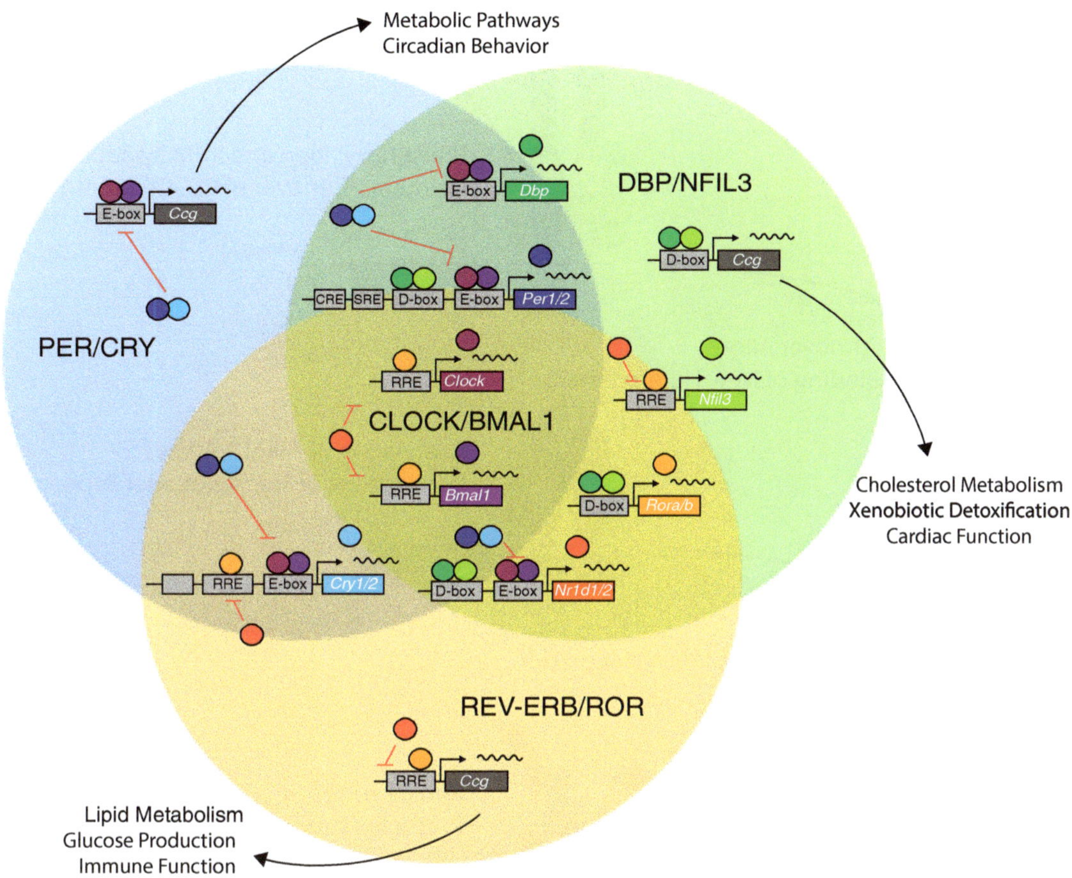

Fig. 1.1 The circadian clock is comprised of pairs of proteins in a triad of transcriptional feedback loops. At the core of the circadian clock are the genes for *Clock* and *Bmal1*, which synthesize CLOCK (burgundy circles) and BMAL1 (purple circles) proteins. These proteins heterodimerize and bind to E-box elements on target gene promoters, both within the core clock and on other clock-controlled genes (*Ccg*). **BLUE**: The first feedback loop is comprised of *Per1/Per2* and *Cry1/Cry2*, target genes of CLOCK:BMAL1. PER (dark blue circles) and CRY (light blue circles) proteins inhibit the binding of CLOCK:BMAL1 heterodimers, both at their own gene promoters as well as at other target genes. Thus, PER and CRY provide negative feedback to the core circadian clock. **ORANGE**: In the second feedback loop, the genes *Nr1d1/Nr1d2* and *Rora/b* produce the REV-ERB (dark orange circles) and retinoic acid-related orphan receptor (ROR, light orange circles) proteins, which bind to retinoic acid response elements (RREs) on gene promoters with opposite effects: REV-ERBs inhibit gene transcription, and RORs promote transcription. REV-ERBs/RORs regulate the expression of *Bmal1*, *Clock*, and *Cry1*; and, importantly, *Nr1d1* is also directly regulated by CLOCK: BMAL1 as well as PER2. Together, these two proteins provide both negative and positive feedback to the core circadian clock and also influence the circadian expression of other clock-controlled genes (*Ccg*). **GREEN**: In the third feedback loop, the genes for *Dbp* and *Nfil3* are under the regulation of CLOCK:BMAL1 and REV-ERBs/RORs, respectively. The DBP (dark green circles) and NFIL3 (light green circles) proteins dimerize and bind to D-box elements on target genes, including *Per1/Per2*, *Nr1d1/Nr1d2*, and *Rora/Rorb*. Therefore, this third feedback loop has many complex interactions with the core clock, as well as regulating additional clock-controlled genes (*Ccg*)

2017; Wang et al. 2018; Mauvoisin 2019), Fig. 1.2. Casein kinase 2α (CK2α) (Tamaru et al. 2009), cyclin-dependent kinase 5 (CDK5) (Kwak et al. 2013), and protein kinase B (AKT) (Luciano et al. 2018) phosphorylate CLOCK at several sites, while glycogen synthase kinase-3β (GSK-3β) and casein kinase 1 ε (CKIε) phosphorylate BMAL1 (Eide et al. 2002; Sahar et al.

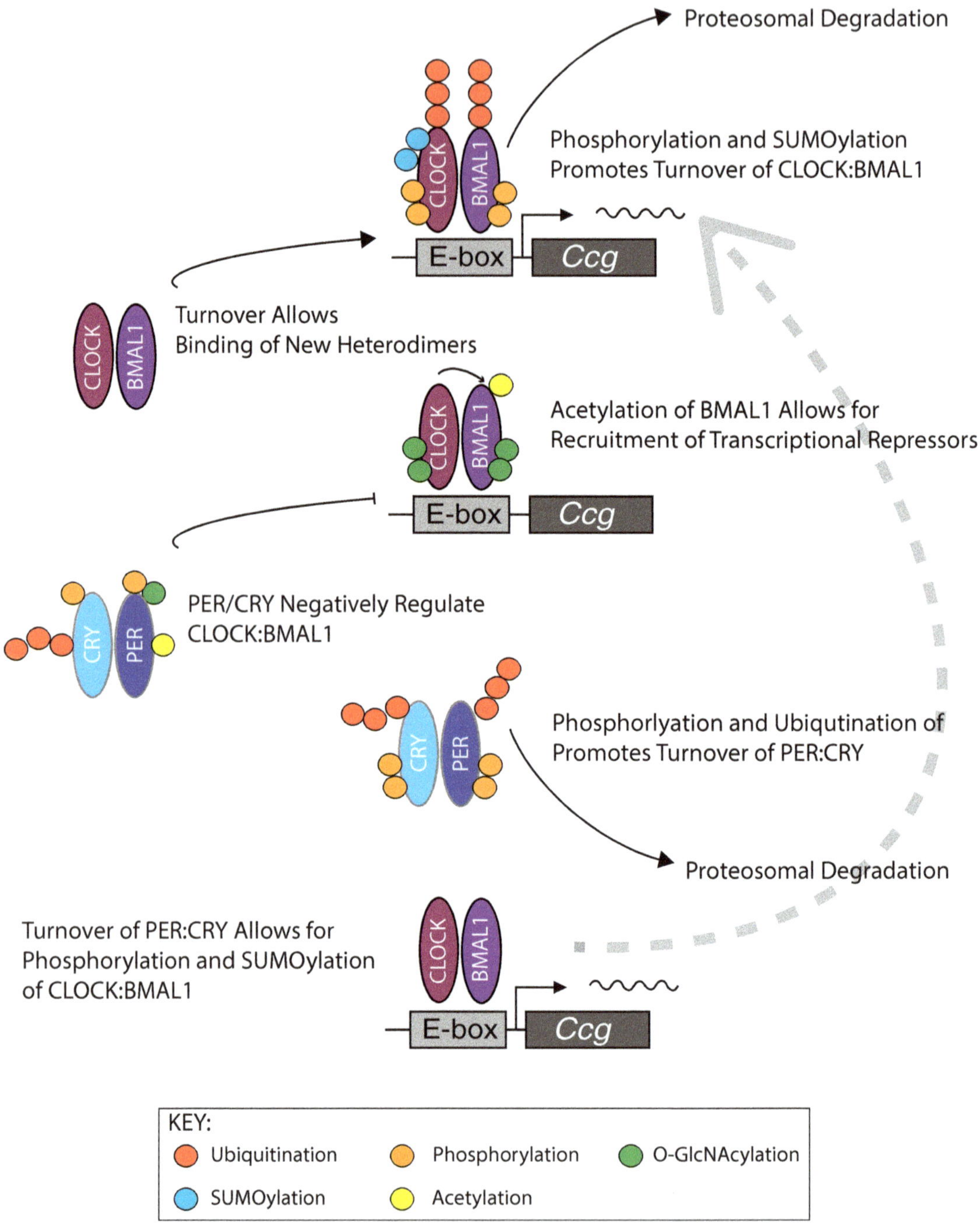

Fig. 1.2 Post-translational regulation of core clock proteins. Phosphorylation of CLOCK and BMAL1 at various sites promotes their degradation, but also allows for their nuclear accumulation and increases their turnover at gene promoters by decreasing their stability. SUMOylation of CLOCK and BMAL1 targets them for ubiquitination. O-GlcNAcylation prevents ubiquitination and degradation of CLOCK and BMAL1 by competing with kinases for phosphorylation sites. CLOCK also directly modulates BMAL1 activity through acetylation, which allows BMAL1 to recruit transcriptional repressors like PER:CRY. Phosphorylation of PER at certain sites regulates PER:CRY nuclear translocation, while Acetylation and O-GlcNAcylation increase PER protein stability, as do the interactions between PER and CRY. Phosphorylation of PER at other sites, as well as CRY, targets them for ubiquitination and proteasomal degradation

2010). Rhythmic phosphorylation of CLOCK and BMAL1 promotes their degradation, but also allows for nuclear accumulation of these proteins and increased turnover at gene promoters, which is important for maintaining the timing of the molecular clock (Spengler et al. 2009; Yoshitane et al. 2009; Sahar et al. 2010; Luciano et al. 2018). In addition to phosphorylation, both CLOCK and BMAL1 are regulated by rhythmic SUMOylation, which targets the proteins for ubiquitination (Cardone et al. 2005; Lee et al. 2008; Li et al. 2013), likely through the E3 ubiquitin ligase UBE3A (Gossan et al. 2014). In contrast, O-GlcNAcylation prevents ubiquitination and degradation of CLOCK and BMAL1 by competing with kinases for phosphorylation sites (Ma et al. 2013). CLOCK also directly modulates BMAL1 activity through its actions as an acetyltransferase, starting a chain of events that cause the termination of transactivation by allowing BMAL1 to recruit transcriptional repressors to CLOCK:BMAL1 complexes (Doi et al. 2006; Hirayama et al. 2007; Hirano et al. 2016a).

CLOCK:BMAL1 binding to DNA is also altered in response to circulating factors, such as the redox cofactors NAD and NADP (Rutter et al. 2001), suggesting another layer of cellular control. In order to influence gene expression, CLOCK:BMAL1 heterodimers form larger complexes with histone modifying enzymes (Katada and Sassone-Corsi 2010; DiTacchio et al. 2011) and transcriptional coactivators such as Sirtuin 1, CBP/p300, and TRAP150 (Nakahata et al. 2008; Nakahata et al. 2009; Ramsey et al. 2009; Lee et al. 2010; Lande-Diner et al. 2013) that foster recruitment to gene promoters (Oishi et al. 2003; Miller et al. 2007). The binding of these CLOCK:BMAL1 co-activator complexes is rhythmic and reliant upon other, often tissue-specific, transcription factors to produce rhythmic chromatin opening, recruitment of RNA polymerase, and gene expression (Panda et al. 2002; Rey et al. 2011; Lande-Diner et al. 2013; Menet et al. 2014; Takahashi 2017; Trott and Menet 2018). In addition to rhythmic binding at gene promoters, recent studies support the idea that CLOCK: BMAL1 can also influence gene expression via rhythmic enhancers and other noncanonical binding sites (Hatanaka et al. 2010; Yoshitane et al. 2014; Beytebiere et al. 2019).

1.3 A Negative Transcriptional Feedback Loop: PER and CRY

CLOCK:BMAL1 regulate the transcription of many genes, including the second set of core clock genes, *Per* and *Cry* (Gekakis et al. 1998; Jin et al. 1999; Kume et al. 1999; Travnickova-Bendova et al. 2002; Yoo et al. 2005), Fig. 1.1. The mammalian *Per* genes (*Per1, Per2,* and *Per3)* were uncovered using homology screens for orthologs of the *Drosophila* circadian gene, *period* (Konopka and Benzer 1971; Bargiello et al. 1984; Albrecht et al. 1997; Shearman et al. 1997; Sun et al. 1997; Tei et al. 1997). These orthologs exhibit circadian expression patterns and are also responsive to light, suggesting that, in mammals as in *Drosophila*, these proteins play central roles in regulating circadian rhythms (Albrecht et al. 1997; Zylka et al. 1998b; Yan et al. 1999; Zheng et al. 1999; Bae et al. 2001; Hamada et al. 2001; Pendergast et al. 2010). Interestingly, while the *Drosophila* PERIOD partners with a protein called, TIMELESS, in mammals, the *Tim* ortholog serves a less prominent function in mediating circadian rhythms (Sangoram et al. 1998; Zylka et al. 1998a; Barnes et al. 2003). Instead, the mammalian PERIOD (PER) proteins heterodimerize with CRYPTOCHROME (CRY) proteins (encoded by *Cry1* and *Cry2*), which are critical for normal circadian function (Griffin et al. 1999; Kume et al. 1999; van der Horst et al. 1999; Vitaterna et al. 1999; Field et al. 2000).

Together, PER and CRY repress their own gene transcription by regulating *Bmal1* expression and inhibiting CLOCK:BMAL1 transcriptional activity (Sangoram et al. 1998; Shearman et al. 2000; Sato et al. 2006; Nangle et al. 2014; Xu et al. 2015), Fig. 1.1. This is accomplished by the negative regulation of CLOCK:BMAL1 phosphorylation via protein phosphatase 5 (Partch et al. 2006; Dardente et al. 2007; Matsumura et al. 2014), which leads to CLOCK:BMAL1 protein stabilization (Kondratov et al. 2006), the recruitment of repressive transcription-termination complex proteins (Brown et al. 2005; Padmanabhan et al. 2012; Michael et al. 2017), and the displacement of CLOCK:BMAL1 dimers from gene promoters (Ye et al. 2014; Chiou et al. 2016). Thus, PER and CRY form a negative autoregulatory feedback loop that allows for rhythmic CLOCK:BMAL1-mediated gene expression, not just for *Per* and *Cry*, but also many other circadian genes, known collectively as clock-controlled genes (Ccgs) (Sangoram et al. 1998; Partch et al. 2014; Takahashi 2017). This negative transcriptional regulation involves the further assembly of large inhibitory complexes that include histone modifying enzymes as well as kinases that directly regulate CLOCK and BMAL1 activity (Duong et al. 2011; Crane and Young 2014; Nangle et al. 2014; Tamaru et al. 2015).

However, the timekeeping mechanism of this feedback loop is more complicated than simply direct transcriptional regulation. Recent evidence suggests that *Per* and *Cry* mRNA expression and translation are dependent on mRNA decay mediated by RNA binding proteins (Kojima et al. 2003; Kwak et al. 2006; Kojima et al. 2007; Woo et al. 2009, 2010; Kojima et al. 2011; Preussner and Heyd 2016). In addition, there is a large literature on post-transcriptional regulation of PER and CRY proteins (Fig. 1.2). PER degradation is regulated by the serine/threonine kinases, casein kinase 1δ (CK1δ), and CK1ϵ, which phosphorylate PERs and target them for proteasomal degradation via the recruitment of the E3 ubiquitin ligases, β-TrCP1 and 2 (Lowrey et al. 2000; Lee et al. 2001; Eide et al. 2002, 2005; Shirogane et al. 2005; Reischl et al. 2007; Meng et al. 2008a; Etchegaray et al. 2009; Lee et al. 2009). Interestingly, the importance of CKI was first demonstrated many years earlier in *tau* mutant hamsters, although its function in the circadian clock at that time was unknown (Ralph and Menaker 1988). In addition to mediating degradation, phosphorylation of PER is also important for regulating PER:CRY nuclear translocation, although the mechanisms for this are not well-understood (Takano et al. 2004; Sakakida et al. 2005). Nonetheless, it is clear that phosphorylation/dephosphorylation of PER proteins is important for regulating CLOCK:BMAL1-mediated gene expression (Lee et al. 2011; Zhou et al. 2015; Narasimamurthy et al. 2018). Furthermore, other modifications, including acetylation and O-GlcNAcylation, also increase PER protein stability (Asher et al. 2008; Nakahata et al. 2008; Kaasik et al. 2013).

The degradation of CRY proteins is regulated post-transcriptionally by the F-box proteins, FBXL3 and FBXL21, which promote and compete for ubiquitination of CRY, respectively (Busino et al. 2007; Godinho et al. 2007; Siepka et al. 2007; Hirano et al. 2013; Yoo et al. 2013). CRY interactions with F-box proteins are mediated by phosphorylation (Harada et al. 2005; Lamia et al. 2009). Importantly, the stability of both PER and CRY is enhanced by their interactions with each other (Hirano et al. 2016a). Once negative transcriptional feedback and post-transcriptional and post-translational regulation of PER and CRY are sufficient to decrease PER/CRY protein levels in the nucleus, repression is relieved and CLOCK:BMAL1 start a new cycle of *Per*/*Cry* gene transcription (Gallego and Virshup 2007; Chen et al. 2009; Lowrey and Takahashi 2011; Buhr and Takahashi 2013; Takahashi 2017). There is also some evidence that PER and CRY regulate circadian expression of genes outside of the core loop (Lamia et al. 2011).

1.4 The Second Loop: Positive Feedback from REV-ERB and ROR

The nuclear receptors REV-ERBα and REV-ERBβ (encoded by the genes *Nr1d1* and *Nr1d2*), along with the retinoic acid-related orphan receptors, RORα, RORβ, and RORγ, provide a second layer of feedback to the core circadian clock (Fig. 1.1). These proteins were found to be involved in circadian rhythms due to the presence of their unique binding sites, called ROR-binding elements (Harding and Lazar 1993; Harding et al. 1997), on the *BMAL1* gene, as well as circadian rhythms in ROR mRNA expression (Sumi et al. 2002). Interestingly, although REV-ERBs and RORs bind to the same sites, these proteins have opposite effects on *BMAL1* transcription, with REV-ERBs providing negative regulation and RORs promoting transcription (Preitner et al. 2002; Ueda et al. 2002; Sato et al. 2004; Akashi and Takumi 2005; Guillaumond et al. 2005). REV-ERBs were later shown to also regulate *CLOCK* gene transcription (Crumbley and Burris 2011), as well as expression of *NPAS2* (Takeda et al. 2011; Matsumura et al. 2013), and *Cry1* (Liu et al. 2008).

Importantly, BMAL1 positively regulates *Nr1d1*, while PER2 represses its expression (Preitner et al. 2002). PER2 has also been shown to interact with REV-ERB at target promoters (Schmutz et al. 2010), and FBXL3 regulates REV-ERB/ROR-mediated transcription by inactivating co-repressor complexes (Shi et al. 2013). Thus, these nuclear receptors interact with the core clock on many different levels (Brown et al. 2012). REV-ERB activity is modulated by heme binding (Meng et al. 2008b), and, like other core clock proteins, REV-ERB protein degradation is regulated by multiple enzymes including GSK-3β (Yin et al. 2006); the E3 ligases, Arf-bp1 and Pam (Yin et al. 2010); and the F-box protein, FBXW7 (Zhao et al. 2016). Overall, REV-ERBα and REV-ERBβ appear to serve largely redundant functions and regulate a considerable number of circadian output genes, particularly genes that are involved in metabolism (Solt et al. 2012; Ikeda et al. 2019).

1.5 The Third Loop: DBP and NFIL3

The D-box binding protein (DBP) is a transcriptional activator that is expressed in a circadian manner. Its gene (*Dbp*) is a direct transcriptional target of CLOCK:BMAL1 (Wuarin et al. 1992; Lopez-Molina et al. 1997; Ripperger et al. 2000; Ripperger and Schibler 2006; Stratmann et al. 2012; Aguilar-Arnal et al. 2013), although CRY1 also regulates its expression (Stratmann et al. 2010). Both DBP and the nuclear factor, interleukin-3 regulated protein (NFIL3, also known as E4BP4), which is a transcriptional repressor, bind to D-box elements on circadian promoters, but with opposite phases of maximal binding (Mitsui et al. 2001; Gachon et al. 2004; Ueda et al. 2005; Ukai-Tadenuma et al. 2008). NFIL3 has also been shown to interact with PER2 to promote transcriptional repression via E-box sites (Ohno et al. 2007a) and also directly represses *Per2* gene expression (Ohno et al. 2007b), while DBP binds to the *Cry1* promoter, influencing the timing of transcription. Thus, this third DBP/NFIL3 loop is also intricately involved in the timing PER/CRY negative feedback (Ukai-Tadenuma et al. 2011). DBP/NFIL3 also influence the circadian transcription of many other target genes and respond to cellular signals, such as PI3 kinase and signal transducer and activator of transcription 3 (STAT3) (Wuarin et al. 1992; Mitsui et al. 2001; Morishita et al. 2016; Wang et al. 2017b).

1.6 Human Mutations in Core Clock Genes

Although most studies of the circadian clock have been performed in animal models, a number of findings in humans have helped solidify the importance of the molecular clock and its role in human disease. One of the first circadian gene mutations was discovered in patients with familial advanced sleep phase syndrome (FASPS), who had a 4-hour phase advance in their sleep, body temperature, and hormonal rhythms that caused them to awaken early (Toh et al. 2001). The mutation, which altered a PER2 phosphorylation site (Fig. 1.2), impaired CK1ϵ binding, and a few years later another cohort of FASPS patients was found with a mutation in *CK1δ* (Xu et al. 2005), corroborating the influence of PER2 phosphorylation in this sleep disorder. Since that time, mutations in *Cry2, Per3, and Tim* have also been associated with FASPS (Hirano et al. 2016b; Zhang et al. 2016; Kurien et al. 2019). In contrast to FASPS, patients with delayed sleep phase disorder (DSPD) have persistent insomnia and delayed sleep onset. So far, only mutations in

Cry1 have been associated with DSPD (Patke et al. 2017).

In addition to investigations into monogenic causes of severe sleep disorders, a number of GWAS studies have also identified genetic variants in core clock genes associated with "early riser" phenotypes (Hu et al. 2016) or self-reported chronotypes (Jones et al. 2016; Lane et al. 2016), including variants in *PER2* and *RGS16* (Kalmbach et al. 2017). More recent GWAS performed on insomnia patients have not yielded variants in known circadian genes, but these cohorts are complicated by other psychological phenotypes (Stein et al. 2018; Jansen et al. 2019). Interestingly, although many neuropsychiatric disorders have associated circadian rhythm phenotypes, to date no variants in known circadian genes have been implicated in these disorders (Jagannath et al. 2017; Lane et al. 2017; McCarthy 2019; Stahl et al. 2019).

1.7 Conclusions and Future Perspectives

The feedback loops that make up the molecular clock are governed by transcriptional (Takahashi 2017), post-transcriptional (Preussner and Heyd 2016), and post-translational (Gallego and Virshup 2007) mechanisms that maintain cellular circadian rhythms. However, as suggested by human genetics studies, the description of the clock provided here is far from finished, and there are likely many other factors that influence the core clock machinery (McCarthy 2019). Moving from the study of the core clock to large-scale studies of circadian gene transcription, we are just beginning to understand how the core clock components alter the transcriptional landscape (Koike et al. 2012; Zhang et al. 2014; Takahashi 2017; Yeung and Naef 2018; Pacheco-Bernal et al. 2019). In addition, recent technological advances in the field of proteomics will no doubt allow for a more complete picture of circadian rhythms in the regulation of protein expression (Narumi et al. 2016; Robles et al. 2017; Wang et al. 2017a; Mauvoisin 2019). Thus, we still have much to learn about the influence of the circadian clock on human disease.

References

Aguilar-Arnal L, Hakim O, Patel VR, Baldi P, Hager GL, Sassone-Corsi P (2013) Cycles in spatial and temporal chromosomal organization driven by the circadian clock. Nat Struct Mol Biol 20(10):1206–1213. https://doi.org/10.1038/nsmb.2667

Akashi M, Takumi T (2005) The orphan nuclear receptor RORalpha regulates circadian transcription of the mammalian core-clock Bmal1. Nat Struct Mol Biol 12(5):441–448. https://doi.org/10.1038/nsmb925

Albrecht U, Sun ZS, Eichele G, Lee CC (1997) A differential response of two putative mammalian circadian regulators, mper1 and mper2, to light. Cell 91 (7):1055–1064

Antoch MP, Song EJ, Chang AM, Vitaterna MH, Zhao Y, Wilsbacher LD, Sangoram AM, King DP, Pinto LH, Takahashi JS (1997) Functional identification of the mouse circadian Clock gene by transgenic BAC rescue. Cell 89(4):655–667. https://doi.org/10.1016/s0092-8674(00)80246-9

Asher G, Gatfield D, Stratmann M, Reinke H, Dibner C, Kreppel F, Mostoslavsky R, Alt FW, Schibler U (2008) SIRT1 regulates circadian clock gene expression through PER2 deacetylation. Cell 134 (2):317–328. https://doi.org/10.1016/j.cell.2008.06.050

Bae K, Jin X, Maywood ES, Hastings MH, Reppert SM, Weaver DR (2001) Differential functions of mPer1, mPer2, and mPer3 in the SCN circadian clock. Neuron 30(2):525–536

Bargiello TA, Jackson FR, Young MW (1984) Restoration of circadian behavioural rhythms by gene transfer in Drosophila. Nature 312(5996):752–754. https://doi.org/10.1038/312752a0

Barnes JW, Tischkau SA, Barnes JA, Mitchell JW, Burgoon PW, Hickok JR, Gillette MU (2003) Requirement of mammalian timeless for circadian rhythmicity. Science 302(5644):439–442. https://doi.org/10.1126/science.1086593

Bertolucci C, Cavallari N, Colognesi I, Aguzzi J, Chen Z, Caruso P, Foa A, Tosini G, Bernardi F, Pinotti M (2008) Evidence for an overlapping role of CLOCK and NPAS2 transcription factors in liver circadian oscillators. Mol Cell Biol 28(9):3070–3075. https://doi.org/10.1128/MCB.01931-07

Beytebiere JR, Trott AJ, Greenwell BJ, Osborne CA, Vitet H, Spence J, Yoo SH, Chen Z, Takahashi JS, Ghaffari N, Menet JS (2019) Tissue-specific BMAL1 cistromes reveal that rhythmic transcription is associated with rhythmic enhancer-enhancer interactions. Genes Dev 33(5-6):294–309. https://doi.org/10.1101/gad.322198.118

Brown SA, Ripperger J, Kadener S, Fleury-Olela F, Vilbois F, Rosbash M, Schibler U (2005) PERIOD1-associated proteins modulate the negative limb of the mammalian circadian oscillator. Science 308 (5722):693–696. https://doi.org/10.1126/science.1107373

Brown SA, Kowalska E, Dallmann R (2012) (Re)-inventing the circadian feedback loop. Dev Cell 22 (3):477–487. https://doi.org/10.1016/j.devcel.2012.02.007

Buhr ED, Takahashi JS (2013) Molecular components of the Mammalian circadian clock. Handb Exp Pharmacol 217:3–27. https://doi.org/10.1007/978-3-642-25950-0_1

Bunger MK, Wilsbacher LD, Moran SM, Clendenin C, Radcliffe LA, Hogenesch JB, Simon MC, Takahashi JS, Bradfield CA (2000) Mop3 is an essential component of the master circadian pacemaker in mammals. Cell 103(7):1009–1017. https://doi.org/10.1016/s0092-8674(00)00205-1

Burbach KM, Poland A, Bradfield CA (1992) Cloning of the Ah-receptor cDNA reveals a distinctive ligand-activated transcription factor. Proc Natl Acad Sci U S A 89(17):8185–8189. https://doi.org/10.1073/pnas.89.17.8185

Busino L, Bassermann F, Maiolica A, Lee C, Nolan PM, Godinho SI, Draetta GF, Pagano M (2007) SCFFbxl3 controls the oscillation of the circadian clock by directing the degradation of cryptochrome proteins. Science 316(5826):900–904. https://doi.org/10.1126/science.1141194

Cardone L, Hirayama J, Giordano F, Tamaru T, Palvimo JJ, Sassone-Corsi P (2005) Circadian clock control by SUMOylation of BMAL1. Science 309 (5739):1390–1394. https://doi.org/10.1126/science.1110689

Chen R, Schirmer A, Lee Y, Lee H, Kumar V, Yoo SH, Takahashi JS, Lee C (2009) Rhythmic PER abundance defines a critical nodal point for negative feedback within the circadian clock mechanism. Mol Cell 36 (3):417–430. https://doi.org/10.1016/j.molcel.2009.10.012

Chiou YY, Yang Y, Rashid N, Ye R, Selby CP, Sancar A (2016) Mammalian Period represses and de-represses transcription by displacing CLOCK-BMAL1 from promoters in a Cryptochrome-dependent manner. Proc Natl Acad Sci U S A 113(41):E6072–E6079. https://doi.org/10.1073/pnas.1612917113

Crane BR, Young MW (2014) Interactive features of proteins composing eukaryotic circadian clocks. Annu Rev Biochem 83:191–219. https://doi.org/10.1146/annurev-biochem-060713-035644

Crumbley C, Burris TP (2011) Direct regulation of CLOCK expression by REV-ERB. PLoS One 6(3): e17290. https://doi.org/10.1371/journal.pone.0017290

Dardente H, Fortier EE, Martineau V, Cermakian N (2007) Cryptochromes impair phosphorylation of transcriptional activators in the clock: a general mechanism for circadian repression. Biochem J 402(3):525–536. https://doi.org/10.1042/BJ20060827

Debruyne JP, Noton E, Lambert CM, Maywood ES, Weaver DR, Reppert SM (2006) A clock shock: mouse CLOCK is not required for circadian oscillator function. Neuron 50(3):465–477. https://doi.org/10.1016/j.neuron.2006.03.041

DeBruyne JP, Weaver DR, Reppert SM (2007a) CLOCK and NPAS2 have overlapping roles in the suprachiasmatic circadian clock. Nat Neurosci 10 (5):543–545. https://doi.org/10.1038/nn1884

DeBruyne JP, Weaver DR, Reppert SM (2007b) Peripheral circadian oscillators require CLOCK. Curr Biol 17 (14):R538–R539. https://doi.org/10.1016/j.cub.2007.05.067

DiTacchio L, Le HD, Vollmers C, Hatori M, Witcher M, Secombe J, Panda S (2011) Histone lysine demethylase JARID1a activates CLOCK-BMAL1 and influences the circadian clock. Science 333 (6051):1881–1885. https://doi.org/10.1126/science.1206022

Doi M, Hirayama J, Sassone-Corsi P (2006) Circadian regulator CLOCK is a histone acetyltransferase. Cell 125(3):497–508. https://doi.org/10.1016/j.cell.2006.03.033

Dudley CA, Erbel-Sieler C, Estill SJ, Reick M, Franken P, Pitts S, McKnight SL (2003) Altered patterns of sleep and behavioral adaptability in NPAS2-deficient mice. Science 301(5631):379–383. https://doi.org/10.1126/science.1082795

Duong HA, Robles MS, Knutti D, Weitz CJ (2011) A molecular mechanism for circadian clock negative feedback. Science 332(6036):1436–1439. https://doi.org/10.1126/science.1196766

Eide EJ, Vielhaber EL, Hinz WA, Virshup DM (2002) The circadian regulatory proteins BMAL1 and cryptochromes are substrates of casein kinase I epsilon. J Biol Chem 277(19):17248–17254. https://doi.org/10.1074/jbc.M111466200

Eide EJ, Kang H, Crapo S, Gallego M, Virshup DM (2005) Casein kinase I in the mammalian circadian clock. Methods Enzymol 393:408–418. https://doi.org/10.1016/S0076-6879(05)93019-X

Etchegaray JP, Machida KK, Noton E, Constance CM, Dallmann R, Di Napoli MN, DeBruyne JP, Lambert CM, Yu EA, Reppert SM, Weaver DR (2009) Casein kinase 1 delta regulates the pace of the mammalian circadian clock. Mol Cell Biol 29(14):3853–3866. https://doi.org/10.1128/MCB.00338-09

Field MD, Maywood ES, O'Brien JA, Weaver DR, Reppert SM, Hastings MH (2000) Analysis of clock proteins in mouse SCN demonstrates phylogenetic divergence of the circadian clockwork and resetting mechanisms. Neuron 25(2):437–447

Gachon F, Fonjallaz P, Damiola F, Gos P, Kodama T, Zakany J, Duboule D, Petit B, Tafti M, Schibler U (2004) The loss of circadian PAR bZip transcription factors results in epilepsy. Genes Dev 18 (12):1397–1412. https://doi.org/10.1101/gad.301404

Gallego M, Virshup DM (2007) Post-translational modifications regulate the ticking of the circadian clock. Nat Rev Mol Cell Biol 8(2):139–148. https://doi.org/10.1038/nrm2106

Gekakis N, Staknis D, Nguyen HB, Davis FC, Wilsbacher LD, King DP, Takahashi JS, Weitz CJ (1998) Role of the CLOCK protein in the mammalian circadian mechanism. Science 280(5369):1564–1569

Godinho SI, Maywood ES, Shaw L, Tucci V, Barnard AR, Busino L, Pagano M, Kendall R, Quwailid MM, Romero MR, O'Neill J, Chesham JE, Brooker D, Lalanne Z, Hastings MH, Nolan PM (2007) The after-hours mutant reveals a role for Fbxl3 in determining mammalian circadian period. Science 316 (5826):897–900. https://doi.org/10.1126/science.1141138

Gossan NC, Zhang F, Guo B, Jin D, Yoshitane H, Yao A, Glossop N, Zhang YQ, Fukada Y, Meng QJ (2014) The E3 ubiquitin ligase UBE3A is an integral component of the molecular circadian clock through regulating the BMAL1 transcription factor. Nucleic Acids Res 42(9):5765–5775. https://doi.org/10.1093/nar/gku225

Griffin EA Jr, Staknis D, Weitz CJ (1999) Light-independent role of CRY1 and CRY2 in the mammalian circadian clock. Science 286(5440):768–771

Guillaumond F, Dardente H, Giguere V, Cermakian N (2005) Differential control of Bmal1 circadian transcription by REV-ERB and ROR nuclear receptors. J Biol Rhythms 20(5):391–403. https://doi.org/10.1177/0748730405277232

Hamada T, LeSauter J, Venuti JM, Silver R (2001) Expression of Period genes: rhythmic and nonrhythmic compartments of the suprachiasmatic nucleus pacemaker. J Neurosci 21(19):7742–7750

Harada Y, Sakai M, Kurabayashi N, Hirota T, Fukada Y (2005) Ser-557-phosphorylated mCRY2 is degraded upon synergistic phosphorylation by glycogen synthase kinase-3 beta. J Biol Chem 280 (36):31714–31721. https://doi.org/10.1074/jbc.M506225200

Harding HP, Lazar MA (1993) The orphan receptor Rev-ErbA alpha activates transcription via a novel response element. Mol Cell Biol 13(5):3113–3121. https://doi.org/10.1128/mcb.13.5.3113

Harding HP, Atkins GB, Jaffe AB, Seo WJ, Lazar MA (1997) Transcriptional activation and repression by RORalpha, an orphan nuclear receptor required for cerebellar development. Mol Endocrinol 11 (11):1737–1746. https://doi.org/10.1210/mend.11.11.0002

Hastings MH, Maywood ES, Brancaccio M (2019) The mammalian circadian timing system and the suprachiasmatic nucleus as its pacemaker. Biology (Basel) 8(1):13. https://doi.org/10.3390/biology8010013

Hatanaka F, Matsubara C, Myung J, Yoritaka T, Kamimura N, Tsutsumi S, Kanai A, Suzuki Y, Sassone-Corsi P, Aburatani H, Sugano S, Takumi T (2010) Genome-wide profiling of the core clock protein BMAL1 targets reveals a strict relationship with metabolism. Mol Cell Biol 30(24):5636–5648. https://doi.org/10.1128/MCB.00781-10

Hirano A, Yumimoto K, Tsunematsu R, Matsumoto M, Oyama M, Kozuka-Hata H, Nakagawa T, Lanjakornsiripan D, Nakayama KI, Fukada Y (2013) FBXL21 regulates oscillation of the circadian clock through ubiquitination and stabilization of cryptochromes. Cell 152(5):1106–1118. https://doi.org/10.1016/j.cell.2013.01.054

Hirano A, Fu YH, Ptacek LJ (2016a) The intricate dance of post-translational modifications in the rhythm of life. Nat Struct Mol Biol 23(12):1053–1060. https://doi.org/10.1038/nsmb.3326

Hirano A, Shi G, Jones CR, Lipzen A, Pennacchio LA, Xu Y, Hallows WC, McMahon T, Yamazaki M, Ptacek LJ, Fu YH (2016b) A Cryptochrome 2 mutation yields advanced sleep phase in humans. Elife 5: e16695. https://doi.org/10.7554/eLife.16695

Hirayama J, Sahar S, Grimaldi B, Tamaru T, Takamatsu K, Nakahata Y, Sassone-Corsi P (2007) CLOCK-mediated acetylation of BMAL1 controls circadian function. Nature 450(7172):1086–1090. https://doi.org/10.1038/nature06394

Hoffman EC, Reyes H, Chu FF, Sander F, Conley LH, Brooks BA, Hankinson O (1991) Cloning of a factor required for activity of the Ah (dioxin) receptor. Science 252(5008):954–958

Hogenesch JB, Gu YZ, Moran SM, Shimomura K, Radcliffe LA, Takahashi JS, Bradfield CA (2000) The basic helix-loop-helix-PAS protein MOP9 is a brain-specific heterodimeric partner of circadian and hypoxia factors. J Neurosci 20(13):RC83

Hu Y, Shmygelska A, Tran D, Eriksson N, Tung JY, Hinds DA (2016) GWAS of 89,283 individuals identifies genetic variants associated with self-reporting of being a morning person. Nat Commun 7:10448. https://doi.org/10.1038/ncomms10448

Ikeda R, Tsuchiya Y, Koike N, Umemura Y, Inokawa H, Ono R, Inoue M, Sasawaki Y, Grieten T, Okubo N, Ikoma K, Fujiwara H, Kubo T, Yagita K (2019) REV-ERBalpha and REV-ERBbeta function as key factors regulating mammalian circadian output. Sci Rep 9(1):10171. https://doi.org/10.1038/s41598-019-46656-0

Jagannath A, Taylor L, Wakaf Z, Vasudevan SR, Foster RG (2017) The genetics of circadian rhythms, sleep and health. Hum Mol Genet 26(R2):R128–R138. https://doi.org/10.1093/hmg/ddx240

Jansen PR, Watanabe K, Stringer S, Skene N, Bryois J, Hammerschlag AR, de Leeuw CA, Benjamins JS, Munoz-Manchado AB, Nagel M, Savage JE, Tiemeier H, White T, Me Research T, Tung JY, Hinds DA, Vacic V, Wang X, Sullivan PF, van der Sluis S, Polderman TJC, Smit AB, Hjerling-Leffler J, Van Someren EJW, Posthuma D (2019) Genome-wide analysis of insomnia in 1,331,010 individuals identifies new risk loci and functional pathways. Nat Genet 51

(3):394–403. https://doi.org/10.1038/s41588-018-0333-3

Jin X, Shearman LP, Weaver DR, Zylka MJ, de Vries GJ, Reppert SM (1999) A molecular mechanism regulating rhythmic output from the suprachiasmatic circadian clock. Cell 96(1):57–68. https://doi.org/10.1016/s0092-8674(00)80959-9

Jones SE, Tyrrell J, Wood AR, Beaumont RN, Ruth KS, Tuke MA, Yaghootkar H, Hu Y, Teder-Laving M, Hayward C, Roenneberg T, Wilson JF, Del Greco F, Hicks AA, Shin C, Yun CH, Lee SK, Metspalu A, Byrne EM, Gehrman PR, Tiemeier H, Allebrandt KV, Freathy RM, Murray A, Hinds DA, Frayling TM, Weedon MN (2016) Genome-wide association analyses in 128,266 individuals identifies new morningness and sleep duration loci. PLoS Genet 12 (8):e1006125. https://doi.org/10.1371/journal.pgen.1006125

Kaasik K, Kivimae S, Allen JJ, Chalkley RJ, Huang Y, Baer K, Kissel H, Burlingame AL, Shokat KM, Ptacek LJ, Fu YH (2013) Glucose sensor O-GlcNAcylation coordinates with phosphorylation to regulate circadian clock. Cell Metab 17(2):291–302. https://doi.org/10.1016/j.cmet.2012.12.017

Kalmbach DA, Schneider LD, Cheung J, Bertrand SJ, Kariharan T, Pack AI, Gehrman PR (2017) Genetic basis of chronotype in humans: insights from three landmark GWAS. Sleep 40(2):zsw048. https://doi.org/10.1093/sleep/zsw048

Katada S, Sassone-Corsi P (2010) The histone methyltransferase MLL1 permits the oscillation of circadian gene expression. Nat Struct Mol Biol 17 (12):1414–1421. https://doi.org/10.1038/nsmb.1961

King DP, Vitaterna MH, Chang AM, Dove WF, Pinto LH, Turek FW, Takahashi JS (1997a) The mouse Clock mutation behaves as an antimorph and maps within the W19H deletion, distal of Kit. Genetics 146 (3):1049–1060

King DP, Zhao Y, Sangoram AM, Wilsbacher LD, Tanaka M, Antoch MP, Steeves TD, Vitaterna MH, Kornhauser JM, Lowrey PL, Turek FW, Takahashi JS (1997b) Positional cloning of the mouse circadian clock gene. Cell 89(4):641–653

Ko CH, Yamada YR, Welsh DK, Buhr ED, Liu AC, Zhang EE, Ralph MR, Kay SA, Forger DB, Takahashi JS (2010) Emergence of noise-induced oscillations in the central circadian pacemaker. PLoS Biol 8(10): e1000513. https://doi.org/10.1371/journal.pbio.1000513

Koike N, Yoo SH, Huang HC, Kumar V, Lee C, Kim TK, Takahashi JS (2012) Transcriptional architecture and chromatin landscape of the core circadian clock in mammals. Science 338(6105):349–354. https://doi.org/10.1126/science.1226339

Kojima S, Hirose M, Tokunaga K, Sakaki Y, Tei H (2003) Structural and functional analysis of 3' untranslated region of mouse Period1 mRNA. Biochem Biophys Res Commun 301(1):1–7. https://doi.org/10.1016/s0006-291x(02)02938-8

Kojima S, Matsumoto K, Hirose M, Shimada M, Nagano M, Shigeyoshi Y, Hoshino S, Ui-Tei K, Saigo K, Green CB, Sakaki Y, Tei H (2007) LARK activates post-transcriptional expression of an essential mammalian clock protein, PERIOD1. Proc Natl Acad Sci U S A 104(6):1859–1864. https://doi.org/10.1073/pnas.0607567104

Kojima S, Shingle DL, Green CB (2011) Post-transcriptional control of circadian rhythms. J Cell Sci 124(Pt 3):311–320. https://doi.org/10.1242/jcs.065771

Kondratov RV, Shamanna RK, Kondratova AA, Gorbacheva VY, Antoch MP (2006) Dual role of the CLOCK/BMAL1 circadian complex in transcriptional regulation. FASEB J 20(3):530–532. https://doi.org/10.1096/fj.05-5321fje

Konopka RJ, Benzer S (1971) Clock mutants of Drosophila melanogaster. Proc Natl Acad Sci U S A 68 (9):2112–2116. https://doi.org/10.1073/pnas.68.9.2112

Kume K, Zylka MJ, Sriram S, Shearman LP, Weaver DR, Jin X, Maywood ES, Hastings MH, Reppert SM (1999) mCRY1 and mCRY2 are essential components of the negative limb of the circadian clock feedback loop. Cell 98(2):193–205

Kurien P, Hsu PK, Leon J, Wu D, McMahon T, Shi G, Xu Y, Lipzen A, Pennacchio LA, Jones CR, Fu YH, Ptacek LJ (2019) TIMELESS mutation alters phase responsiveness and causes advanced sleep phase. Proc Natl Acad Sci U S A 116(24):12045–12053. https://doi.org/10.1073/pnas.1819110116

Kwak E, Kim TD, Kim KT (2006) Essential role of 3'--untranslated region-mediated mRNA decay in circadian oscillations of mouse Period3 mRNA. J Biol Chem 281(28):19100–19106. https://doi.org/10.1074/jbc.M511927200

Kwak Y, Jeong J, Lee S, Park YU, Lee SA, Han DH, Kim JH, Ohshima T, Mikoshiba K, Suh YH, Cho S, Park SK (2013) Cyclin-dependent kinase 5 (Cdk5) regulates the function of CLOCK protein by direct phosphorylation. J Biol Chem 288(52):36878–36889. https://doi.org/10.1074/jbc.M113.494856

Lamia KA, Sachdeva UM, DiTacchio L, Williams EC, Alvarez JG, Egan DF, Vasquez DS, Juguilon H, Panda S, Shaw RJ, Thompson CB, Evans RM (2009) AMPK regulates the circadian clock by cryptochrome phosphorylation and degradation. Science 326 (5951):437–440. https://doi.org/10.1126/science.1172156

Lamia KA, Papp SJ, Yu RT, Barish GD, Uhlenhaut NH, Jonker JW, Downes M, Evans RM (2011) Cryptochromes mediate rhythmic repression of the glucocorticoid receptor. Nature 480(7378):552–556. https://doi.org/10.1038/nature10700

Lande-Diner L, Boyault C, Kim JY, Weitz CJ (2013) A positive feedback loop links circadian clock factor CLOCK-BMAL1 to the basic transcriptional machinery. Proc Natl Acad Sci U S A 110(40):16021–16026. https://doi.org/10.1073/pnas.1305980110

Landgraf D, Wang LL, Diemer T, Welsh DK (2016) NPAS2 compensates for loss of CLOCK in peripheral circadian oscillators. PLoS Genet 12(2):e1005882. https://doi.org/10.1371/journal.pgen.1005882

Lane JM, Vlasac I, Anderson SG, Kyle SD, Dixon WG, Bechtold DA, Gill S, Little MA, Luik A, Loudon A, Emsley R, Scheer FA, Lawlor DA, Redline S, Ray DW, Rutter MK, Saxena R (2016) Genome-wide association analysis identifies novel loci for chronotype in 100,420 individuals from the UK Biobank. Nat Commun 7:10889. https://doi.org/10.1038/ncomms10889

Lane JM, Liang J, Vlasac I, Anderson SG, Bechtold DA, Bowden J, Emsley R, Gill S, Little MA, Luik AI, Loudon A, Scheer FA, Purcell SM, Kyle SD, Lawlor DA, Zhu X, Redline S, Ray DW, Rutter MK, Saxena R (2017) Genome-wide association analyses of sleep disturbance traits identify new loci and highlight shared genetics with neuropsychiatric and metabolic traits. Nat Genet 49(2):274–281. https://doi.org/10.1038/ng.3749

Lee C, Etchegaray JP, Cagampang FR, Loudon AS, Reppert SM (2001) Post-translational mechanisms regulate the mammalian circadian clock. Cell 107 (7):855–867

Lee J, Lee Y, Lee MJ, Park E, Kang SH, Chung CH, Lee KH, Kim K (2008) Dual modification of BMAL1 by SUMO2/3 and ubiquitin promotes circadian activation of the CLOCK/BMAL1 complex. Mol Cell Biol 28 (19):6056–6065. https://doi.org/10.1128/MCB.00583-08

Lee H, Chen R, Lee Y, Yoo S, Lee C (2009) Essential roles of CKIdelta and CKIepsilon in the mammalian circadian clock. Proc Natl Acad Sci U S A 106 (50):21359–21364. https://doi.org/10.1073/pnas.0906651106

Lee Y, Lee J, Kwon I, Nakajima Y, Ohmiya Y, Son GH, Lee KH, Kim K (2010) Coactivation of the CLOCK-BMAL1 complex by CBP mediates resetting of the circadian clock. J Cell Sci 123(Pt 20):3547–3557. https://doi.org/10.1242/jcs.070300

Lee HM, Chen R, Kim H, Etchegaray JP, Weaver DR, Lee C (2011) The period of the circadian oscillator is primarily determined by the balance between casein kinase 1 and protein phosphatase 1. Proc Natl Acad Sci U S A 108(39):16451–16456. https://doi.org/10.1073/pnas.1107178108

Li S, Wang M, Ao X, Chang AK, Yang C, Zhao F, Bi H, Liu Y, Xiao L, Wu H (2013) CLOCK is a substrate of SUMO and sumoylation of CLOCK upregulates the transcriptional activity of estrogen receptor-alpha. Oncogene 32(41):4883–4891. https://doi.org/10.1038/onc.2012.518

Lin ST, Zhang L, Lin X, Zhang LC, Garcia VE, Tsai CW, Ptacek L, Fu YH (2014) Nuclear envelope protein MAN1 regulates clock through BMAL1. Elife 3: e02981. https://doi.org/10.7554/eLife.02981

Liu AC, Tran HG, Zhang EE, Priest AA, Welsh DK, Kay SA (2008) Redundant function of REV-ERBalpha and beta and non-essential role for Bmal1 cycling in transcriptional regulation of intracellular circadian rhythms. PLoS Genet 4(2):e1000023. https://doi.org/10.1371/journal.pgen.1000023

Lopez-Molina L, Conquet F, Dubois-Dauphin M, Schibler U (1997) The DBP gene is expressed according to a circadian rhythm in the suprachiasmatic nucleus and influences circadian behavior. EMBO J 16 (22):6762–6771. https://doi.org/10.1093/emboj/16.22.6762

Lowrey PL, Takahashi JS (2011) Genetics of circadian rhythms in Mammalian model organisms. Adv Genet 74:175–230. https://doi.org/10.1016/B978-0-12-387690-4.00006-4

Lowrey PL, Shimomura K, Antoch MP, Yamazaki S, Zemenides PD, Ralph MR, Menaker M, Takahashi JS (2000) Positional syntenic cloning and functional characterization of the mammalian circadian mutation tau. Science 288(5465):483–492. https://doi.org/10.1126/science.288.5465.483

Luciano AK, Zhou W, Santana JM, Kyriakides C, Velazquez H, Sessa WC (2018) CLOCK phosphorylation by AKT regulates its nuclear accumulation and circadian gene expression in peripheral tissues. J Biol Chem 293(23):9126–9136. https://doi.org/10.1074/jbc.RA117.000773

Ma YT, Luo H, Guan WJ, Zhang H, Chen C, Wang Z, Li JD (2013) O-GlcNAcylation of BMAL1 regulates circadian rhythms in NIH3T3 fibroblasts. Biochem Biophys Res Commun 431(3):382–387. https://doi.org/10.1016/j.bbrc.2013.01.043

Masri S, Patel VR, Eckel-Mahan KL, Peleg S, Forne I, Ladurner AG, Baldi P, Imhof A, Sassone-Corsi P (2013) Circadian acetylome reveals regulation of mitochondrial metabolic pathways. Proc Natl Acad Sci U S A 110(9):3339–3344. https://doi.org/10.1073/pnas.1217632110

Matsumura R, Matsubara C, Node K, Takumi T, Akashi M (2013) Nuclear receptor-mediated cell-autonomous oscillatory expression of the circadian transcription factor, neuronal PAS domain protein 2 (NPAS2). J Biol Chem 288(51):36548–36553. https://doi.org/10.1074/jbc.M113.517235

Matsumura R, Tsuchiya Y, Tokuda I, Matsuo T, Sato M, Node K, Nishida E, Akashi M (2014) The mammalian circadian clock protein period counteracts cryptochrome in phosphorylation dynamics of circadian locomotor output cycles kaput (CLOCK). J Biol Chem 289(46):32064–32072. https://doi.org/10.1074/jbc.M114.578278

Mauvoisin D (2019) Circadian rhythms and proteomics: it's all about post-translational modifications! Wiley Interdiscip Rev Syst Biol Med 11(5):e1450. https://doi.org/10.1002/wsbm.1450

Maywood ES, O'Brien JA, Hastings MH (2003) Expression of mCLOCK and other circadian clock-relevant proteins in the mouse suprachiasmatic nuclei. J Neuroendocrinol 15(4):329–334

McCarthy MJ (2019) Missing a beat: assessment of circadian rhythm abnormalities in bipolar disorder in the genomic era. Psychiatr Genet 29(2):29–36. https://doi.org/10.1097/YPG.0000000000000215
McDearmon EL, Patel KN, Ko CH, Walisser JA, Schook AC, Chong JL, Wilsbacher LD, Song EJ, Hong HK, Bradfield CA, Takahashi JS (2006) Dissecting the functions of the mammalian clock protein BMAL1 by tissue-specific rescue in mice. Science 314 (5803):1304–1308. https://doi.org/10.1126/science.1132430
Menet JS, Pescatore S, Rosbash M (2014) CLOCK: BMAL1 is a pioneer-like transcription factor. Genes Dev 28(1):8–13. https://doi.org/10.1101/gad.228536.113
Meng QJ, Logunova L, Maywood ES, Gallego M, Lebiecki J, Brown TM, Sladek M, Semikhodskii AS, Glossop NRJ, Piggins HD, Chesham JE, Bechtold DA, Yoo SH, Takahashi JS, Virshup DM, Boot-Handford RP, Hastings MH, Loudon ASI (2008a) Setting clock speed in mammals: the CK1 epsilon tau mutation in mice accelerates circadian pacemakers by selectively destabilizing PERIOD proteins. Neuron 58(1):78–88. https://doi.org/10.1016/j.neuron.2008.01.019
Meng QJ, McMaster A, Beesley S, Lu WQ, Gibbs J, Parks D, Collins J, Farrow S, Donn R, Ray D, Loudon A (2008b) Ligand modulation of REV-ERBalpha function resets the peripheral circadian clock in a phasic manner. J Cell Sci 121(Pt 21):3629–3635. https://doi.org/10.1242/jcs.035048
Michael AK, Fribourgh JL, Chelliah Y, Sandate CR, Hura GL, Schneidman-Duhovny D, Tripathi SM, Takahashi JS, Partch CL (2017) Formation of a repressive complex in the mammalian circadian clock is mediated by the secondary pocket of CRY1. Proc Natl Acad Sci U S A 114(7):1560–1565. https://doi.org/10.1073/pnas.1615310114
Miller BH, McDearmon EL, Panda S, Hayes KR, Zhang J, Andrews JL, Antoch MP, Walker JR, Esser KA, Hogenesch JB, Takahashi JS (2007) Circadian and CLOCK-controlled regulation of the mouse transcriptome and cell proliferation. Proc Natl Acad Sci U S A 104(9):3342–3347. https://doi.org/10.1073/pnas.0611724104
Mitsui S, Yamaguchi S, Matsuo T, Ishida Y, Okamura H (2001) Antagonistic role of E4BP4 and PAR proteins in the circadian oscillatory mechanism. Genes Dev 15 (8):995–1006. https://doi.org/10.1101/gad.873501
Morishita Y, Miura D, Kida S (2016) PI3K regulates BMAL1/CLOCK-mediated circadian transcription from the Dbp promoter. Biosci Biotechnol Biochem 80(6):1131–1140. https://doi.org/10.1080/09168451.2015.1136885
Murre C, McCaw PS, Baltimore D (1989) A new DNA binding and dimerization motif in immunoglobulin enhancer binding, daughterless, MyoD, and myc proteins. Cell 56(5):777–783
Nakahata Y, Kaluzova M, Grimaldi B, Sahar S, Hirayama J, Chen D, Guarente LP, Sassone-Corsi P (2008) The NAD+-dependent deacetylase SIRT1 modulates CLOCK-mediated chromatin remodeling and circadian control. Cell 134(2):329–340. https://doi.org/10.1016/j.cell.2008.07.002
Nakahata Y, Sahar S, Astarita G, Kaluzova M, Sassone-Corsi P (2009) Circadian control of the NAD+ salvage pathway by CLOCK-SIRT1. Science 324 (5927):654–657. https://doi.org/10.1126/science.1170803
Nambu JR, Lewis JO, Wharton KA Jr, Crews ST (1991) The Drosophila single-minded gene encodes a helix-loop-helix protein that acts as a master regulator of CNS midline development. Cell 67(6):1157–1167
Nangle SN, Rosensweig C, Koike N, Tei H, Takahashi JS, Green CB, Zheng N (2014) Molecular assembly of the period-cryptochrome circadian transcriptional repressor complex. Elife 3:e03674. https://doi.org/10.7554/eLife.03674
Narasimamurthy R, Hunt SR, Lu Y, Fustin JM, Okamura H, Partch CL, Forger DB, Kim JK, Virshup DM (2018) CK1delta/epsilon protein kinase primes the PER2 circadian phosphoswitch. Proc Natl Acad Sci U S A 115(23):5986–5991. https://doi.org/10.1073/pnas.1721076115
Narumi R, Shimizu Y, Ukai-Tadenuma M, Ode KL, Kanda GN, Shinohara Y, Sato A, Matsumoto K, Ueda HR (2016) Mass spectrometry-based absolute quantification reveals rhythmic variation of mouse circadian clock proteins. Proc Natl Acad Sci U S A 113 (24):E3461–E3467. https://doi.org/10.1073/pnas.1603799113
Noshiro M, Furukawa M, Honma S, Kawamoto T, Hamada T, Honma K, Kato Y (2005) Tissue-specific disruption of rhythmic expression of Dec1 and Dec2 in clock mutant mice. J Biol Rhythms 20(5):404–418. https://doi.org/10.1177/0748730405280195
Ohno T, Onishi Y, Ishida N (2007a) The negative transcription factor E4BP4 is associated with circadian clock protein PERIOD2. Biochem Biophys Res Commun 354(4):1010–1015. https://doi.org/10.1016/j.bbrc.2007.01.084
Ohno T, Onishi Y, Ishida N (2007b) A novel E4BP4 element drives circadian expression of mPeriod2. Nucleic Acids Res 35(2):648–655. https://doi.org/10.1093/nar/gkl868
Oishi K, Fukui H, Ishida N (2000) Rhythmic expression of BMAL1 mRNA is altered in Clock mutant mice: differential regulation in the suprachiasmatic nucleus and peripheral tissues. Biochem Biophys Res Commun 268(1):164–171. https://doi.org/10.1006/bbrc.1999.2054
Oishi K, Miyazaki K, Kadota K, Kikuno R, Nagase T, Atsumi G, Ohkura N, Azama T, Mesaki M, Yukimasa S, Kobayashi H, Iitaka C, Umehara T, Horikoshi M, Kudo T, Shimizu Y, Yano M, Monden M, Machida K, Matsuda J, Horie S, Todo T, Ishida N (2003) Genome-wide expression analysis of mouse liver reveals CLOCK-regulated circadian

output genes. J Biol Chem 278(42):41519–41527. https://doi.org/10.1074/jbc.M304564200

Pacheco-Bernal I, Becerril-Perez F, Aguilar-Arnal L (2019) Circadian rhythms in the three-dimensional genome: implications of chromatin interactions for cyclic transcription. Clin Epigenetics 11(1):79. https://doi.org/10.1186/s13148-019-0677-2

Padmanabhan K, Robles MS, Westerling T, Weitz CJ (2012) Feedback regulation of transcriptional termination by the mammalian circadian clock PERIOD complex. Science 337(6094):599–602. https://doi.org/10.1126/science.1221592

Panda S, Antoch MP, Miller BH, Su AI, Schook AB, Straume M, Schultz PG, Kay SA, Takahashi JS, Hogenesch JB (2002) Coordinated transcription of key pathways in the mouse by the circadian clock. Cell 109(3):307–320. https://doi.org/10.1016/s0092-8674(02)00722-5

Partch CL, Shields KF, Thompson CL, Selby CP, Sancar A (2006) Post-translational regulation of the mammalian circadian clock by cryptochrome and protein phosphatase 5. Proc Natl Acad Sci U S A 103(27):10467–10472. https://doi.org/10.1073/pnas.0604138103

Partch CL, Green CB, Takahashi JS (2014) Molecular architecture of the mammalian circadian clock. Trends Cell Biol 24(2):90–99. https://doi.org/10.1016/j.tcb.2013.07.002

Patke A, Murphy PJ, Onat OE, Krieger AC, Ozcelik T, Campbell SS, Young MW (2017) Mutation of the human circadian clock gene CRY1 in familial delayed sleep phase disorder. Cell 169(2):203–215. e213. https://doi.org/10.1016/j.cell.2017.03.027

Pendergast JS, Friday RC, Yamazaki S (2010) Distinct functions of Period2 and Period3 in the mouse circadian system revealed by in vitro analysis. PLoS One 5(1):e8552. https://doi.org/10.1371/journal.pone.0008552

Preitner N, Damiola F, Lopez-Molina L, Zakany J, Duboule D, Albrecht U, Schibler U (2002) The orphan nuclear receptor REV-ERBalpha controls circadian transcription within the positive limb of the mammalian circadian oscillator. Cell 110(2):251–260

Preussner M, Heyd F (2016) Post-transcriptional control of the mammalian circadian clock: implications for health and disease. Pflugers Arch 468(6):983–991. https://doi.org/10.1007/s00424-016-1820-y

Ralph MR, Menaker M (1988) A mutation of the circadian system in golden hamsters. Science 241(4870):1225–1227. https://doi.org/10.1126/science.3413487

Ramsey KM, Yoshino J, Brace CS, Abrassart D, Kobayashi Y, Marcheva B, Hong HK, Chong JL, Buhr ED, Lee C, Takahashi JS, Imai S, Bass J (2009) Circadian clock feedback cycle through NAMPT-mediated NAD+ biosynthesis. Science 324(5927):651–654. https://doi.org/10.1126/science.1171641

Reddy AB, Karp NA, Maywood ES, Sage EA, Deery M, O'Neill JS, Wong GK, Chesham J, Odell M, Lilley KS, Kyriacou CP, Hastings MH (2006) Circadian orchestration of the hepatic proteome. Curr Biol 16(11):1107–1115. https://doi.org/10.1016/j.cub.2006.04.026

Reick M, Garcia JA, Dudley C, McKnight SL (2001) NPAS2: an analog of clock operative in the mammalian forebrain. Science 293(5529):506–509. https://doi.org/10.1126/science.1060699

Reischl S, Kramer A (2011) Kinases and phosphatases in the mammalian circadian clock. FEBS Lett 585(10):1393–1399. https://doi.org/10.1016/j.febslet.2011.02.038

Reischl S, Vanselow K, Westermark PO, Thierfelder N, Maier B, Herzel H, Kramer A (2007) Beta-TrCP1-mediated degradation of PERIOD2 is essential for circadian dynamics. J Biol Rhythms 22(5):375–386. https://doi.org/10.1177/0748730407303926

Rey G, Cesbron F, Rougemont J, Reinke H, Brunner M, Naef F (2011) Genome-wide and phase-specific DNA-binding rhythms of BMAL1 control circadian output functions in mouse liver. PLoS Biol 9(2):e1000595. https://doi.org/10.1371/journal.pbio.1000595

Ripperger JA, Schibler U (2006) Rhythmic CLOCK-BMAL1 binding to multiple E-box motifs drives circadian Dbp transcription and chromatin transitions. Nat Genet 38(3):369–374. https://doi.org/10.1038/ng1738

Ripperger JA, Shearman LP, Reppert SM, Schibler U (2000) CLOCK, an essential pacemaker component, controls expression of the circadian transcription factor DBP. Genes Dev 14(6):679–689

Robles MS, Humphrey SJ, Mann M (2017) Phosphorylation is a central mechanism for circadian control of metabolism and physiology. Cell Metab 25(1):118–127. https://doi.org/10.1016/j.cmet.2016.10.004

Rutter J, Reick M, Wu LC, McKnight SL (2001) Regulation of clock and NPAS2 DNA binding by the redox state of NAD cofactors. Science 293(5529):510–514. https://doi.org/10.1126/science.1060698

Sahar S, Zocchi L, Kinoshita C, Borrelli E, Sassone-Corsi P (2010) Regulation of BMAL1 protein stability and circadian function by GSK3beta-mediated phosphorylation. PLoS One 5(1):e8561. https://doi.org/10.1371/journal.pone.0008561

Sakakida Y, Miyamoto Y, Nagoshi E, Akashi M, Nakamura TJ, Mamine T, Kasahara M, Minami Y, Yoneda Y, Takumi T (2005) Importin alpha/beta mediates nuclear transport of a mammalian circadian clock component, mCRY2, together with mPER2, through a bipartite nuclear localization signal. J Biol Chem 280(14):13272–13278. https://doi.org/10.1074/jbc.M413236200

Sangoram AM, Saez L, Antoch MP, Gekakis N, Staknis D, Whiteley A, Fruechte EM, Vitaterna MH, Shimomura K, King DP, Young MW, Weitz CJ, Takahashi JS (1998) Mammalian circadian autoregulatory loop: a timeless ortholog and mPer1 interact and negatively regulate CLOCK-BMAL1-induced transcription. Neuron 21(5):1101–1113

Sato TK, Panda S, Miraglia LJ, Reyes TM, Rudic RD, McNamara P, Naik KA, FitzGerald GA, Kay SA, Hogenesch JB (2004) A functional genomics strategy reveals Rora as a component of the mammalian circadian clock. Neuron 43(4):527–537. https://doi.org/10.1016/j.neuron.2004.07.018

Sato TK, Yamada RG, Ukai H, Baggs JE, Miraglia LJ, Kobayashi TJ, Welsh DK, Kay SA, Ueda HR, Hogenesch JB (2006) Feedback repression is required for mammalian circadian clock function. Nat Genet 38 (3):312–319. https://doi.org/10.1038/ng1745

Schmutz I, Ripperger JA, Baeriswyl-Aebischer S, Albrecht U (2010) The mammalian clock component PERIOD2 coordinates circadian output by interaction with nuclear receptors. Genes Dev 24(4):345–357. https://doi.org/10.1101/gad.564110

Shearman LP, Zylka MJ, Weaver DR, Kolakowski LF Jr, Reppert SM (1997) Two period homologs: circadian expression and photic regulation in the suprachiasmatic nuclei. Neuron 19(6):1261–1269

Shearman LP, Sriram S, Weaver DR, Maywood ES, Chaves I, Zheng B, Kume K, Lee CC, van der Horst GT, Hastings MH, Reppert SM (2000) Interacting molecular loops in the mammalian circadian clock. Science 288(5468):1013–1019. https://doi.org/10.1126/science.288.5468.1013

Shi G, Xing L, Liu Z, Qu Z, Wu X, Dong Z, Wang X, Gao X, Huang M, Yan J, Yang L, Liu Y, Ptacek LJ, Xu Y (2013) Dual roles of FBXL3 in the mammalian circadian feedback loops are important for period determination and robustness of the clock. Proc Natl Acad Sci U S A 110(12):4750–4755. https://doi.org/10.1073/pnas.1302560110

Shirogane T, Jin J, Ang XL, Harper JW (2005) SCFbeta-TRCP controls clock-dependent transcription via casein kinase 1-dependent degradation of the mammalian period-1 (Per1) protein. J Biol Chem 280 (29):26863–26872. https://doi.org/10.1074/jbc.M502862200

Siepka SM, Yoo SH, Park J, Song W, Kumar V, Hu Y, Lee C, Takahashi JS (2007) Circadian mutant Overtime reveals F-box protein FBXL3 regulation of cryptochrome and period gene expression. Cell 129 (5):1011–1023. https://doi.org/10.1016/j.cell.2007.04.030

Solt LA, Wang Y, Banerjee S, Hughes T, Kojetin DJ, Lundasen T, Shin Y, Liu J, Cameron MD, Noel R, Yoo SH, Takahashi JS, Butler AA, Kamenecka TM, Burris TP (2012) Regulation of circadian behaviour and metabolism by synthetic REV-ERB agonists. Nature 485(7396):62–68. https://doi.org/10.1038/nature11030

Spengler ML, Kuropatwinski KK, Schumer M, Antoch MP (2009) A serine cluster mediates BMAL1-dependent CLOCK phosphorylation and degradation. Cell Cycle 8(24):4138–4146. https://doi.org/10.4161/cc.8.24.10273

Stahl EA, Breen G, Forstner AJ, McQuillin A, Ripke S, Trubetskoy V, Mattheisen M, Wang Y, Coleman JRI, Gaspar HA, de Leeuw CA, Steinberg S, Pavlides JMW, Trzaskowski M, Byrne EM, Pers TH, Holmans PA, Richards AL, Abbott L, Agerbo E, Akil H, Albani D, Alliey-Rodriguez N, Als TD, Anjorin A, Antilla V, Awasthi S, Badner JA, Baekvad-Hansen M, Barchas JD, Bass N, Bauer M, Belliveau R, Bergen SE, Pedersen CB, Boen E, Boks MP, Boocock J, Budde M, Bunney W, Burmeister M, Bybjerg-Grauholm J, Byerley W, Casas M, Cerrato F, Cervantes P, Chambert K, Charney AW, Chen D, Churchhouse C, Clarke TK, Coryell W, Craig DW, Cruceanu C, Curtis D, Czerski PM, Dale AM, de Jong S, Degenhardt F, Del-Favero J, DePaulo JR, Djurovic S, Dobbyn AL, Dumont A, Elvsashagen T, Escott-Price-V, Fan CC, Fischer SB, Flickinger M, Foroud TM, Forty L, Frank J, Fraser C, Freimer NB, Frisen L, Gade K, Gage D, Garnham J, Giambartolomei C, Pedersen MG, Goldstein J, Gordon SD, Gordon-Smith K, Green EK, Green MJ, Greenwood TA, Grove J, Guan W, Guzman-Parra J, Hamshere ML, Hautzinger M, Heilbronner U, Herms S, Hipolito M, Hoffmann P, Holland D, Huckins L, Jamain S, Johnson JS, Jureus A, Kandaswamy R, Karlsson R, Kennedy JL, Kittel-Schneider S, Knowles JA, Kogevinas M, Koller AC, Kupka R, Lavebratt C, Lawrence J, Lawson WB, Leber M, Lee PH, Levy SE, Li JZ, Liu C, Lucae S, Maaser A, MacIntyre DJ, Mahon PB, Maier W, Martinsson L, McCarroll S, McGuffin P, McInnis MG, McKay JD, Medeiros H, Medland SE, Meng F, Milani L, Montgomery GW, Morris DW, Muhleisen TW, Mullins N, Nguyen H, Nievergelt CM, Adolfsson AN, Nwulia EA, O'Donovan C, Loohuis LMO, Ori APS, Oruc L, Osby U, Perlis RH, Perry A, Pfennig A, Potash JB, Purcell SM, Regeer EJ, Reif A, Reinbold CS, Rice JP, Rivas F, Rivera M, Roussos P, Ruderfer DM, Ryu E, Sanchez-Mora C, Schatzberg AF, Scheftner WA, Schork NJ, Shannon Weickert C, Shehktman T, Shilling PD, Sigurdsson E, Slaney C, Smeland OB, Sobell JL, Soholm Hansen C, Spijker AT, St Clair D, Steffens M, Strauss JS, Streit F, Strohmaier J, Szelinger S, Thompson RC, Thorgeirsson TE, Treutlein J, Vedder H, Wang W, Watson SJ, Weickert TW, Witt SH, Xi S, Xu W, Young AH, Zandi P, Zhang P, Zollner S, eQTLGen Consortium, Adolfsson R, Agartz I, Alda M, Backlund L, Baune BT, Bellivier F, Berrettini WH, Biernacka JM, DHR B, Boehnke M, Borglum AD, Corvin A, Craddock N, Daly MJ, Dannlowski U, Esko T, Etain B, Frye M, Fullerton JM, Gershon ES, Gill M, Goes F, Grigoroiu-Serbanescu M, Hauser J, Hougaard DM, Hultman CM, Jones I, Jones LA, Kahn RS, Kirov G, Landen M, Leboyer M, Lewis CM, Li QS, Lissowska J, Martin NG, Mayoral F, SL ME, AM MI, FJ MM, Melle I, Metspalu A, Mitchell PB, Morken G, Mors O,

Mortensen PB, Muller-Myhsok B, Myers RM, Neale BM, Nimgaonkar V, Nordentoft M, Nothen MM, O'Donovan MC, Oedegaard KJ, Owen MJ, Paciga SA, Pato C, Pato MT, Posthuma D, Ramos-Quiroga JA, Ribases M, Rietschel M, Rouleau GA, Schalling M, Schofield PR, Schulze TG, Serretti A, Smoller JW, Stefansson H, Stefansson K, Stordal E, Sullivan PF, Turecki G, Vaaler AE, Vieta E, Vincent JB, Werge T, Nurnberger JI, Wray NR, Di Florio A, Edenberg HJ, Cichon S, Ophoff RA, Scott LJ, Andreassen OA, Kelsoe J, Sklar P, Bipolar Disorder Working Group of the Psychiatric Genomics C (2019) Genome-wide association study identifies 30 loci associated with bipolar disorder. Nat Genet 51 (5):793–803. https://doi.org/10.1038/s41588-019-0397-8

Steeves TD, King DP, Zhao Y, Sangoram AM, Du F, Bowcock AM, Moore RY, Takahashi JS (1999) Molecular cloning and characterization of the human CLOCK gene: expression in the suprachiasmatic nuclei. Genomics 57(2):189–200. https://doi.org/10.1006/geno.1998.5675

Stein MB, McCarthy MJ, Chen CY, Jain S, Gelernter J, He F, Heeringa SG, Kessler RC, Nock MK, Ripke S, Sun X, Wynn GH, Smoller JW, Ursano RJ (2018) Genome-wide analysis of insomnia disorder. Mol Psychiatry 23(11):2238–2250. https://doi.org/10.1038/s41380-018-0033-5

Stratmann M, Stadler F, Tamanini F, van der Horst GT, Ripperger JA (2010) Flexible phase adjustment of circadian albumin D site-binding protein (DBP) gene expression by CRYPTOCHROME1. Genes Dev 24 (12):1317–1328. https://doi.org/10.1101/gad.578810

Stratmann M, Suter DM, Molina N, Naef F, Schibler U (2012) Circadian Dbp transcription relies on highly dynamic BMAL1-CLOCK interaction with E boxes and requires the proteasome. Mol Cell 48 (2):277–287. https://doi.org/10.1016/j.molcel.2012.08.012

Sumi Y, Yagita K, Yamaguchi S, Ishida Y, Kuroda Y, Okamura H (2002) Rhythmic expression of ROR beta mRNA in the mice suprachiasmatic nucleus. Neurosci Lett 320(1–2):13–16. https://doi.org/10.1016/s0304-3940(02)00011-3

Sun ZS, Albrecht U, Zhuchenko O, Bailey J, Eichele G, Lee CC (1997) RIGUI, a putative mammalian ortholog of the Drosophila period gene. Cell 90(6):1003–1011. https://doi.org/10.1016/s0092-8674(00)80366-9

Takahashi JS (2017) Transcriptional architecture of the mammalian circadian clock. Nat Rev Genet 18 (3):164–179. https://doi.org/10.1038/nrg.2016.150

Takano A, Isojima Y, Nagai K (2004) Identification of mPer1 phosphorylation sites responsible for the nuclear entry. J Biol Chem 279(31):32578–32585. https://doi.org/10.1074/jbc.M403433200

Takeda Y, Kang HS, Angers M, Jetten AM (2011) Retinoic acid-related orphan receptor gamma directly regulates neuronal PAS domain protein 2 transcription in vivo. Nucleic Acids Res 39(11):4769–4782. https://doi.org/10.1093/nar/gkq1335

Tamaru T, Hirayama J, Isojima Y, Nagai K, Norioka S, Takamatsu K, Sassone-Corsi P (2009) CK2alpha phosphorylates BMAL1 to regulate the mammalian clock. Nat Struct Mol Biol 16(4):446–448. https://doi.org/10.1038/nsmb.1578

Tamaru T, Hattori M, Honda K, Nakahata Y, Sassone-Corsi P, van der Horst GT, Ozawa T, Takamatsu K (2015) CRY Drives Cyclic CK2-Mediated BMAL1 phosphorylation to control the mammalian circadian clock. PLoS Biol 13(11):e1002293. https://doi.org/10.1371/journal.pbio.1002293

Tei H, Okamura H, Shigeyoshi Y, Fukuhara C, Ozawa R, Hirose M, Sakaki Y (1997) Circadian oscillation of a mammalian homologue of the Drosophila period gene. Nature 389(6650):512–516. https://doi.org/10.1038/39086

Toh KL, Jones CR, He Y, Eide EJ, Hinz WA, Virshup DM, Ptacek LJ, Fu YH (2001) An hPer2 phosphorylation site mutation in familial advanced sleep phase syndrome. Science 291(5506):1040–1043. https://doi.org/10.1126/science.1057499

Travnickova-Bendova Z, Cermakian N, Reppert SM, Sassone-Corsi P (2002) Bimodal regulation of mPeriod promoters by CREB-dependent signaling and CLOCK/BMAL1 activity. Proc Natl Acad Sci U S A 99(11):7728–7733. https://doi.org/10.1073/pnas.102075599

Trott AJ, Menet JS (2018) Regulation of circadian clock transcriptional output by CLOCK:BMAL1. PLoS Genet 14(1):e1007156. https://doi.org/10.1371/journal.pgen.1007156

Ueda HR, Chen W, Adachi A, Wakamatsu H, Hayashi S, Takasugi T, Nagano M, Nakahama K, Suzuki Y, Sugano S, Iino M, Shigeyoshi Y, Hashimoto S (2002) A transcription factor response element for gene expression during circadian night. Nature 418 (6897):534–539. https://doi.org/10.1038/nature00906

Ueda HR, Hayashi S, Chen W, Sano M, Machida M, Shigeyoshi Y, Iino M, Hashimoto S (2005) System-level identification of transcriptional circuits underlying mammalian circadian clocks. Nat Genet 37 (2):187–192. https://doi.org/10.1038/ng1504

Ukai-Tadenuma M, Kasukawa T, Ueda HR (2008) Proof-by-synthesis of the transcriptional logic of mammalian circadian clocks. Nat Cell Biol 10(10):1154–1163. https://doi.org/10.1038/ncb1775

Ukai-Tadenuma M, Yamada RG, Xu H, Ripperger JA, Liu AC, Ueda HR (2011) Delay in feedback repression by cryptochrome 1 is required for circadian clock function. Cell 144(2):268–281. https://doi.org/10.1016/j.cell.2010.12.019

van der Horst GT, Muijtjens M, Kobayashi K, Takano R, Kanno S, Takao M, de Wit J, Verkerk A, Eker AP, van Leenen D, Buijs R, Bootsma D, Hoeijmakers JH, Yasui A (1999) Mammalian Cry1 and Cry2 are essential for maintenance of circadian rhythms. Nature 398 (6728):627–630. https://doi.org/10.1038/19323

Vitaterna MH, King DP, Chang AM, Kornhauser JM, Lowrey PL, McDonald JD, Dove WF, Pinto LH, Turek FW, Takahashi JS (1994) Mutagenesis and mapping of a mouse gene, Clock, essential for circadian behavior. Science 264(5159):719–725

Vitaterna MH, Selby CP, Todo T, Niwa H, Thompson C, Fruechte EM, Hitomi K, Thresher RJ, Ishikawa T, Miyazaki J, Takahashi JS, Sancar A (1999) Differential regulation of mammalian period genes and circadian rhythmicity by cryptochromes 1 and 2. Proc Natl Acad Sci U S A 96(21):12114–12119. https://doi.org/10.1073/pnas.96.21.12114

Wang J, Mauvoisin D, Martin E, Atger F, Galindo AN, Dayon L, Sizzano F, Palini A, Kussmann M, Waridel P, Quadroni M, Dulic V, Naef F, Gachon F (2017a) Nuclear proteomics uncovers diurnal regulatory landscapes in mouse liver. Cell Metab 25 (1):102–117. https://doi.org/10.1016/j.cmet.2016.10.003

Wang Y, Kuang Z, Yu X, Ruhn KA, Kubo M, Hooper LV (2017b) The intestinal microbiota regulates body composition through NFIL3 and the circadian clock. Science 357(6354):912–916. https://doi.org/10.1126/science.aan0677

Wang Y, Song L, Liu M, Ge R, Zhou Q, Liu W, Li R, Qie J, Zhen B, Wang Y, He F, Qin J, Ding C (2018) A proteomics landscape of circadian clock in mouse liver. Nat Commun 9(1):1553. https://doi.org/10.1038/s41467-018-03898-2

Wilsbacher LD, Sangoram AM, Antoch MP, Takahashi JS (2000) The mouse Clock locus: sequence and comparative analysis of 204 kb from mouse chromosome 5. Genome Res 10(12):1928–1940. https://doi.org/10.1101/gr.10.12.1928

Woo KC, Kim TD, Lee KH, Kim DY, Kim W, Lee KY, Kim KT (2009) Mouse period 2 mRNA circadian oscillation is modulated by PTB-mediated rhythmic mRNA degradation. Nucleic Acids Res 37(1):26–37. https://doi.org/10.1093/nar/gkn893

Woo KC, Ha DC, Lee KH, Kim DY, Kim TD, Kim KT (2010) Circadian amplitude of cryptochrome 1 is modulated by mRNA stability regulation via cytoplasmic hnRNP D oscillation. Mol Cell Biol 30 (1):197–205. https://doi.org/10.1128/MCB.01154-09

Wuarin J, Falvey E, Lavery D, Talbot D, Schmidt E, Ossipow V, Fonjallaz P, Schibler U (1992) The role of the transcriptional activator protein DBP in circadian liver gene expression. J Cell Sci Suppl 16:123–127

Xu Y, Padiath QS, Shapiro RE, Jones CR, Wu SC, Saigoh N, Saigoh K, Ptacek LJ, Fu YH (2005) Functional consequences of a CKIdelta mutation causing familial advanced sleep phase syndrome. Nature 434 (7033):640–644. https://doi.org/10.1038/nature03453

Xu H, Gustafson CL, Sammons PJ, Khan SK, Parsley NC, Ramanathan C, Lee HW, Liu AC, Partch CL (2015) Cryptochrome 1 regulates the circadian clock through dynamic interactions with the BMAL1 C terminus. Nat Struct Mol Biol 22(6):476–484. https://doi.org/10.1038/nsmb.3018

Yan L, Takekida S, Shigeyoshi Y, Okamura H (1999) Per1 and Per2 gene expression in the rat suprachiasmatic nucleus: circadian profile and the compartment-specific response to light. Neuroscience 94(1):141–150

Ye R, Selby CP, Chiou YY, Ozkan-Dagliyan I, Gaddameedhi S, Sancar A (2014) Dual modes of CLOCK:BMAL1 inhibition mediated by Cryptochrome and Period proteins in the mammalian circadian clock. Genes Dev 28(18):1989–1998. https://doi.org/10.1101/gad.249417.114

Yeung J, Naef F (2018) Rhythms of the genome: circadian dynamics from chromatin topology, tissue-specific gene expression, to behavior. Trends Genet 34 (12):915–926. https://doi.org/10.1016/j.tig.2018.09.005

Yin L, Wang J, Klein PS, Lazar MA (2006) Nuclear receptor Rev-erbalpha is a critical lithium-sensitive component of the circadian clock. Science 311 (5763):1002–1005. https://doi.org/10.1126/science.1121613

Yin L, Joshi S, Wu N, Tong X, Lazar MA (2010) E3 ligases Arf-bp1 and Pam mediate lithium-stimulated degradation of the circadian heme receptor Rev-erb alpha. Proc Natl Acad Sci U S A 107 (25):11614–11619. https://doi.org/10.1073/pnas.1000438107

Yoo SH, Ko CH, Lowrey PL, Buhr ED, Song EJ, Chang S, Yoo OJ, Yamazaki S, Lee C, Takahashi JS (2005) A noncanonical E-box enhancer drives mouse Period2 circadian oscillations in vivo. Proc Natl Acad Sci U S A 102(7):2608–2613. https://doi.org/10.1073/pnas.0409763102

Yoo SH, Mohawk JA, Siepka SM, Shan Y, Huh SK, Hong HK, Kornblum I, Kumar V, Koike N, Xu M, Nussbaum J, Liu X, Chen Z, Chen ZJ, Green CB, Takahashi JS (2013) Competing E3 ubiquitin ligases govern circadian periodicity by degradation of CRY in nucleus and cytoplasm. Cell 152(5):1091–1105. https://doi.org/10.1016/j.cell.2013.01.055

Yoshitane H, Takao T, Satomi Y, Du NH, Okano T, Fukada Y (2009) Roles of CLOCK phosphorylation in suppression of E-box-dependent transcription. Mol Cell Biol 29(13):3675–3686. https://doi.org/10.1128/MCB.01864-08

Yoshitane H, Ozaki H, Terajima H, Du NH, Suzuki Y, Fujimori T, Kosaka N, Shimba S, Sugano S, Takagi T, Iwasaki W, Fukada Y (2014) CLOCK-controlled polyphonic regulation of circadian rhythms through canonical and noncanonical E-boxes. Mol Cell Biol 34 (10):1776–1787. https://doi.org/10.1128/MCB.01465-13

Zhang R, Lahens NF, Ballance HI, Hughes ME, Hogenesch JB (2014) A circadian gene expression atlas in mammals: implications for biology and medicine. Proc Natl Acad Sci U S A 111(45):16219–16224. https://doi.org/10.1073/pnas.1408886111

Zhang L, Hirano A, Hsu PK, Jones CR, Sakai N, Okuro M, McMahon T, Yamazaki M, Xu Y, Saigoh N, Saigoh K, Lin ST, Kaasik K, Nishino S, Ptacek LJ, Fu YH (2016) A PERIOD3 variant causes a circadian phenotype and is associated with a seasonal mood trait. Proc Natl Acad Sci U S A 113(11):E1536–E1544. https://doi.org/10.1073/pnas.1600039113

Zhao X, Hirota T, Han X, Cho H, Chong LW, Lamia K, Liu S, Atkins AR, Banayo E, Liddle C, Yu RT, Yates JR 3rd, Kay SA, Downes M, Evans RM (2016) Circadian amplitude regulation via FBXW7-targeted REV-ERBalpha degradation. Cell 165(7):1644–1657. https://doi.org/10.1016/j.cell.2016.05.012

Zheng B, Larkin DW, Albrecht U, Sun ZS, Sage M, Eichele G, Lee CC, Bradley A (1999) The mPer2 gene encodes a functional component of the mammalian circadian clock. Nature 400 (6740):169–173. https://doi.org/10.1038/22118

Zhou M, Kim JK, Eng GW, Forger DB, Virshup DM (2015) A Period2 phosphoswitch regulates and temperature compensates circadian period. Mol Cell 60 (1):77–88. https://doi.org/10.1016/j.molcel.2015.08.022

Zylka MJ, Shearman LP, Levine JD, Jin X, Weaver DR, Reppert SM (1998a) Molecular analysis of mammalian timeless. Neuron 21(5):1115–1122

Zylka MJ, Shearman LP, Weaver DR, Reppert SM (1998b) Three period homologs in mammals: differential light responses in the suprachiasmatic circadian clock and oscillating transcripts outside of brain. Neuron 20(6):1103–1110

Circadian Clocks, Sleep, and Metabolism 2

Nora Nowak, Audrey Rawleigh, and Steven A. Brown

Abstract

A molecular circadian clock exists not only in the brain, but also in most cells of the body. Research over the past two decades has demonstrated that it directs daily rhythmicity of nearly every aspect of metabolism. It also consolidates sleep-wake behavior each day into an activity/feeding period and a sleep/fasting period. Otherwise, sleep-wake states are mostly controlled by hypothalamic and thalamic regulatory circuits in the brain that direct overall brain state. Recent evidence suggests that hypothalamic control of appetite and metabolism may be concomitant with sleep-wake regulation, and even share the same control centers. Thus, circadian control of metabolic pathways might be overlaid by sleep-wake control of the same pathways, providing a flexible and redundant system to modify metabolism according to both activity and environment.

Nora Nowak and Audrey Rawleigh contributed equally with all other contributors.

N. Nowak
ETH Zurich, Department of Chemistry and Applied Biosciences, Zurich, Switzerland

University of Zurich, Institute of Pharmacology and Toxicology, Zurich, Switzerland

A. Rawleigh · S. A. Brown (✉)
University of Zurich, Institute of Pharmacology and Toxicology, Zurich, Switzerland
e-mail: steven.brown@pharma.uzh.ch

Keywords

Sleep · Circadian · Metabolism · Hypothalamus · Homeostasis · Sleep disorder

2.1 Introduction

Modern society has allowed humans easy access to light, food, transportation, and entertainment, twenty-four hours per day. Perhaps, increased productivity and hedonistic pleasure have resulted, but also increased shift work, social jetlag, sleep loss, and an epidemic of obesity, type 2 diabetes (T2D), and associated metabolic syndrome (Roenneberg et al. 2012; Coomans et al. 2015; Kelly et al. 2018). At the same time, there has also been an increase in other aspects of pathology such as psychological disorders, cancers, immune system dysregulation, and gastrointestinal diseases (Sahar and Sassone-Corsi 2009; Asarnow et al. 2013; Castanon-Cervantes et al. 2010; Codoñer-Franch and Gombert 2018). In this chapter, we focus on the relationship between circadian clocks, sleep, and metabolism, and the consequences of these connections for modern health. The circadian clockwork, the sleep homeostat, and their regulatory networks have been studied considerably throughout the past decades, and there has also been extensive work tying each of these processes to metabolism. Despite this, very few studies exist that focus on the interplay between circadian and sleep

O. Engmann, M. Brancaccio (eds.), *Circadian Clock in Brain Health and Disease*, Advances in Experimental Medicine and Biology 1344, https://doi.org/10.1007/978-3-030-81147-1_2

processes in terms of metabolic regulation, function, and pathology. We propose that by looking through the lenses of both chronobiology and sleep science together, fresh insights may be found which will further the understanding and development of novel strategies for metabolic health.

2.2 The Circadian Clock

2.2.1 Clock Molecules and Circuits

2.2.1.1 The Molecular Pacemaker

A circadian clock temporally controls virtually every aspect of the cellular function, including energy balance, macromolecular synthesis, and signaling. Circadian mechanisms and outputs have been reviewed elsewhere previously (Brown 2016; O'Neill et al. 2013), and we outline briefly only the main transcriptional mechanism for this review. At the core of the molecular clock in mammals are two transcriptional-translational feedback loops in which the protein products inhibit their own transcription. Conventionally, this mechanism is separated into two "limbs". In the positive limb, circadian locomotor output cycles kaput protein (CLOCK) and aryl hydrocarbon receptor nuclear translocator-like protein 1 (BMAL1), form heterodimers that bind to cis-acting DNA elements (E-boxes) to activate transcription of *Period (Per1-3)* and *Cryptochrome (Cry1-2)* genes, which are translated into proteins in the cytoplasm. These proteins form the negative limb, in which PER and CRY proteins then are transported from the cytoplasm to the nucleus and inhibit their own transcription. These positive and negative limbs are further interwoven by a connected molecular circuit in which the gene encoding the nuclear orphan receptor REV-ERBα is induced by CLOCK/BMAL1, and REVERBα itself is a transcriptional repressor of *Bmal1* while other nuclear receptors binding the same site (RORs) activate it.

2.2.1.2 The Master Pacemaker

It is thought that virtually every mammalian cell possesses a clock of its own (Balsalobre et al. 1998). In mammals, a master pacemaker is necessary to keep all of these clocks synchronously tracking geophysical time; this is the job of the suprachiasmatic nucleus (SCN). The SCN are bilateral nuclei of about 20,000 neurons located in the hypothalamus of the brain. They integrate time of day and environmental lighting cues by input received from the eyes via the retinohypothalamic tract (Moore and Lenn 1972), and via a web of direct and indirect cues, they then coordinate all cellular clocks in the body (Buijs and Kalsbeek 2001). Each cell of the SCN has an autonomous clock, and these clocks are coupled together via synaptic connections, gap junctions, and neuropeptidergic signaling to create a precisely oscillating entity (Brown 2016). In turn, the SCN synchronizes body clocks through electrical and endocrine mechanisms, food timing and body temperature, and various other signaling pathways to peripheral cells (Albrecht 2012). Studies have shown that the sleep-wake cycle, activity, and feeding behavior are under the control of the SCN, as well as hormone secretions and many other aspects of physiology (Albrecht and Eichele 2003).

2.2.1.3 Peripheral Clocks

Peripheral clocks refer to any cellular or organ clock outside of the SCN, including elsewhere in the brain. Peripheral clocks are capable of integrating phase information with respect to a stimulus (either at a cellular or organ level), and integrating this information to time output—generally at a clock phase slightly later than in SCN, as demonstrated nicely in young male baboons (Mure et al. 2018). One example of this is glucocorticoid signaling. Via the hypothalamic-pituitary-adrenal axis, the SCN triggers the release of glucocorticoids from the adrenal glands in a time-of-day-dependent manner (Ramamoorthy and Cidlowski 2016). Once released into the blood stream, glucocorticoids act as a Zeitgeber (German for "time giver") to peripheral clocks by inducing clock gene expression, which in turn alters oscillations in these cells. Importantly, the SCN is not the only Zeitgeber for peripheral clocks. For example, transcription in the liver is not only driven by the clock, but also by the timing of food intake (Vollmers et al. 2009). When food is consistently given at an abnormal time, the peripheral clocks become uncoupled from the central ones (Damiola et al. 2000). It is likely that a number of similar cues exist, providing avenues by which different signals such as metabolic and immune state might influence circadian timing. Thus, although the circadian clock circuitry itself is cell-autonomous and robustly self-sustained, nevertheless it is systemically coordinated across cells and tissues. Even in the absence of the SCN, experimental models show that clocks in different organs retain coherence via signals that are currently unknown (Saini et al. 2013).

2.2.2 The Circadian Clock and Metabolism

One of the primary functional outputs of the circadian clock is metabolic regulation: virtually all aspects of metabolism exhibit daily oscillations, which persist even in constant environmental conditions (Brown 2016). Recently, there has been a growing interest about the interplay between metabolism and the circadian clock, and for compelling reasons. The circadian clock exercises its metabolic control at every level of physiology from individual cells to the organism as a whole, but at the same time metabolic state can influence circadian timing. This regulation may be thought of as the clock anticipating and preparing the body so that it can respond in a timely manner to predictable stimuli, such as feeding. In turn, the timing of such stimuli can feed back to affect clock time and refine such prediction. In this section, the ways in which the circadian clock impacts metabolic processes and regulation will be broken up into four main categories: cellular control, system-wide or organ-specific control, neuronal or central control, and behavioral control.

2.2.2.1 Cellular Control

There is an overwhelming amount of evidence that supports the coordination of cellular energy metabolism by the circadian clock. Some examples of how the clock exercises its metabolic control include, but are not limited to: regulation of transcription and metabolite levels (Eckel-Mahan et al. 2012; Nakahata et al. 2009), (Zhang et al. 2015) integration of nutrient sensors and nuclear receptors with the circadian clock (Schmutz et al. 2010; Asher et al. 2008; Lee and Kim 2013), and mitochondrial respiration (Schmitt et al. 2018). A diversity of mechanisms has been elucidated to explain this extensive control.

At the level of transcription, it has been reported that around 10% of transcripts in any given tissue are regulated in a circadian manner. A significant portion of these in all tissues relate to metabolic function (Mure et al. 2018). In individual cellular compartments, this proportion can be even higher. For example, 67% of synaptic mRNA display biphasic circadian oscillations, and the peak preceding dawn is entirely related to metabolism and mitochondrial function (Noya et al. 2019).

An example of a post-translational modification under clock control is seen in mitochondrial bioenergetics and morphology. Daily rhythms in mitochondrial fission are dependent on the circadian phosphorylation of DRP1 and this in turn controls cellular oxygen consumption and ATP levels, as they correlate with the morphological state of mitochondria (Schmitt et al. 2018). Phosphorylation control of circadian metabolism is by no means limited to mitochondrial function. A study of liver circadian phosphoproteomics revealed control of most major cellular signaling pathways (Robles et al. 2017), thereby regulating both cellular energetics and xenobiotic metabolism in rhythmic fashion. Lipid levels also show dynamic circadian regulation. One study looked at temporal dynamics in membrane lipids of the nucleus and mitochondria of the mouse liver, and found that lipid species regulation is driven by feeding time and the circadian clock (Aviram et al. 2016). Another study demonstrated that circadian variations in lipid metabolites are independent of feeding, because they also occur in cultured myotubes. This suggests cell-autonomous regulation in diurnal lipid profiles, and this control was dependent upon the local circadian clock in these cells (Loizides-Mangold et al. 2017).

Broadly speaking, multiple cellular circuits have been characterized that at least partially explain widespread circadian control of metabolism. Glucocorticoid hormone-dependent gene expression certainly explains a portion of this control. It was also found that CRY1 and 2 repress glucocorticoid receptor activation, adding an additional layer of complexity in this regulation (Lamia et al. 2011).

Another major axis is the level of redox cofactors such as NAD^+ and NADH. This control occurs through circadian clock regulation of the rate-limiting enzyme nicotinamide phosphoribosyltransferase, a key enzyme within the NAD^+ salvage pathway (Nakahata et al. 2009). An additional important pathway in the circadian regulation of metabolism centers on AMP-Dependent Protein Kinase (AMPK), which regulates ATP production (Jordan and Lamia 2013). Additionally, the circadian clock protein BMAL1 not only plays a crucial role in circadian transcription, but also in translation and coupling the mTOR signaling pathway to the circadian clock (Lipton et al. 2015). Finally, clock proteins can also act together with factors that modify chromatin, thereby globally orchestrating transcriptional activation and repression across families of related genes. These chromatin modifying factors include histone acetylases and deacetylases, methyltransferases and demethylases, and nucleosome remodeling complexes, among others (Brown 2016; Masri et al. 2015). For example, the sirtuin class of histone deacetylases is regulated in circadian fashion via its NAD cofactor (Eckel-Mahan et al. 2012; Asher et al. 2008), and circadian transcriptional regulation of the deacetylase HDAC3 and the REV-ERB and ROR transcription factors controls a large swath of fatty acid metabolism (Feng et al. 2011). Another clock protein, PER2, coregulates nuclear receptor-mediated transcription through its interaction with PPARα and other nuclear hormone receptors (Schmutz et al. 2010). Importantly, for each of these mechanisms, feedback control from metabolic state to circadian clock function has also been characterized. Levels of NAD+ also control the deacetylation of the clock protein PER2 (Asher et al. 2008), and AMPK is able to phosphorylate the cryptochrome circadian proteins (Jordan and Lamia 2013). Thus, redox and cellular energetics can directly influence the core transcription-translation feedback loop that drives the circadian clock.

A second, more speculative class of control occurs via possible "non-canonical" circadian timing mechanisms based entirely on post-translational idioms. Although a post-translational circadian clock mechanism is well-established in bacteria (Nakajima et al. 2005), so far it has been documented in mammalian cells for only a limited family of redox enzymes (peroxiredoxins), and operates via an unknown mechanism (O'Neill and Reddy 2011; Kil et al. 2015). Nevertheless, given recently reported widespread circadian control of metabolism in mammalian cells and tissues genetically modified to destroy the "canonical" mechanism, it is possible that post-translational clock mechanisms provide important additional layers of control.

2.2.2.2 Organ-Specific Control

The circadian clock coordinates peripheral tissues so that they are able to carry out appropriate metabolic responses. One way this is carried out is through clock-controlled genes (CCGs), which regulate various tissue-specific functions in different tissues or organs (Korenčič et al. 2014). For example, gluconeogenesis and glycogenolysis are promoted in the liver during sleep (fasting time) and glycogen/cholesterol synthesis is promoted during wake (feeding time). Examining the liver metabolome, amino acids, carbohydrates, nucleotides, lipids, cofactors, vitamins, and xenobiotics, they all display rhythmicity under the control of the circadian clock transcriptional machinery (Eckel-Mahan et al. 2012). Similarly, the circadian clock is of imperative importance in regulating glucose sensing and insulin secretion in the pancreas, and loss of these clocks (even specifically in beta-cells) leads to glucose intolerance (Perelis et al. 2015; Sadacca et al. 2011). Another important peripheral clock for mammalian metabolism is adipose tissue. Adipocytes store energy as triglycerides when the body has an excess. They also act as a regulator of triglycerides in the blood stream, and this regulation is compromised when adipocytes lack a functional clock and results in obesity and a defect in the adipocyte-hypothalamic axis (Paschos et al. 2012). Finally, circadian rhythms in human skeletal myotubes have been reported to have self-sustained rhythms *in vitro*, and skeletal muscle in general plays a massive role in whole body glucose homeostasis as it is the principal organ responsive to insulin. The circadian clock might also be implicated in muscle myokine secretion, which is also important in glucose homeostasis (Perrin et al. 2015).

As mentioned earlier, for circadian rhythms in all of these peripheral tissues, the SCN is not the only Zeitgeber. Timing of food intake can also entrain peripheral clocks, and in mammals is even dominant to light after about a week of timed feeding. Indeed, the majority of oscillating mouse liver gene expression was recently found to be controlled by rhythmic food intake (Greenwell et al. 2019). When the timing of food intake was arrhythmic, more than 70% of cycling genes were lost its rhythmicity (Greenwell et al. 2019). The pancreas also shifts its clock time along with the liver, heart, kidney, and muscle in response to feeding in the inactive period (Damiola et al. 2000).

The mechanisms by which food or feeding entrains peripheral circadian clocks have not yet been fully elucidated, but likely include a variety of cues including temperature (Brown et al. 2002) and redox state (Reinke and Asher 2019). Post-transcriptional mechanisms likely play an important role: for example, the circadian-implicated RNA-binding protein non-POU domain-containing octamer-binding protein (NONO), which regulates the pre-mRNA processing of liver circadian genes in response to glucose (Benegiamo et al. 2018); or poly-ADP ribosyltransferase, which modifies clock factors in NAD+-dependent manner (Asher et al. 2010).

2.2.2.3 Central Control by Activity and Behavior

At the level of the whole organism, glucocorticoid hormones, insulin, and appetite hormones play key roles in the regulation of rhythms in activity and behavior (Brown 2016). It has been proposed that the molecular circadian clock acts as a metabolic rheostat, and circadian regulation of glucose metabolism has been extensively reviewed (Qian and Scheer 2016). More broadly, however, the control of feeding behavior and metabolism by hypothalamic circuits themselves plays an essential part in the circadian control of metabolism (Kalsbeek and Fliers 2013). Glucocorticoid hormones, whose secretion is commanded by the hypothalamic-pituitary-adrenal axis, have been discussed earlier in this chapter. Glucocorticoids show daily oscillations, ultradian rhythms, and are also secreted in response to acute stress (Leliavski et al. 2015), playing a major circadian role in anticipating metabolic requirements imposed by food intake (Oster et al. 2017). An additional facet is provided by appetite itself, which is also clock-controlled and in humans, peaks in the evening before sleep and fasting, as opposed to in the morning following an extended fasting period (Scheer et al. 2013). Some of the main hormones at play here include leptin, ghrelin, cholecystokinin, and insulin. Leptin is released mainly from adipocytes and binds to its receptor in the hypothalamus; this signaling is essential for this hormone's suppressive effects on feeding (Hsuchou et al. 2013). It has been proposed that obesity and leptin resistance can disrupt circadian regulation, as well as the reverse (Hsuchou et al. 2013). It was recently established in mice that the clock of energy-sensing AgRP neurons mediates transcriptional responses to leptin to help align appetite behaviors to the sleep-wake cycle (Cedernaes et al. 2019a). Ghrelin, an orexigenic peptide hormone essential for appetite stimulation, has recently been found to oscillate in humans (Qian et al. 2019). This hints at a neuroendocrine-mediated circadian variation in hunger, perhaps involving the entrainment of the stomach cells which secrete ghrelin (Qian et al. 2019). Further information regarding the circadian regulation of appetite behavior with respects to nutrient state and sleep-wake behavior has recently been reviewed (Cedernaes et al. 2019b).

2.2.3 Pathophysiology Related to the Circadian Clock and Metabolism

Disruption to the circadian clock—for example via sleep deprivation, jetlag, or diet—can have numerous negative effects on the circadian coordination of metabolic systems (Bass and Takahashi 2010). Some examples relevant to human disease include, but are certainly not limited to cancer and tumor development, obesity and related comorbidities, death rate from cardiovascular disease and stroke, disrupted menstrual cycles, night time asthma, abnormal cortisol rhythm in Cushing's syndrome, and some psychiatric disorders (Arendt 2010; Akerstedt 1990; Shilts et al. 2018; Ramos-Lopez et al. 2018). Metabolic syndrome describes an increased risk of diabetes, stroke, and heart disease due to a collection of risk factors such as obesity, high blood sugar, cholesterol, and high blood pressure (Brown 2016). The link between metabolic syndrome and associated metabolic diseases with circadian dysfunction has been evident since it was found that mice with a deficient clock gene have obesity and metabolic syndrome including hyperleptinemia, hyperlipidemia, hepatic steatosis, hyperglycemia, and hypoinsulinemia (Turek et al. 2005). The interplay and cross-talk between the circadian clock and metabolism have been reviewed recently and often, and it is important to note that it is a reciprocal relationship (Reinke and Asher 2019).

A simple example to illustrate the importance of the clock to metabolism was demonstrated in a study where high-fat food was fed to mice at either a normal or an inappropriate circadian time. The mice who received food at an inappropriate time (the inactive phase) gained significantly more fat than mice fed at a normal time (Arble et al. 2009). Conversely, when feeding is restricted to a normal time (the active phase), it promotes synchrony with circadian rhythms and actually prevents obesity (Hatori et al. 2012). In normal or homeostatic conditions, metabolic physiology is driven by the clock (Bass and Takahashi 2010). However, when shift work or high-fat feeding for example disrupts either system, this disruption in metabolic pathways leads to dampening and lengthening of circadian oscillations (Kohsaka et al. 2007). There is also evidence that the correct timing of eating applies to humans, and has recently been reviewed (Beccuti et al. 2017). Such approaches might help in the prevention or treatment of obesity, diabetes, metabolic syndrome, and many other metabolic dysregulations, although more long-term and large-scale clinical trials are necessary to clarify and optimize this treatment potential.

Research over the past decade has placed circadian dysfunction as a strong possible contributor to diabetic pathology. Normally, pancreatic β-cell located in the islets of Langerhans operates to secrete insulin in response to food intake, and abnormalities such as a reduction in β-cell mass is considered to be the main cause of T2D (Sun and Han 2019). However, defects in islet function are also linked to circadian clock perturbations, since the β-cell clock coordinates transcription and eventual insulin release (Perelis et al. 2015). The intrinsic clock regulates many cellular processes that are crucial to normal β-cell function including glucose-sensing, substrate metabolism, mitochondrial function, stress response, and insulin secretion via exocytosis and proliferation (Lee et al. 2018). In reverse, circadian period length in cells from human diabetic subjects is inversely correlated with HbA_{1c} values, a measure of chronic blood sugar levels and hence diabetic severity (Sinturel et al. 2019).

Long-term epidemiological studies have shown that prolonged desynchrony between circadian clock and environment is demonstrably deleterious not only to metabolic syndrome and diabetes (Pan et al. 2011), but also to many other aspects of health. Chronic jet lag is associated with increased risk of cancer (Shilts et al. 2018; Greene 2012). Shift work in nursing is one of the most prevalent examples of circadian misalignment and internal desynchrony. It is known that shift work is associated with metabolic syndrome and cancer (Brum et al. 2015; Schernhammer et al. 2006). Mechanistically, night shift work affected gene expression in peripheral blood mononucleated cells and circadian alignment in core body temperature, peak cortisol, and melatonin onset compared to day shift work (Resuehr et al. 2019), suggesting that shiftwork might lead to circadian desynchrony among internal organs. Metabolomics studies of simulated shiftwork have provided further evidence for this idea (Kervezee et al. 2019; Skene et al. 2018). It has also been shown that the gut microbiota play a key role in this equilibrium. When gut microbiota were eradicated via antibiotics, these mice did not develop obesity or glucose intolerance (Thaiss et al. 2014), suggesting that they were spared at least some aspects of metabolic syndrome. Thus, the ill effects of circadian desynchrony might also be a problem of dysbiosis.

Recent studies have suggested that even simple chronotype—an individual's timing in their sleep-wake schedules and circadian physiology—may affect metabolic health in fundamental ways. Morning-types have earlier timing and evening-types have a later timing in their circadian and sleep-wake physiology, and most people fall somewhere in between these two groups. Surprisingly, one study found that evening types are more prone to diabetes, metabolic syndrome, and sarcopenia (the loss of skeletal muscle mass and strength with aging) (Yu et al. 2015). A second study found that an evening chronotype is associated with diabetes and also a greater all-cause mortality and cardiovascular disease mortality (Knutson and von Schantz 2018). Even for "normal" chronotypes, weekend schedules often differ significantly from weekday ones due to social activities and obligations, a phenomenon called "social jetlag". This creates a shift every week, which disrupts both the circadian and sleep systems. In rats, it was found that social jet lag altered cholesterol, elevating the risk of metabolic syndrome and increasing appetite for fat-rich and carbohydrate heavy food (Espitia-Bautista et al. 2017). It has also been suggested that people with evening chronotypes who work regular hours during the week are at an increased risk of social jet lag and T2D since their endogenous schedule is later (Stenvers et al. 2019; Vetter et al. 2015). Thus, awareness of one's chronotype could be one strategy to preventively combat metabolic disorders, for example by adjusting daily schedules.

Overall, much has been discovered in the past few decades about how the circadian clock might contribute to health and disease. Above, we have discussed extensively how circadian rhythms might themselves be important for health. Equally important, however, and beyond the scope of this review are circadian effects upon drug delivery, due either to circadian pharmacokinetic effects (daily changes in drug metabolism and excretion) or circadian pharmacodynamics (daily changes in target susceptibility). The same drug may be more effective when taken at one time of day, regardless if this was considered during the development of the drug, and many examples are included in another review (Cederroth et al. 2019).

2.3 Sleep

Sleep is both one of the major outputs of the circadian clock and an important recuperative neurobiological process independently regulated and essential for health and well-being. However, distinct functions of sleep are still poorly understood and the question "Why do we need to sleep?" is difficult to answer. Nevertheless, there are several hypotheses about the functions of sleep. Apart from sleep acting as an important mechanism for brain plasticity and cognitive functions (Gorgoni et al. 2013; Puentes-Mestril et al. 2019; Tononi and Cirelli 2020), there are clear indications that sleep has a fundamental impact on metabolism.

2.3.1 *Sleep Architecture and Regulation*

In brief, mammalian sleep is categorized into different sleep stages based on types of cortical neural oscillations, and consists of cycles of alternating rapid-eye movement (REM) sleep and nonrapid eye movement (NREM) sleep. REM sleep is characterized not only by rapid eye movements, but also by a very low muscle tonus throughout the body. In contrast, brain activity during REM sleep is comparable with wakefulness, showing high frequency and low voltage waves. REM sleep occurs primarily during the second half of the night and it is associated with dreaming. NREM sleep, in contrast, occurs predominantly during the first half of the night and is characterized mainly by brain waves of lower frequency (Rechtschaffen and Kales 1968). It is therefore also called slow wave sleep (SWS). Dreaming may also occur during NREM sleep. In humans, NREM sleep is further divided into N1, N2, and N3 sleep. N3 sleep is specified by high-amplitude brain waves of 0–3 Hz and is commonly referred to as deep sleep; N1 and N2 sleep are gradual transitions from wakefulness to deep sleep.

How does the brain control sleep and wakefulness? In broad theoretical terms, sleep-wake cycles are known to be driven by two main "processes": a homeostatic process and a circadian one. Sleep propensity grows with increasing time awake. When it reaches an upper threshold, sleep onset occurs and at a lower threshold, awakening is induced. This hourglass-like mechanism defines the homeostatic process. However, the propensity levels sufficient to trigger wake and sleep vary with time of day: these thresholds are under circadian control. In this way, both circadian and homeostatic influences can contribute additively to sleep duration and intensity. SWS is well predicted by duration of wakefulness at all circadian phases, leading some to suggest that SWS is determined mostly by homeostatic factors (Borbély 1982; Borbély et al. 2016). Nevertheless, both REM and SWS are altered in mice lacking core clock genes, adding confusion to this picture (Laposky et al. 2005; Mang et al. 2016).

2.3.2 *Sleep Molecules and Circuits*

At a molecular level, in contrast to the circadian clock relatively little is known about the workings of the sleep homeostat. Early studies hypothesized that specific molecules ("somnogens") might accumulate with time awake and thereby drive sleep (Rosenbaum 1892). Interestingly, one of these molecules, adenosine, is also a metabolic byproduct -- a topic that we discuss below as a connection between sleep and metabolism. Although adenosine is well established to promote sleep and may accumulate in the brain during sleep deprivation (Leenaars et al. 2018), nevertheless its possible role as the principal molecular "currency" of sleep need remains ambiguous. Beyond adenosine, various sleep-promoting as well as wake-promoting neurotransmitters have been identified. Examples of sleep-promoting molecules are gamma-aminobutyric acid (GABA), galanin, growth hormone releasing hormone (GHRH), and also cytokines. Wakefulness is promoted for example by acetylcholine, norepinephrine (Berridge et al. 2012), glutamate, histamine, serotonin, and orexins (Berridge et al. 2012; Oh et al. 2019).

Recent studies have suggested that a general increase in phosphorylation of specific synaptic proteins might serve the same somnogenic function (Diering et al. 2017; Wang et al. 2018), though this simple idea is complicated by the fact that other phosphorylations in the same proteins are also driven oppositely and in circadian fashion (Bruning et al. 2019a). Other recent studies have focused upon the molecular determinants of sleep oscillations such as slow waves, and concluded that cortical potassium channels play key roles (Muheim et al. 2019; Tatsuki et al. 2016). Transcriptional regulation, perhaps controlled via MAP kinase signaling, is likely also implicated (Mikhail et al. 2017). Other yet unelucidated pathways likely exist.

At a circuit level, a considerable amount is known about the control of sleep and wake. Origins of major sleep oscillations have been proposed. For example, although not the only mechanism to generate slow waves, a thalamocortical circuit certainly plays a major role (Crunelli and Hughes 2010). More broadly, several brain nuclei mostly in the hypothalamus are involved in the regulation of sleep and wakefulness (Kalia 2006). In the lateral hypothalamus, so-called arousal centers such as the tuberomammillary nucleus (TMN) and raphe nuclei send neurotransmitters to the cerebral cortex, promoting wakefulness. Closely connected to the arousal centers is the ventrolateral preoptic nucleus (VLPO), which counteracts them and thus promotes sleep. Switching between inactivation and activation of wake- and sleep-promoting nuclei regulates sleep and wakefulness (McGinty and Szymusiak 2000; Lu et al. 2006). A similar "flip-flop" mechanism has been proposed for the switch between REM and NREM sleep involving the sublaterodorsal nucleus (SLD) and the ventrolateral part of the periaqueductal gray matter (vlPAG) in the brainstem (Lu et al. 2006). A recent study suggests that two groups of neurons in the dorsomedial nucleus of the hypothalamus (DMH) project to the preoptic area and to the raphe pallidus area are involved in the REM sleep switch as well (Chen et al. 2018).

2.3.3 Sleep and Metabolism

Many of the molecules mentioned above as somnogens are also involved in the regulation of metabolic functions such as energy homeostasis, hormone regulation, and immune response. Brain regions controlling metabolic functions are also located close to sleep and arousal centers. This spatial arrangement makes it appear likely that neuronal circuits connecting both are an important link between sleep and its metabolic functions, a subject to which we turn next.

2.3.3.1 Cellular Control: Restoration of Brain Energy

It has been suggested that sleep plays an important role in the restoration of brain energy. In particular, cerebral energy metabolism and its relation to sleep have been reviewed recently. According to these arguments, during wakefulness glucose is the main cerebral fuel and brain metabolism is mainly glycolytic (Aalling et al. 2018). (N.B. Whether neurons are directly using this glucose, or rather burning lactate provided to them by astrocytes, is an interesting question that has also been a subject of recent discussion, though not yet in the context of sleep (Machler et al. 2016)). During sleep, brain levels of glucose increase and lactate levels decrease. Correspondingly, metabolic rates of glucose drop significantly during sleep. The metabolic cost of sleep for the brain is probably almost the same during sleep as quiet wakefulness—a mere 5% difference in whole-body respiratory quotient in humans (DiNuzzo and Nedergaard 2017). These changes therefore indicate a transition from glycolysis to oxidative metabolism during sleep. Moreover, enhanced lactate efflux from the brain during sleep has also been measured (Aalling et al. 2018). Interestingly, glucose and lactate are both involved in sleep regulation in the brain as well. Extracellular glucose levels, for example, have been shown to promote sleep by inhibiting orexinergic neurons in the lateral hypothalamus (Burdakov et al. 2006) and by activating sleep-promoting GABAergic neurons in VLPO

(Varin et al. 2015a, b). They also act by inhibiting wake-promoting orexinergic neurons (Yamanaka et al. 2003). In contrast, an association between elevated extracellular levels of lactate and the activation of noradrenergic neurons in the locus coeruleus, promoting wakefulness, has been reported (Tang et al. 2014).

In addition to these potentially direct connections between metabolism and sleep, various studies suggest that other neurotransmitters and signaling molecules might coordinate both processes in synchrony. For example, norepinephrine (NE) not only stimulates arousal centers, but also promotes aerobic glycolysis (Dienel and Cruz 2016). NE might thus represent a link between sleep regulation and cerebral energy metabolism. Furthermore, there is evidence for specific biosynthetic pathways to be up-regulated in the brain during sleep. Several studies reported alterations in glycogen storage, and gene expression experiments in rat brain revealed enhanced biosynthesis of lipids and proteins during sleep (Cirelli et al. 2004; Petit et al. 2002). AMPK might serve as a switch between anabolic (energy-consuming) and catabolic (energy-producing) processes in order to maintain sleep homeostasis (Chikahisa and Séi 2011).

Although multiple studies including those above have suggested general links between metabolism and sleep, knowledge about sleep-stage specific cerebral metabolism remains limited. However, some studies report differences between REM and NREM sleep, such as higher glucose utilization during REM sleep compared to NREM sleep (Maquet et al. 1990; Boyle et al. 1994; Maquet 1995).

2.3.3.2 Organ-Specific Control: Energy Homeostasis

At the level of the entire organism, one major function attributed to sleep is the maintenance of energy homeostasis (Berger and Phillips 1995). Sleep is the most energy-efficient human behavior and metabolic rate during sleep is reduced compared to resting during wakefulness (White et al. 1985). In mammals, sleep duration decreases with increasing size of the animal (Elgar et al. 1988), perhaps suggesting that a higher metabolic rate requires more sleep to keep the energy balance. In humans, inter-individual differences in sleep duration have been associated with genetic polymorphism of the SUR2 subunit of ATP sensitive potassium channels, which sense the state of cellular energy metabolism (Allebrandt et al. 2013). In addition to this, a genetic link between sleep duration and lipid levels in blood has been found with TRIB1 (Ollila et al. 2012).

Despite overnight fasting, blood glucose levels remain stable during the night with a small increase toward the end of the night. Under constant glucose infusion, an increase in glucose levels is observed with sleep onset, independent of time of day (Van Cauter et al. 1991). This decreased glucose tolerance is caused by reduced glucose utilization by brain and muscles, but also due to decreased insulin sensitivity (Knutson 2007). Thus, not only circadian influences but direct sleep-dependent mechanisms likely regulate circulating glucose levels.

Findings about differences in whole-body energy expenditure across different sleep stages are contradictory. Whereas some studies report lower energy expenditure during SWS compared to REM sleep (Brebbia and Altshuler 1965; Fontvieille et al. 1994), others did not confirm this finding (Webb and Hiestand 1975; Haskell et al. 1981; Palca et al. 1986; Jung et al. 2011). A recent study using whole-room calorimetry has found differences in respiration quotients across different stages of sleep. Carbohydrate oxidation was lowest during NREM sleep, which was explained with decreased glucose consumption by the brain during NREM sleep (Kayaba et al. 2017).

2.3.3.3 Central Control: Hormone Regulation

Appetite Regulating Hormones

Leptin and ghrelin are hormones that regulate hunger and appetite as a response to changes in energy balance. It has been shown that there is a link between these hormones and sleep (Sharma and Kavuru 2010). Ghrelin is released in the stomach and acts rapidly in response to caloric shortage or fasting by promoting hunger and appetite (Kojima et al. 1999). It also acts as a sleep-promoting factor and can induce SWS (Weikel et al. 2003). Leptin, in contrast suppresses appetite and is produced in adipose tissue (Gale et al. 2004). Both of these hormones increase during sleep and decrease in the morning. In the first part of the night, it is thought that leptin masks the effect of rising ghrelin levels in order to prevent arousals due to hunger (Sharma and Kavuru 2010). Leptin is also under circadian control and food intake is a confounding factor. However, under continuous enteral nutrition and during daytime sleep, increased leptin levels are observed with sleep onset (Simon et al. 1998). Moreover, animal studies have given evidence for leptin reducing REM sleep and modulating SWS. Leptin-deficient mice have more arousals and mice with mutated leptin receptors show increased total sleep time, but more fragmented sleep as well as a decrease in compensatory response to acute sleep deprivation (Laposky et al. 2006). Hence, these appetite regulating hormones might be an important link between sleep, circadian rhythms, and metabolism.

Orexins

Orexin A and B (hypocretins) could represent another major part of the link between hormonal control of metabolism and sleep. These excitatory neuropeptide hormones are expressed by neurons in the hypothalamus where energy homeostasis is regulated (Siegel 2004). They are influenced by peripheral hormones like ghrelin and leptin and also by glucose (Sharma and Kavuru 2010). Orexin administration has effects on sleep regulation as well as metabolism. It induces wakefulness, which comes along with increased energy expenditure and increased food intake (Yamanaka et al. 2003). However, orexin-deficient mice also show reduced energy expenditure regardless of sleep duration (Teske and Mavanji 2012). This suggests that there is a direct link between orexins and metabolism, and metabolic changes are not just a secondary effect of orexins regulating sleep-wake time.

Pituitary Hormones

The pituitary hormones, growth hormone (Sassin et al. 1969) and prolactin (Schmid et al. 2006) are both secreted upon sleep onset and reach a maximum 2 hours later. The extent of this hormonal release is associated with delta activity during NREM sleep (Latta et al. 2005). Also, levels of posterior pituitary hormones, such as plasma arginine vasopressin and oxytocin, are increased during sleep. These hormones profoundly regulate different aspects of metabolism, ranging from protein anabolism and triglyceride breakdown to milk production. Moreover, these hormones are also involved in sleep regulation and their administration is associated with alterations in sleep (Gómez-González et al. 2012). Therefore, also here, the connection between sleep and metabolism is bidirectional.

2.3.3.4 Immune Function

Another function associated with sleep is the immune response. Similar to hormonal regulation, there is a bidirectional communication between the immune system and the central nervous system and therefore sleep. Cytokines are main messenger molecules involved in immune responses, which are produced and released by the central nervous system with highest levels during sleep. Examples are interleukins (ILs) and tumor necrosis factors (TNF) (Besedovsky et al. 2011). Cytokines are also involved in sleep-wake regulation (Imeri and Opp 2009). Immune function also broadly regulates metabolism, especially adipocyte function (Brestoff and Artis 2015), making immune modulation a possible further route by which sleep influences metabolism.

2.3.4 Pathophysiological Consequences of Impaired Sleep

When sleep is impaired, the negative consequences for health and metabolic as well as cognitive functions are well established. Typical experiments to investigate these negative effects in healthy individuals are sleep restriction, partial sleep deprivation, and total sleep deprivation studies. Metabolic alterations in patients with sleep-related diseases, metabolic diseases, and their comorbidities can also be studied.

From these experiments, a relatively homogenous picture emerges. Insufficient sleep across several days results in a 5% increase of daily energy expenditure (Markwald et al. 2013). Acute sleep deprivation has also been shown to increase energy expenditure, supporting the hypothesis that energy conservation is a function of sleep (Jung et al. 2011). Under controlled conditions of caloric intake and physical activity prolonged wake can artificially provoke a negative energy balance. However, this does not correspond to real-life situations in modern society, where food shortage is no longer an issue. It has been shown that short sleep promotes snacking behavior (Nedeltcheva et al. 2009) and reduces physical activity (Schmid et al. 2009). With *ad libitum* feeding, an increased energy intake during wakefulness was observed, especially after dinner, resulting in a positive energy balance (Markwald et al. 2013). Overeating occurred despite proper signaling of leptin and ghrelin, indicating that it is not just due to a longer period of food availability, but also physiological adaptation: energy intake is increased to sustain prolonged wakefulness (Penev 2007). Nonhomeostatic food intake is likely to be driven by brain mechanisms similar to those by which mood and comfort regulate feeding (Spiegel et al. 2004). This imbalance between food intake and energy expenditure due to a lack of sleep might partly explain the association between short and fragmented sleep and an increased risk for metabolic diseases such as T2D and obesity, which has been found in various epidemiologic studies (Cappuccio et al. 2010). Importantly, recent studies suggest that even unlimited "recovery sleep" on weekends is insufficient to compensate for metabolic dysregulation incurred during weekday sleep restriction (Depner et al. 2019).

2.3.4.1 Obesity, T2D, and Sleep

Sleep restriction has been associated with reduced insulin sensitivity, indicating that impaired sleep alters glucose metabolism (Buxton et al. 2010). Similarly, large epidemiological studies have related insufficient sleep and disturbed sleep to obesity and T2D (Anothaisintawee et al. 2016; Cappuccio et al. 2008). Mechanistically, appetite and metabolic hormones—the same that we describe above as capable of altering sleep per se—are believed to play a strong role in this pathology. Leptin, ghrelin, endocannabinoids, and other appetite peptides have all been shown to be dysregulated by sleep loss, restriction, or disturbance, and the direction of dysregulation is consistent with increased caloric intake and decreased glucose clearance. This topic has been reviewed recently (Reutrakul and Van Cauter 2018).

2.3.4.2 Obesity, T2D, and Obstructive Sleep Apnea

Obesity is considered as one of the most important risk factors for obstructive sleep apnea (OSA). In turn, OSA was also found to promote weight gain. Causal relationships are still unclear and it is hypothesized as a vicious cycle (Carter and Watenpaugh 2008). Both physiopathologies are linked genetically, and worsen each other. Adipose tissue deposits in obese patients lead to reduced ventilatory stability and promote the development of OSA. OSA often goes along with physical inactivity, dysregulated appetite hormones, and insulin resistance, thereby increasing the risk for obesity. Dysregulated appetite hormones are also likely contributors, since OSA patients have increased ghrelin levels (Harsch et al. 2003).

Apart from obesity, a link between OSA and T2D has been found (Mallon et al. 2005; West et al. 2006), and especially amongst obese T2D patients there is a high prevalence of OSA (Foster et al. 2009). One suggested mechanism for the link between T2D and OSA is that OSA causes sympathetic activation, which inhibits leptin secretion and promotes HPA axis stimulation.

This leads to increased cortisol secretion resulting in impaired glucose homeostasis (Barone and Menna-Barreto 2011). Several studies have shown that treatment of OSA patients with continuous positive airway pressure (CPAP) also improved insulin sensitivity, corroborating the hypothesis that impaired sleep is promoting T2D (Pallayova et al. 2008). However, other studies suggest the opposite: no effect of CPAP therapy on glucose metabolism or T2D (Smurra et al. 2001; Hecht et al. 2011). The problem here is that obesity acts as a confounding factor, since obesity is considered as an important risk factor for OSA and occurs often together with T2D (Pillar and Shehadeh 2008). In order to elucidate causal relationships, nonobese OSA patients with and without T2D would need to be investigated.

OSA is not the only sleep disorder linked to metabolic dysfunction. Narcolepsy, a REM sleep disorder resulting from a deficiency in orexigenic neurons, is associated with excessive daytime sleepiness and poor sleep quality, abnormalities in REM sleep and orexin deficiency (Dauvilliers et al. 2007), and has been linked to obesity (Dahmen et al. 2001).

2.3.4.3 Inflammatory Response to Impaired Sleep

Sleep deprivation is associated with altered immune responses due to an increase of proinflammatory markers (Yehuda et al. 2009). This is supported by increased systemic levels of TNF-α in OSA, narcolepsy, and insomnia patients (Gómez-González et al. 2012). Moreover, CPAP treatment can decrease TNF-α levels (Ryan et al. 2006). Furthermore, sleep deprivation results in impaired host defense against pathogens (Everson and Toth 2000) and many autoimmune diseases are associated with sleep disruption, daytime sleepiness, and an increased risk for sleep disorders (Gómez-González et al. 2012).

2.3.4.4 Alzheimer's Disease and Sleep

There is evidence of a link between sleep, T2D, and Alzheimer's disease (AD). These interactions suggest that sleep impairment and metabolic dysregulation promote the progression of AD (Carroll and Macauley 2019). AD patients often show sleep impairments (Peter-Derex et al. 2015) and recently, reduced amounts of SWS have been associated with tau pathophysiology of AD (Lucey et al. 2019). Additionally, cerebrospinal fluid (CSF) levels of several AD biomarkers have been found to correlate with sleep-wake cycles (Lucey et al. 2017). Moreover, elevated CSF levels of orexin A are reported in AD patients (Gabelle et al. 2017). Links between T2D, AD, and sleep further suggest that impaired glucose metabolism might be a key player in interactions between sleep impairment and cognitive dysfunction in AD (Holingue et al. 2018). An interesting alternative is that connections between AD and sleep impairment might relate to glymphatic flow—the "waste clearance" system of the brain -- which increases during sleep and contributes to Alzheimer-associated peptide (Aβ) clearance from the brain (Xie et al. 2013).

2.4 Clocks or Sleep: Future Directions

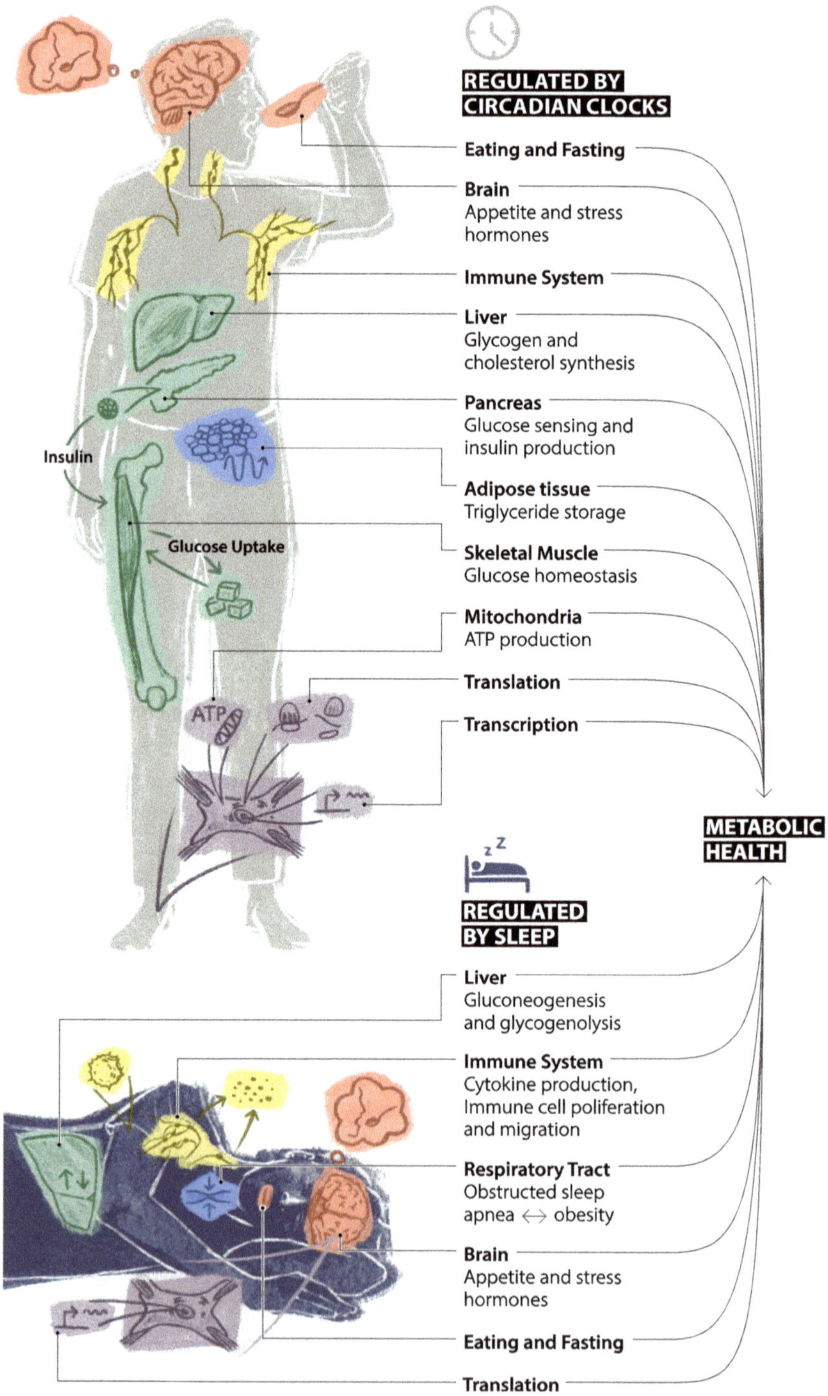

Circadian clocks and sleep: two related factors in the multifaceted regulation of metabolism

Overlapping aspects of metabolism have been shown to be regulated by circadian clocks (standing person) and by sleep-wake (sleeping person). Behaviorally, they both regulate eating and fasting across the day, as well as hormones related to feeding (red). Other co-regulated aspect of physiology includes immune function (yellow), glucose and carbohydrate levels in multiple organs (green), and adiposity (blue). At a cellular level (purple), evidence exists for coregulation of transcription, translation, and mitochondrial output.

It is clear that modern society has increasingly intruded upon natural circadian rhythms in humans, possibly leading to profound metabolic consequences. Classically, the phase desynchrony between central and peripheral clocks is thought of as the main contributor. However, one direct test of this assumption has failed: it was shown that a 6-hour phase misalignment between central and peripheral clocks was not sufficient to cause obesity and glucose intolerance in mice (van der Vinne et al. 2018). Thus, we favor the hypothesis that metabolic consequences of circadian disruption arise via multiple mechanisms, rather than solely from internal desynchrony. In humans, it is unlikely that central and peripheral clocks would have such a large phase misalignment as those artificially tested in animals outside situations of shiftwork. Thus, if circadian disruptions such as social jetlag result in metabolic dysfunction, it is likely that other factors are at work, and sleep disturbance is a prime possibility.

Equally, sleep restriction is an omnipresent issue in contemporary society, with adverse effects on health and metabolism. However, these metabolic considerations are mostly based on measures of total energy expenditure rather than pathway-specific investigations. It is likely that advances in metabolomics techniques using high resolution mass spectrometry that have been made in recent years have great potential for novel insights. The sensitivity of some methods is sufficient to identify thousands of compounds in a single human breath, making them powerful noninvasive techniques to overcome limitations in sampling rate (Martinez-Lozano Sinues et al. 2014; Sinues et al. 2013).

Strikingly, most aspects of metabolism regulated by sleep and circadian clocks are shared (FIGURE). Molecular mechanisms are continually discovered about single aspects of the relationship between the circadian clock or sleep and the downstream physiology they control, but often these studies look at a single facet. More recent studies that compare effects of circadian disruption and sleep disruption demonstrate that each might play a role in cellular physiology. For example, considering brain, individual transcripts might be regulated by either circadian or sleep influences, or both (Hor et al. 2019). Even more locally, it has been suggested that at synapses, RNA abundance is primarily clock-driven, whereas translation and phosphorylation of proteins are mostly controlled by sleep-wake cycles (Noya et al. 2019; Bruning et al. 2019b).

These studies, as well as this review, have mostly addressed the idea that different aspects of physiology might be controlled by circadian clocks or sleep. However, molecular pathway-specific investigations are lacking. Indeed, circadian rhythm sleep disorder is often misdiagnosed as insomnia (Kim et al. 2013; Dagan and Ayalon 2005). Experimentally, circadian clock gene disruptions in mice also affect sleep consolidation (Laposky et al. 2005; Wisor et al. 2002). Thus, the classical paradigm of a clock gene deletion as a way to ascertain that a process is directly clock controlled contains an unavoidable flaw. Similarly, most epidemiological studies—and many laboratory ones—examining effects of sleep disruption are in fact examining unknown degrees of circadian disruption as well. The only real solution to this conundrum is a detailed mechanistic understanding of the regulatory processes involved. Without a doubt, such an understanding will lead to improved therapies as well.

References

Aalling NN, Nedergaard M, DiNuzzo M (2018) Cerebral metabolic changes during sleep. Curr Neurol Neurosci Rep 18(9):57

Akerstedt T (1990) Psychological and psychophysiological effects of shift work. Scand J Work Environ Health. 16(Suppl 1):67–73

Albrecht U (2012) Timing to perfection: the biology of central and peripheral circadian clocks. Neuron. 74 (2):246–260

Albrecht U, Eichele G (2003) The mammalian circadian clock. Curr Opin Genet Dev. 13(3):271–277

Allebrandt KV, Amin N, Müller-Myhsok B, Esko T, Teder-Laving M, Azevedo RVDM et al (2013) A KATP channel gene effect on sleep duration: from genome-wide association studies to function in Drosophila. Mol Psychiatry. 18(1):122–132

Anothaisintawee T, Reutrakul S, Van Cauter E, Thakkinstian A (2016) Sleep disturbances compared to traditional risk factors for diabetes development: systematic review and meta-analysis. Sleep Med Rev. 30:11–24

Arble DM, Bass J, Laposky AD, Vitaterna MH, Turek FW (2009) Circadian timing of food intake contributes to weight gain. Obesity (Silver Spring). 17 (11):2100–2102

Arendt J (2010) Shift work: coping with the biological clock. Occup Med (Lond). 60(1):10–20

Asarnow LD, Soehner AM, Harvey AG (2013) Circadian rhythms and psychiatric illness. Curr Opin Psychiatry. 26(6):566–571

Asher G, Gatfield D, Stratmann M, Reinke H, Dibner C, Kreppel F et al (2008) SIRT1 regulates circadian clock gene expression through PER2 deacetylation. Cell. 134 (2):317–328

Asher G, Reinke H, Altmeyer M, Gutierrez-Arcelus M, Hottiger MO, Schibler U (2010) Poly(ADP-ribose) polymerase 1 participates in the phase entrainment of circadian clocks to feeding. Cell. 142(6):943–953

Aviram R, Manella G, Kopelman N, Neufeld-Cohen A, Zwighaft Z, Elimelech M et al (2016) Lipidomics analyses reveal temporal and spatial lipid organization and uncover daily oscillations in intracellular organelles. Mol Cell. 62(4):636–648

Balsalobre A, Damiola F, Schibler U (1998) A serum shock induces circadian gene expression in mammalian tissue culture cells. Cell. 93(6):929–937

Barone MTU, Menna-Barreto L (2011) Diabetes and sleep: a complex cause-and-effect relationship. Diabetes Res Clin Pract 91(2):129–137

Bass J, Takahashi JS (2010) Circadian integration of metabolism and energetics. Science. 330 (6009):1349–1354

Beccuti G, Monagheddu C, Evangelista A, Ciccone G, Broglio F, Soldati L et al (2017) Timing of food intake: Sounding the alarm about metabolic impairments? A systematic review. Pharmacol Res 125(Pt B):132–141

Benegiamo G, Mure LS, Erikson G, Le HD, Moriggi E, Brown SA et al (2018) The RNA-binding protein NONO coordinates hepatic adaptation to feeding. Cell Metab 27(2):404–418

Berger RJ, Phillips NH (1995) Energy conservation and sleep. Behav Brain Res 69(1):65–73

Berridge CW, Schmeichel BE, España RA (2012) Noradrenergic modulation of wakefulness/arousal. Sleep Med Rev 16(2):187–197

Besedovsky L, Lange T, Born J (2011) Sleep and immune function. Pflügers Archiv—Eur J Physiol 463 (1):121–137

Borbély AA (1982) A two process model of sleep regulation. Hum Neurobiol 1(3):195–204

Borbély AA, Daan S, Wirz-Justice A, Deboer T (2016) The two-process model of sleep regulation: a reappraisal. J Sleep Res 25(2):131–143

Boyle PJ, Scott JC, Krentz AJ, Nagy RJ, Comstock E, Hoffman C (1994) Diminished brain glucose metabolism is a significant determinant for falling rates of systemic glucose utilization during sleep in normal humans. J Clin Invest 93(2):529–535

Brebbia DR, Altshuler KZ (1965) Oxygen consumption rate and electroencephalographic stage of sleep. Science 150(3703):1621–1623

Brestoff JR, Artis D (2015) Immune regulation of metabolic homeostasis in health and disease. Cell. 161 (1):146–160

Brown SA (2016) Circadian metabolism: from mechanisms to metabolomics and medicine. Trends Endocrinol Metab. 27(6):415–426

Brown SA, Zumbrunn G, Fleury-Olela F, Preitner N, Schibler U (2002) Rhythms of mammalian body temperature can sustain peripheral circadian clocks. Curr Biol. 12(18):1574–1583

Brum MC, Filho FF, Schnorr CC, Bottega GB, Rodrigues TC (2015) Shift work and its association with metabolic disorders. Diabetol Metab Syndr. 7:45

Bruning F, Noya SB, Bange T, Koutsouli S, Rudolph JD, Tyagarajan SK et al (2019a) Sleep-wake cycles drive daily dynamics of synaptic phosphorylation. Science 366(6462):eaav3617

Bruning F, Noya SB, Bange T, Koutsouli S, Rudolph JD, Tyagarajan S et al (2019b) Sleep-wake cycles drive daily dynamics of synaptic phosphorylation. Science 366(6462):201

Buijs RM, Kalsbeek A (2001) Hypothalamic integration of central and peripheral clocks. Nat Rev Neurosci. 2 (7):521–526

Burdakov D, Jensen LT, Alexopoulos H, Williams RH, Fearon IM, O'Kelly I et al (2006) Tandem-pore K+ channels mediate inhibition of orexin neurons by glucose. Neuron. 50(5):711–722

Buxton OM, Pavlova M, Reid EW, Wang W, Simonson DC, Adler GK (2010) Sleep restriction for 1 week reduces insulin sensitivity in healthy men. Diabetes. 59(9):2126

Cappuccio FP, Taggart FM, Kandala NB, Currie A, Peile E, Stranges S et al (2008) Meta-analysis of short sleep duration and obesity in children and adults. Sleep. 31(5):619–626

Cappuccio FP, Elia L, Strazzullo P, Miller MA (2010) Quantity and quality of sleep and incidence of type 2 diabetes. Diabetes Care. 33(2):414

Carroll CM, Macauley SL (2019) The interaction between sleep and metabolism in Alzheimer's disease: cause or consequence of disease? Front Aging Neurosci 11:258

Carter R, Watenpaugh DE (2008) Obesity and obstructive sleep apnea: or is it OSA and obesity? Pathophysiology. 15(2):71–77

Castanon-Cervantes O, Wu M, Ehlen JC, Paul K, Gamble KL, Johnson RL et al (2010) Dysregulation of inflammatory responses by chronic circadian disruption. J Immunol. 185(10):5796–5805

Cedernaes J, Huang W, Ramsey KM, Waldeck N, Cheng L, Marcheva B et al (2019a) Transcriptional basis for rhythmic control of hunger and metabolism within the AgRP neuron. Cell Metab. 29 (5):1078–1091. e5

Cedernaes J, Waldeck N, Bass J (2019b) Neurogenetic basis for circadian regulation of metabolism by the hypothalamus. Genes Dev. 33(17-18):1136–1158

Cederroth CR, Albrecht U, Bass J, Brown SA, Dyhrfjeld-Johnsen J, Gachon F et al (2019) Medicine in the fourth dimension. Cell Metab. 30(2):238–250

Chen K-S, Xu M, Zhang Z, Chang W-C, Gaj T, Schaffer DV et al (2018) A hypothalamic switch for REM and non-REM sleep. Neuron 97(5):1168–76.e4

Chikahisa S, Séi H (2011) The role of ATP in sleep regulation. Front Neurol 2:87

Cirelli C, Gutierrez CM, Tononi G (2004) Extensive and divergent effects of sleep and wakefulness on brain gene expression. Neuron. 41(1):35–43

Codoñer-Franch P, Gombert M (2018) Circadian rhythms in the pathogenesis of gastrointestinal diseases. World J Gastroenterol. 24(38):4297–4303

Coomans CP, Lucassen EA, Kooijman S, Fifel K, Deboer T, Rensen PC et al (2015) Plasticity of circadian clocks and consequences for metabolism. Diabetes Obes Metab. 17(Suppl 1):65–75

Crunelli V, Hughes SW (2010) The slow (<1 Hz) rhythm of non-REM sleep: a dialogue between three cardinal oscillators. Nat Neurosci. 13(1):9–17

Dagan Y, Ayalon L (2005) Case study: psychiatric misdiagnosis of non-24-hours sleep-wake schedule disorder resolved by melatonin. J Am Acad Child Adolesc Psychiatry 44(12):1271–1275

Dahmen N, Bierbrauer J, Kasten M (2001) Increased prevalence of obesity in narcoleptic patients and relatives. Eur Arch Psychiatry Clin Neurosci 251 (2):85–89

Damiola F, Le Minh N, Preitner N, Kornmann B, Fleury-Olela F, Schibler U (2000) Restricted feeding uncouples circadian oscillators in peripheral tissues from the central pacemaker in the suprachiasmatic nucleus. Genes Dev. 14(23):2950–2961

Dauvilliers Y, Rompré S, Gagnon J-F, Vendette M, Petit D, Montplaisir J (2007) REM sleep characteristics in narcolepsy and REM sleep behavior disorder. Sleep. 30(7):844–849

Depner CM, Melanson EL, Eckel RH, Snell-Bergeon JK, Perreault L, Bergman BC et al (2019) Ad libitum weekend recovery sleep fails to prevent metabolic dysregulation during a repeating pattern of insufficient sleep and weekend recovery sleep. Curr Biol. 29 (6):957–967. e4

Dienel GA, Cruz NF (2016) Aerobic glycolysis during brain activation: adrenergic regulation and influence of norepinephrine on astrocytic metabolism. J Neurochemistry. 138(1):14–52

Diering GH, Nirujogi RS, Roth RH, Worley PF, Pandey A, Huganir RL (2017) Homer1a drives homeostatic scaling-down of excitatory synapses during sleep. Science. 355(6324):511–515

DiNuzzo M, Nedergaard M (2017) Brain energetics during the sleep–wake cycle. Curr Opin Neurobiol 47:65–72

Eckel-Mahan KL, Patel VR, Mohney RP, Vignola KS, Baldi P, Sassone-Corsi P (2012) Coordination of the transcriptome and metabolome by the circadian clock. Proc Natl Acad Sci U S A. 109(14):5541–5546

Elgar MA, Pagel MD, Harvey PH (1988) Sleep in mammals. Anim Behav 36(5):1407–1419

Espitia-Bautista E, Velasco-Ramos M, Osnaya-Ramírez I, Ángeles-Castellanos M, Buijs RM, Escobar C (2017) Social jet-lag potentiates obesity and metabolic syndrome when combined with cafeteria diet in rats. Metabolism. 72:83–93

Everson CA, Toth LA (2000) Systemic bacterial invasion induced by sleep deprivation. Am J Physiol Regul Integr Comp Physiol 278(4):R905–RR16

Feng D, Liu T, Sun Z, Bugge A, Mullican SE, Alenghat T et al (2011) A circadian rhythm orchestrated by histone deacetylase 3 controls hepatic lipid metabolism. Science. 331(6022):1315–1319

Fontvieille AM, Rising R, Spraul M, Larson DE, Ravussin E (1994) Relationship between sleep stages and metabolic rate in humans. Am J Physiol Endocrinol Metab 267(5):E732–E737. https://doi.org/10.1152/ajpendo19942675E732

Foster GD, Sanders MH, Millman R, Zammit G, Borradaile KE, Newman AB et al (2009) Obstructive sleep apnea among obese patients with type 2 diabetes. Diabetes Care. 32(6):1017

Gabelle A, Jaussent I, Hirtz C, Vialaret J, Navucet S, Grasselli C et al (2017) Cerebrospinal fluid levels of orexin-A and histamine, and sleep profile within the Alzheimer process. Neurobiology of Aging. 53:59–66

Gale SM, Castracane VD, Mantzoros CS (2004) Energy homeostasis, obesity and eating disorders: recent advances in endocrinology. J Nutr 134(2):295–298

Gómez-González B, Domínguez-Salazar E, Hurtado-Alvarado G, Esqueda-Leon E, Santana-Miranda R, Rojas-Zamorano JA et al (2012) Role of sleep in the regulation of the immune system and the pituitary hormones. Annals New York Acad Sci 1261 (1):97–106

Gorgoni M, D'Atri A, Lauri G, Rossini PM, Ferlazzo F, De Gennaro L (2013) Is sleep essential for neural plasticity in humans, and how does it affect motor and cognitive recovery? Neural Plast 2013:13

Greene MW (2012) Circadian rhythms and tumor growth. Cancer Lett. 318(2):115–123

Greenwell BJ, Trott AJ, Beytebiere JR, Pao S, Bosley A, Beach E et al (2019) Rhythmic food intake drives rhythmic gene expression more potently than the hepatic circadian clock in mice. Cell Rep 27 (3):649–57.e5

Harsch IA, Konturek PC, Koebnick C, Kuehnlein PP, Fuchs FS, Pour Schahin S et al (2003) Leptin and ghrelin levels in patients with obstructive sleep apnoea: effect of CPAP treatment. Eur Respirat J 22(2):251

Haskell EH, Palca JW, Walker JM, Berger RJ, Heller HC (1981) Metabolism and thermoregulation during stages of sleep in humans exposed to heat and cold. J Appl Phys 51(4):948–954. https://doi.org/10.1152/jappl1981514948

Hatori M, Vollmers C, Zarrinpar A, DiTacchio L, Bushong EA, Gill S et al (2012) Time-restricted feeding without reducing caloric intake prevents metabolic diseases in mice fed a high-fat diet. Cell Metab. 15 (6):848–860

Hecht L, Möhler R, Meyer G (2011) Effects of CPAP-respiration on markers of glucose metabolism in patients with obstructive sleep apnoea syndrome: a systematic review and meta-analysis. Ger Med Sci 9: Doc20-Doc

Holingue C, Wennberg A, Berger S, Polotsky VY, Spira AP (2018) Disturbed sleep and diabetes: a potential nexus of dementia risk. Metabolism. 84:85–93

Hor CN, Yeung J, Jan M, Emmenegger Y, Hubbard J, Xenarios I et al (2019) Sleep-wake-driven and circadian contributions to daily rhythms in gene expression and chromatin accessibility in the murine cortex. Proc Natl Acad Sci U S A. 116(51):25773–25783

Hsuchou H, Wang Y, Cornelissen-Guillaume GG, Kastin AJ, Jang E, Halberg F et al (2013) Diminished leptin signaling can alter circadian rhythm of metabolic activity and feeding. J Appl Physiol (1985) 115 (7):995–1003

Imeri L, Opp MR (2009) How (and why) the immune system makes us sleep. Nature Rev Neurosci 10 (3):199–210

Jordan SD, Lamia KA (2013) AMPK at the crossroads of circadian clocks and metabolism. Mol Cell Endocrinol. 366(2):163–169

Jung CM, Melanson EL, Frydendall EJ, Perreault L, Eckel RH, Wright KP (2011) Energy expenditure during sleep, sleep deprivation and sleep following sleep deprivation in adult humans. J Physiol 589(1):235–244

Kalia M (2006) Neurobiology of sleep. Metabolism. 55: S2–S6

Kalsbeek A, Fliers E (2013) Daily regulation of hormone profiles. Handb Exp Pharmacol. 217:185–226

Kayaba M, Park I, Iwayama K, Seya Y, Ogata H, Yajima K et al (2017) Energy metabolism differs between sleep stages and begins to increase prior to awakening. Metabolism. 69:14–23

Kelly RM, Healy U, Sreenan S, McDermott JH, Coogan AN (2018) Clocks in the clinic: circadian rhythms in health and disease. Postgrad Med J. 94(1117):653–658

Kervezee L, Cermakian N, Boivin DB (2019) Individual metabolomic signatures of circadian misalignment during simulated night shifts in humans. PLoS Biol. 17(6): e3000303

Kil IS, Ryu KW, Lee SK, Kim JY, Chu SY, Kim JH et al (2015) Circadian oscillation of sulfiredoxin in the mitochondria. Mol Cell. 59(4):651–663

Kim MJ, Lee JH, Duffy JF (2013) Circadian rhythm sleep disorders. J Clin Outcomes Manag. 20(11):513–528

Knutson KL (2007) Impact of sleep and sleep loss on glucose homeostasis and appetite regulation. Sleep Med Clin. 2(2):187–197

Knutson KL, von Schantz M (2018) Associations between chronotype, morbidity and mortality in the UK Biobank cohort. Chronobiol Int. 35(8):1045–1053

Kohsaka A, Laposky AD, Ramsey KM, Estrada C, Joshu C, Kobayashi Y et al (2007) High-fat diet disrupts behavioral and molecular circadian rhythms in mice. Cell Metab. 6(5):414–421

Kojima M, Hosoda H, Date Y, Nakazato M, Matsuo H, Kangawa K (1999) Ghrelin is a growth-hormone-releasing acylated peptide from stomach. Nature. 402 (6762):656–660

Korenčič A, Košir R, Bordyugov G, Lehmann R, Rozman D, Herzel H (2014) Timing of circadian genes in mammalian tissues. Sci Rep. 4:5782

Lamia KA, Papp SJ, Yu RT, Barish GD, Uhlenhaut NH, Jonker JW et al (2011) Cryptochromes mediate rhythmic repression of the glucocorticoid receptor. Nature. 480(7378):552–556

Laposky A, Easton A, Dugovic C, Walisser J, Bradfield C, Turek F (2005) Deletion of the mammalian circadian clock gene BMAL1/Mop3 alters baseline sleep architecture and the response to sleep deprivation. Sleep. 28 (4):395–409

Laposky AD, Shelton J, Bass J, Dugovic C, Perrino N, Turek FW (2006) Altered sleep regulation in leptin-deficient mice. Am J Physiol Regul Integr Comp Phys 290(4):R894–R903

Latta F, Leproult R, Tasali E, Hofmann E, L'Hermite-Balériaux M, Copinschi G et al (2005) Sex differences in nocturnal growth hormone and prolactin secretion in healthy older adults: relationships with sleep EEG variables. Sleep. 28(12):1519–1524

Lee Y, Kim EK (2013) AMP-activated protein kinase as a key molecular link between metabolism and clockwork. Exp Mol Med. 45:e33

Lee J, Ma K, Moulik M, Yechoor V (2018) Untimely oxidative stress in β-cells leads to diabetes—role of circadian clock in β-cell function. Free Radic Biol Med. 119:69–74

Leenaars CHC, Savelyev SA, Van der Mierden S, Joosten R, Dematteis M, Porkka-Heiskanen T et al (2018) Intracerebral adenosine during sleep deprivation: a meta-analysis and new experimental data. J Circadian Rhythms. 16:11

Leliavski A, Dumbell R, Ott V, Oster H (2015) Adrenal clocks and the role of adrenal hormones in the regulation of circadian physiology. J Biol Rhythms. 30 (1):20–34

Lipton JO, Yuan ED, Boyle LM, Ebrahimi-Fakhari D, Kwiatkowski E, Nathan A et al (2015) The circadian protein BMAL1 regulates translation in response to S6K1-mediated phosphorylation. Cell. 161 (5):1138–1151

Loizides-Mangold U, Perrin L, Vandereycken B, Betts JA, Walhin JP, Templeman I et al (2017) Lipidomics reveals diurnal lipid oscillations in human skeletal muscle persisting in cellular myotubes cultured in vitro. Proc Natl Acad Sci U S A. 114(41):E8565–E8E74

Lu J, Sherman D, Devor M, Saper CB (2006) A putative flip–flop switch for control of REM sleep. Nature. 441 (7093):589–594

Lucey BP, Fagan AM, Holtzman DM, Morris JC, Bateman RJ (2017) Diurnal oscillation of CSF Aβ and other AD biomarkers. Mol Neurodegener 12(1):36

Lucey BP, McCullough A, Landsness EC, Toedebusch CD, McLeland JS, Zaza AM et al (2019) Reduced non–rapid eye movement sleep is associated with tau pathology in early Alzheimer's disease. Sci Transl Med 11(474):eaau6550

Machler P, Wyss MT, Elsayed M, Stobart J, Gutierrez R, von Faber-Castell A et al (2016) In vivo evidence for a lactate gradient from astrocytes to neurons. Cell Metab. 23(1):94–102

Mallon L, Broman J-E, Hetta J (2005) High incidence of diabetes in men with sleep complaints or short sleep duration. Diabetes Care. 28(11):2762

Mang GM, La Spada F, Emmenegger Y, Chappuis S, Ripperger JA, Albrecht U et al (2016) Altered sleep homeostasis in rev-erbalpha knockout mice. Sleep. 39 (3):589–601

Maquet P (1995) Sleep function(s) and cerebral metabolism. Behav Brain Res 69(1):75–83

Maquet P, Dive D, Salmon E, Sadzot B, Franco G, Poirrier R et al (1990) Cerebral glucose utilization during sleep-wake cycle in man determined by positron emission tomography and [18F]2-fluoro-2-deoxy-d-glucose method. Brain Res 513(1):136–143

Markwald RR, Melanson EL, Smith MR, Higgins J, Perreault L, Eckel RH et al (2013) Impact of insufficient sleep on total daily energy expenditure, food intake, and weight gain. Proc Nat Acad Sci 110 (14):5695–5700

Martinez-Lozano Sinues P, Tarokh L, Li X, Kohler M, Brown SA, Zenobi R et al (2014) Circadian variation of the human metabolome captured by real-time breath analysis. PLoS One. 9(12):e114422

Masri S, Orozco-Solis R, Aguilar-Arnal L, Cervantes M, Sassone-Corsi P (2015) Coupling circadian rhythms of metabolism and chromatin remodelling. Diabetes Obes Metab. 17(Suppl 1):17–22

McGinty D, Szymusiak R (2000) The sleep–wake switch: a neuronal alarm clock. Nature Med 6(5):510–511

Mikhail C, Vaucher A, Jimenez S, Tafti M (2017) ERK signaling pathway regulates sleep duration through activity-induced gene expression during wakefulness. Sci Signal 10(463):eaai9219

Moore RY, Lenn NJ (1972) A retinohypothalamic projection in the rat. J Comp Neurol. 146(1):1–14

Muheim CM, Spinnler A, Sartorius T, Durr R, Huber R, Kabagema C et al (2019) Dynamic- and frequency-specific regulation of sleep oscillations by cortical potassium channels. Curr Biol. 29(18):2983–2992. e3

Mure LS, Le HD, Benegiamo G, Chang MW, Rios L, Jillani N et al (2018) Diurnal transcriptome atlas of a primate across major neural and peripheral tissues. Science 359(6381):eaao0318

Nakahata Y, Sahar S, Astarita G, Kaluzova M, Sassone-Corsi P (2009) Circadian control of the NAD+ salvage pathway by CLOCK-SIRT1. Science. 324 (5927):654–657

Nakajima M, Imai K, Ito H, Nishiwaki T, Murayama Y, Iwasaki H et al (2005) Reconstitution of circadian oscillation of cyanobacterial KaiC phosphorylation in vitro. Science. 308(5720):414–415

Nedeltcheva AV, Kilkus JM, Imperial J, Kasza K, Schoeller DA, Penev PD (2009) Sleep curtailment is accompanied by increased intake of calories from snacks. Am J Clin Nutr. 89(1):126–133

Noya SB, Colameo D, Bruning F, Spinnler A, Mircsof D, Opitz L et al (2019) The forebrain synaptic transcriptome is organized by clocks but its proteome is driven by sleep. Science 366(6462):eaav2642

Oh J, Petersen C, Walsh CM, Bittencourt JC, Neylan TC, Grinberg LT (2019) The role of co-neurotransmitters in sleep and wake regulation. Mol Psychiatry. 24 (9):1284–1295

Ollila HM, Utge S, Kronholm E, Aho V, Leeuwen WV, Silander K et al (2012) TRIB1 constitutes a molecular link between regulation of sleep and lipid metabolism in humans. Translat Psychiatry 2(3):e97

O'Neill JS, Reddy AB (2011) Circadian clocks in human red blood cells. Nature. 469(7331):498–503

O'Neill JS, Maywood ES, Hastings MH (2013) Cellular mechanisms of circadian pacemaking: beyond transcriptional loops. Handb Exp Pharmacol. 217:67–103

Oster H, Challet E, Ott V, Arvat E, de Kloet ER, Dijk DJ et al (2017) The functional and clinical significance of the 24-hour rhythm of circulating glucocorticoids. Endocr Rev. 38(1):3–45

Palca JW, Walker JM, Berger RJ (1986) Thermoregulation, metabolism, and stages of sleep in cold-exposed men. J Appl Phys 61(3):940–947. https://doi.org/10.1152/jappl1986613940

Pallayova M, Donic V, Tomori Z (2008) Beneficial effects of severe sleep apnea therapy on nocturnal glucose control in persons with type 2 diabetes mellitus. Diabetes Res Clin Pract 81(1):e8–e11

Pan A, Schernhammer ES, Sun Q, Hu FB (2011) Rotating night shift work and risk of type 2 diabetes: two

prospective cohort studies in women. PLoS Med. 8 (12):e1001141
Paschos GK, Ibrahim S, Song WL, Kunieda T, Grant G, Reyes TM et al (2012) Obesity in mice with adipocyte-specific deletion of clock component Arntl. Nat Med. 18(12):1768–1777
Penev PD (2007) Sleep deprivation and energy metabolism: to sleep, perchance to eat? Curr Opin Endocrinol Diabetes Obes 14(5):374–381
Perelis M, Marcheva B, Ramsey KM, Schipma MJ, Hutchison AL, Taguchi A et al (2015) Pancreatic β cell enhancers regulate rhythmic transcription of genes controlling insulin secretion. Science 350(6261): aac4250
Perrin L, Loizides-Mangold U, Skarupelova S, Pulimeno P, Chanon S, Robert M et al (2015) Human skeletal myotubes display a cell-autonomous circadian clock implicated in basal myokine secretion. Mol Metab. 4(11):834–845
Peter-Derex L, Yammine P, Bastuji H, Croisile B (2015) Sleep and Alzheimer's disease. Sleep Med Rev 19:29–38
Petit JM, Tobler I, Allaman I, Borbély AA, Magistretti PJ (2002) Sleep deprivation modulates brain mRNAs encoding genes of glycogen metabolism. Eur J Neurosci. 16(6):1163–1167
Pillar G, Shehadeh N (2008) Abdominal fat and sleep Apnea. Diabetes Care 31(Supplement 2):S303
Puentes-Mestril C, Roach J, Niethard N, Zochowski M, Aton SJ (2019) How rhythms of the sleeping brain tune memory and synaptic plasticity. Sleep 42(7):zsz095
Qian J, Scheer FAJL (2016) Circadian system and glucose metabolism: implications for physiology and disease. Trends Endocrinol Metab. 27(5):282–293
Qian J, Morris CJ, Caputo R, Garaulet M, Scheer FAJL (2019) Ghrelin is impacted by the endogenous circadian system and by circadian misalignment in humans. Int J Obes (Lond). 43(8):1644–1649
Ramamoorthy S, Cidlowski JA (2016) Corticosteroids: mechanisms of action in health and disease. Rheum Dis Clin North Am. 42(1):15–31. vii
Ramos-Lopez O, Samblas M, Milagro FI, Riezu-Boj JI, Crujeiras AB, Martinez JA et al (2018) Circadian gene methylation profiles are associated with obesity, metabolic disturbances and carbohydrate intake. Chronobiol Int. 35(7):969–981. https://doi.org/10.1080/07420528.2018.1446021
Rechtschaffen A, Kales A (1968) A manual of standardized terminology, technique and scoring system for sleep stages of human sleep. Brain Information Service, Los Angeles
Reinke H, Asher G (2019) Crosstalk between metabolism and circadian clocks. Nat Rev Mol Cell Biol. 20 (4):227–241
Resuehr D, Wu G, Johnson RL Jr, Young ME, Hogenesch JB, Gamble KL (2019) Shift work disrupts circadian regulation of the transcriptome in hospital nurses. J Biol Rhythms. 34(2):167–177
Reutrakul S, Van Cauter E (2018) Sleep influences on obesity, insulin resistance, and risk of type 2 diabetes. Metabolism. 84:56–66
Robles MS, Humphrey SJ, Mann M (2017) Phosphorylation is a central mechanism for circadian control of metabolism and physiology. Cell Metab. 25 (1):118–127
Roenneberg T, Allebrandt KV, Merrow M, Vetter C (2012) Social jetlag and obesity. Curr Biol. 22 (10):939–943
Rosenbaum E (1892) Warum müssen wir schlafen? Eine neue Theorie des Schlafes. August Hirschwald, Berlin
Ryan S, Taylor CT, McNicholas WT (2006) Predictors of elevated nuclear factor-κB–dependent genes in obstructive sleep Apnea Syndrome. Am J Respir Crit Care Med 174(7):824–830
Sadacca LA, Lamia KA, de Lemos AS, Blum B, Weitz CJ (2011) An intrinsic circadian clock of the pancreas is required for normal insulin release and glucose homeostasis in mice. Diabetologia. 54(1):120–124
Sahar S, Sassone-Corsi P (2009) Metabolism and cancer: the circadian clock connection. Nat Rev Cancer. 9 (12):886–896
Saini C, Liani A, Curie T, Gos P, Kreppel F, Emmenegger Y et al (2013) Real-time recording of circadian liver gene expression in freely moving mice reveals the phase-setting behavior of hepatocyte clocks. Genes Dev. 27(13):1526–1536
Sassin JF, Parker DC, Mace JW, Gotlin RW, Johnson LC, Rossman LG (1969) Human growth hormone release: relation to slow-wave sleep and sleep-waking cycles. Science 165(3892):513–515
Scheer FA, Morris CJ, Shea SA (2013) The internal circadian clock increases hunger and appetite in the evening independent of food intake and other behaviors. Obesity (Silver Spring). 21(3):421–423
Schernhammer ES, Kroenke CH, Laden F, Hankinson SE (2006) Night work and risk of breast cancer. Epidemiology. 17(1):108–111
Schmid DA, Wichniak A, Uhr M, Ising M, Brunner H, Held K et al (2006) Changes of sleep architecture, spectral composition of sleep EEG, the nocturnal secretion of cortisol, ACTH, GH, Prolactin, Melatonin, Ghrelin, and Leptin, and the DEX-CRH test in depressed patients during treatment with mirtazapine. Neuropsychopharmacology. 31(4):832–844
Schmid SM, Hallschmid M, Jauch-Chara K, Wilms B, Benedict C, Lehnert H et al (2009) Short-term sleep loss decreases physical activity under free-living conditions but does not increase food intake under time-deprived laboratory conditions in healthy men. Am J Clin Nutr 90(6):1476–1482
Schmitt K, Grimm A, Dallmann R, Oettinghaus B, Restelli LM, Witzig M et al (2018) Circadian control of DRP1 activity regulates mitochondrial dynamics and bioenergetics. Cell Metab 27(3):657–66.e5
Schmutz I, Ripperger JA, Baeriswyl-Aebischer S, Albrecht U (2010) The mammalian clock component

PERIOD2 coordinates circadian output by interaction with nuclear receptors. Genes Dev. 24(4):345–357
Sharma S, Kavuru M (2010) Sleep and metabolism: an overview. Int J Endocrinol. 2010:270832
Shilts J, Chen G, Hughey JJ (2018) Evidence for widespread dysregulation of circadian clock progression in human cancer. PeerJ. 6:e4327
Siegel JM (2004) Hypocretin (OREXIN): role in normal behavior and neuropathology. Annual Rev Psychol 55 (1):125–148
Simon C, Gronfier C, Schlienger JL, Brandenberger G (1998) Circadian and ultradian variations of leptin in normal man under continuous enteral nutrition: relationship to sleep and body temperature. J Clin Endocrinol Metab. 83(6):1893–1899
Sinturel F, Makhlouf AM, Meyer P, Tran C, Pataky Z, Golay A et al (2019) Cellular circadian period length inversely correlates with HbA. Diabetologia. 62 (8):1453–1462
Sinues PM, Kohler M, Zenobi R (2013) Monitoring diurnal changes in exhaled human breath. Anal Chem. 85 (1):369–373
Skene DJ, Skornyakov E, Chowdhury NR, Gajula RP, Middleton B, Satterfield BC et al (2018) Separation of circadian- and behavior-driven metabolite rhythms in humans provides a window on peripheral oscillators and metabolism. Proc Natl Acad Sci U S A. 115 (30):7825–7830
Smurra M, Philip P, Taillard J, Guilleminault C, Bioulac B, Gin H (2001) CPAP treatment does not affect glucose–insulin metabolism in sleep apneic patients. Sleep Med 2(3):207–213
Spiegel K, Tasali E, Penev P, Van Cauter E (2004) Brief communication: Sleep curtailment in healthy young men is associated with decreased leptin levels, elevated ghrelin levels, and increased hunger and appetite. Ann Intern Med. 141(11):846–850
Stenvers DJ, Scheer F, Schrauwen P, la Fleur SE, Kalsbeek A (2019) Circadian clocks and insulin resistance. Nat Rev Endocrinol. 15(2):75–89
Sun T, Han X (2019) Death versus dedifferentiation: The molecular bases of beta cell mass reduction in type 2 diabetes. Semin Cell Dev Biol. 103:76–82
Tang F, Lane S, Korsak A, Paton JFR, Gourine AV, Kasparov S et al (2014) Lactate-mediated glia-neuronal signalling in the mammalian brain. Nat Commun 5 (1):1–14
Tatsuki F, Sunagawa GA, Shi S, Susaki EA, Yukinaga H, Perrin D et al (2016) Involvement of Ca(2+)-dependent hyperpolarization in sleep duration in mammals. Neuron. 90(1):70–85
Teske JA, Mavanji V (2012) Energy expenditure: role of orexin. Vitam Horm. 89:91–109
Thaiss CA, Zeevi D, Levy M, Zilberman-Schapira G, Suez J, Tengeler AC et al (2014) Transkingdom control of microbiota diurnal oscillations promotes metabolic homeostasis. Cell. 159(3):514–529
Tononi G, Cirelli C (2020) Sleep and synaptic downselection. Eur J Neurosci. 51(1):413–421
Turek FW, Joshu C, Kohsaka A, Lin E, Ivanova G, McDearmon E et al (2005) Obesity and metabolic syndrome in circadian Clock mutant mice. Science. 308(5724):1043–1045
Van Cauter E, Blackman JD, Roland D, Spire JP, Refetoff S, Polonsky KS (1991) Modulation of glucose regulation and insulin secretion by circadian rhythmicity and sleep. J Clin Invest 88(3):934–942
van der Vinne V, Swoap SJ, Vajtay TJ, Weaver DR (2018) Desynchrony between brain and peripheral clocks caused by CK1δ/ε disruption in GABA neurons does not lead to adverse metabolic outcomes. Proc Natl Acad Sci U S A. 115(10):E2437–E2E46
Varin C, Rancillac A, Geoffroy H, Arthaud S, Fort P, Gallopin T (2015a) Glucose induces slow-wave sleep by exciting the sleep-promoting neurons in the ventrolateral preoptic nucleus: a new link between sleep and metabolism. J Neurosci 35(27):9900–9911
Varin C, Rancillac A, Geoffroy H, Arthaud S, Fort P, Gallopin T (2015b) Glucose induces slow-wave sleep by exciting the sleep-promoting neurons in the ventrolateral preoptic nucleus: a new link between sleep and metabolism. J Neurosci 35(27):9900–9911
Vetter C, Devore EE, Ramin CA, Speizer FE, Willett WC, Schernhammer ES (2015) Mismatch of sleep and work timing and risk of type 2 diabetes. Diabetes Care. 38 (9):1707–1713
Vollmers C, Gill S, DiTacchio L, Pulivarthy SR, Le HD, Panda S (2009) Time of feeding and the intrinsic circadian clock drive rhythms in hepatic gene expression. Proc Natl Acad Sci U S A. 106(50):21453–21458
Wang Z, Ma J, Miyoshi C, Li Y, Sato M, Ogawa Y et al (2018) Quantitative phosphoproteomic analysis of the molecular substrates of sleep need. Nature. 558 (7710):435–439
Webb P, Hiestand M (1975) Sleep metabolism and age. J Appl Phys 38(2):257–262. https://doi.org/10.1152/jappl1975382257
Weikel JC, Wichniak A, Ising M, Brunner H, Friess E, Held K et al (2003) Ghrelin promotes slow-wave sleep in humans. Am J Physiol Endocrinol Metab 284(2): E407–EE15
West SD, Nicoll DJ, Stradling JR (2006) Prevalence of obstructive sleep apnoea in men with type 2 diabetes. Thorax. 61(11):945
White DP, Weil JV, Zwillich CW (1985) Metabolic rate and breathing during sleep. J Appl Phys 59 (2):384–391. https://doi.org/10.1152/jappl1985592384
Wisor JP, O'Hara BF, Terao A, Selby CP, Kilduff TS, Sancar A et al (2002) A role for cryptochromes in sleep regulation. BMC Neurosci. 3:20
Xie L, Kang H, Xu Q, Chen MJ, Liao Y, Thiyagarajan M et al (2013) Sleep drives metabolite clearance from the adult brain. Science. 342(6156):373–377
Yamanaka A, Beuckmann CT, Willie JT, Hara J, Tsujino N, Mieda M et al (2003) Hypothalamic orexin neurons regulate arousal according to energy balance in mice. Neuron. 38(5):701–713

Yehuda S, Sredni B, Carasso RL, Kenigsbuch-Sredni D (2009) REM sleep deprivation in rats results in inflammation and interleukin-17 elevation. J Interf Cytokine Res 29(7):393–398

Yu JH, Yun CH, Ahn JH, Suh S, Cho HJ, Lee SK et al (2015) Evening chronotype is associated with metabolic disorders and body composition in middle-aged adults. J Clin Endocrinol Metab. 100 (4):1494–1502

Zhang Y, Fang B, Emmett MJ, Damle M, Sun Z, Feng D et al (2015) Gene regulation. Discrete functions of nuclear receptor Rev-erbalpha couple metabolism to the clock. Science. 348(6242):1488–1492

Linking Depression to Epigenetics: Role of the Circadian Clock

3

Shogo Sato and Paolo Sassone-Corsi

Abstract

The circadian clock governs multiple biological functions at the molecular level and plays an essential role in providing temporal diversity of behavior and physiology including neuronal activity. Studies spanning the past two decades have deciphered the molecular mechanisms of the circadian clock, which appears to operate as an essential interface in linking cellular metabolism to epigenetic control. Accumulating evidence illustrates that disruption of circadian rhythms through jet lag, shift work, and temporary irregular life-style could lead to depression-like symptoms. Remarkably, abnormal neuronal activity and depression-like behavior appear in animals lacking elements of the molecular clock. Recent studies demonstrate that neuronal and synaptic gene induction is under epigenetic control, and robust epigenetic remodeling is observed under depression and related psychiatric disorders. Thus, the intertwined links between the circadian clock and epigenetics may point to novel approaches for antidepressant treatments, epigenetic therapy, and chronotherapy. In this chapter we summarize how the circadian clock is involved in neuronal functions and depressive-like behavior and propose that potential strategies for antidepressant therapy by incorporating circadian genomic and epigenetic rewiring of neuronal signaling pathways.

Keywords

Epigenetics · Clock · Depression · Metabolism · Chronotherapy

3.1 Highlights

1. Disruption of circadian rhythms is critically detrimental for neuronal functions and mental health.
2. Neuronal and synaptic activities are associated with metabolism and under epigenetic gene regulation.
3. Depression disrupts circadian clockwork and remodels chromatin structure.
4. The circadian clock intimately modulates epigenetic gene regulation and vice versa.
5. The link between metabolism, epigenetics, and the circadian clock may help developing antidepressant therapy.

S. Sato (✉)
Center for Epigenetics and Metabolism, INSERM U1233, Department of Biological Chemistry, School of Medicine, University of California, Irvine, USA

Department of Biology, College of Science, Texas A&M University, College Station, TX, USA
e-mail: shogo.sato@bio.tamu.edu

P. Sassone-Corsi
Center for Epigenetics and Metabolism, INSERM U1233, Department of Biological Chemistry, School of Medicine, University of California, Irvine, USA

O. Engmann, M. Brancaccio (eds.), *Circadian Clock in Brain Health and Disease*, Advances in Experimental Medicine and Biology 1344, https://doi.org/10.1007/978-3-030-81147-1_3

3.2 Introduction

Timed oscillations in physiology, metabolism, and neuronal activity, which control biological daily rhythms of multiple functions and behaviors, are under the control of a so-called circadian clock (Green et al. 2008; Panda 2016; Top and Young 2017). The word circadian derives from Latin (*circa diem*, about a day) and relates to events that follow a cycle of ~24 hours. Disruption of circadian rhythms through jet lag, shift work, and sleep disturbance increases the risk of mental health problems (Lyall et al. 2018; Zaki et al. 2018). Notably, depressed patients may have abnormal sleep-wake cycles, temperature rhythms, hormonal secretion, and mood (Bunney and Bunney 2013). However, the molecular mechanisms of the disharmony between the circadian clock, behaviors, and psychiatric disorders such as depression are not fully understood.

Epigenetic control, and specifically chromatin remodeling, has been associated to depression-like disorders (Deussing and Jakovcevski 2017; Mahgoub and Monteggia 2013; Nagy et al. 2018; Sato and Sassone-Corsi 2019). Of note, histone modifications display robust daily oscillation elicited by clock-controlled epigenetic enzymes as well as the cyclic availability of metabolites involved in epigenetic modifications (Berger and Sassone-Corsi 2016; Koike et al. 2012). An interplay thereby exists between circadian control, depression, and epigenetic modifications. While the molecular mechanisms of this connection remain unexplored, it is conceivable that disruption of clock-controlled rhythmic histone modifications may lead to dysfunctions of neuronal physiology and activity.

In this chapter we discuss: (1) the relationship between depression and circadian rhythms; (2) how the circadian clock plays a key role in connecting cellular metabolism and epigenetic control; (3) the possible mechanisms by which epigenetic modifications underlie depression; and (4) the future of the use of chronotherapy to tackle depression-related behaviors.

3.3 Link of the Circadian Clock to Depression-Like Behavior

A remarkable array of physiological and behavioral functions follows circadian cycles. These are governed by a molecular system that, through the regulation of thousands of genes (Greco and Sassone-Corsi 2018; Masri and Sassone-Corsi 2014; Takahashi 2017), controls daily rhythmicity of the feeding behavior, immune responses, metabolism, mood, and sleep-wake cycles (Green et al. 2008; Masri and Sassone-Corsi 2018; Top and Young 2017). The central clock in mammals is located in the suprachiasmatic nucleus (SCN) of the hypothalamus. Disruption of SCN function has been linked to depression-like behaviors. Indeed, circadian behaviors are orchestrated by a subset of SCN neurons expressing the neuropeptide neuromedin S (NMS). Abolishing the molecular clock of NMS neurons disrupts in vivo circadian behavior (Lee et al. 2015). Also, depression-like behaviors such as helplessness, behavioral despair, and anxiety-like behavior have been associated to genetic disruption of SCN circadian rhythms in mice (Landgraf et al. 2016).

Environmental cues influence the free-running function of the clock. These are called *zeitgebers*, light being the most powerful for the central clock (LeGates et al. 2014). The effect of light on circadian behavior could explain why irregular light/dark cycles, such as those generated by shift work and jet lag, may lead to mental and behavioral disorders. Different light schedules have been applied to rodents, establishing models for experimentation (LeGates et al. 2014). Epidemiological studies indicate that disruption of circadian rhythms in humans increases the risk for psychiatric problems (LeGates et al. 2014; Lyall et al. 2018; Zaki et al. 2018). A robust association of disrupted circadian rhythmicity with mood disorders has been revealed in a cross-sectional study of 91,105 participants (Lyall et al. 2018). An estimated 300 million people worldwide suffer from depression and suicide accounts for >800,000 deaths per year. Notably, the expression of core clock genes is significantly dysregulated across six brain regions in patients compared with nonpsychiatric controls (Li et al. 2013). In addition to the circadian genomic signatures, major depressive disorder (MDD) patients often exhibit abnormal blood concentration of hormones, such as cortisol and melatonin, and pronounced circadian phase shift in the hormonal secretion (Germain and Kupfer 2008; LeGates et al. 2014). Thus, development of depressive-like behaviors appears to be intimately linked to alteration of circadian rhythms.

3.4 The Circadian Clock: A Molecular Link to Antidepressant Action

The molecular organization of the mammalian clock is based on the transcriptional activators CLOCK and BMAL1 that dimerize to activate thousands of genes through E-box promoter elements (Takahashi 2017). The CLOCK/BMAL1 complex drives circadian gene expression and contributes to transcriptional/translational feedback loops (TTFLs) composed of additional core clock components (Fig. 3.1). Among the clock-controlled genes (CCGs), there are genes encoding the circadian repressors PERs and CRYs (Takahashi 2017). We have investigated the impact of antidepressant treatments on clock gene response. Specifically, ketamine, a noncompetitive high-affinity N-methyl-D-aspartate (NMDA) receptor antagonist, impacts CLOCK/BMAL1 function, leading to altered circadian gene expression in neuronal cells cultures (Bellet et al. 2011). An additional study compared the effects of two rapid-acting antidepressant treatments, low-dose ketamine, and sleep deprivation (SD) on the transcriptome in the anterior cingulate cortex of mice (Orozco-Solis et al. 2017). Notably, both treatments elicit common transcriptional responses related to neuronal plasticity and the circadian clock. Most recently, the circadian transcription factor BMAL1 has been demonstrated to be essential for gene induction of *Homer1a*, a stress-responsible antidepressant gene in response to SD stress (Sato et al. 2020), underscoring the molecular implication of the circadian clock in the regulation of neuronal activity, depressive behavior, and the action of antidepressant. These findings point to the interaction between depression, antidepressant treatments, and the circadian clock.

3.5 The Circadian Clock: A Critical Interface Between Metabolism and Epigenetics

The importance of a physiologically timed and well-regulated feeding lifestyle on the relationship between the circadian clock and metabolism has been underscored (Asher and Sassone-Corsi 2015). Also, through the remodeling of chromatin structure, a link exists between metabolic alterations and gene regulation (Berger and Sassone-Corsi 2016). Small metabolites, such as acetyl-coenzyme A (CoA), S-adenosyl methionine (SAM), and nicotinamide adenine dinucleotide (NAD^+) play a regulatory role in the modification of epigenetic histone and nonhistone proteins (Berger and Sassone-Corsi 2016). These metabolites often serve as substrates or upstream intermediates that ultimately modulate chromatin marks. Thus, cellular metabolism and epigenetic control are intimately linked. Importantly, the circadian clock controls the synthesis and usage of several metabolites. This places the clock in a strategic position to bridge metabolism and epigenetic function.

An important link between cellular metabolism and epigenetic control relates to the synthesis of NAD^+. Levels of NAD^+ exhibit robust daily oscillation driven by the rhythmic expression of nicotinamide phosphoribosyltransferase (NAMPT), the rate-limiting enzyme in the salvage pathway of NAD^+ biosynthesis. The expression of the *Nampt* gene is directly controlled by the circadian clock, in virtue of E-box elements in its promoter which bind the CLOCK-BMAL1 complex (Nakahata et al. 2009; Ramsey et al. 2009). This regulation leads to the rhythmic activity of NAD^+-dependent SIRT1, a class III histone deacetylase (HDAC). SIRT1 regulates the cyclic deacetylation of nonhistone BMAL1 and histone

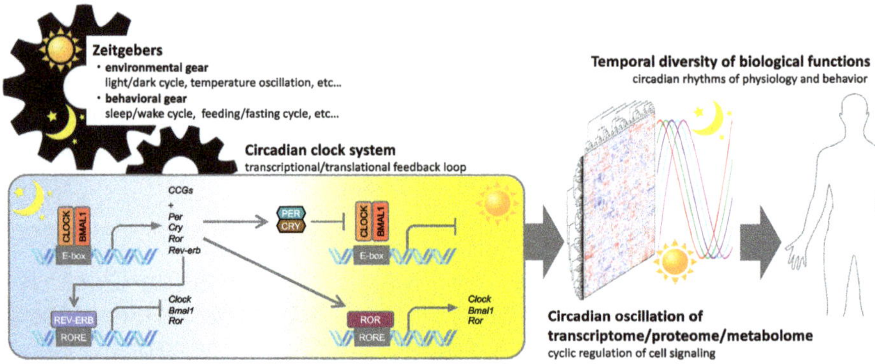

Fig. 3.1 Transcriptional/translational feedback loops (TTFLs) composed of molecular core clock components and their regulation of rhythmic biological functions. Rhythmic behavior and circadian biological homeostasis are sustained by the cooperation from external environmental cues, including light/dark cycles and feeding/fasting cycles, and the internal molecular cogs, the circadian clock. The core transcriptional factors CLOCK and BMAL1 heterodimer activates the gene expression of clock-controlled genes (CCGs) through the binding to E-box elements in the promoter region. CLOCK-BMAL1 also transactivates the expression of *Period* (*Per*) and *Cryptochrome* (*Cry*) genes. The accumulated PERs and CRYs proteins can translocate into nuclear to repress CLOCK-BMAL1-driven transcription. The other key players in the core circadian TTFLs are the orphan nuclear receptors RAR-related orphan receptors (RORs) and REV-ERBs which activate and repress *Bmal1* transcription, respectively, through access to the specific response elements (ROREs). The TTFLs play an essential role in the operation of daily oscillation of physiology and behavior. At the molecular level, the TTFLs directly contribute to cyclic fluctuation of gene/protein expression, epigenetic modification, and metabolites abundance

H3 lysine9/lysine14 at circadian gene promoters, resulting in rhythmic gene expression (Nakahata et al. 2008).

Another key epigenetic player in circadian control is mixed-lineage leukemia (MLL1). This enzyme operates as an H3 lysine4 tri-methyltransferase and is directly linked to activation of gene expression. It has been demonstrated that MLL1 contributes to efficient and time-controlled expression of circadian genes by directing the cyclic recruiting of the CLOCK/BMAL1 complex to target promoters at chromatin (Katada and Sassone-Corsi 2010). Importantly, conditional ablation of the gene encoding this enzyme in the postnatal forebrain is associated with excessive nocturnal activity and with absent or blunted responses to stimulant and dopaminergic agonist drugs, in conjunction with near-complete loss of spike-timing-dependent long-term potentiation in medium spiny neurons (MSNs). In addition, as MLL1 contributes to the transcriptional activation through the positive mark H3 lysine4 mono- and tri-methylation, its deletion causes transcriptional repression of a number of genes, including the 5HTR2A serotonin receptor, strongly associated with anxiety- and depression-related disorders in human and animal models (Shen et al. 2016).

It was also demonstrated that the NAD^+-SIRT1-dependent cyclic histone H3 lysine4 trimethylation was driven by circadian deacetylation of MLL1 (Aguilar-Arnal et al. 2015). In addition, a time-of-day-dependent landscape of rhythmic epigenome in mouse liver shows distinct phase distribution of cyclic modifications of several histone marks (Koike et al. 2012). Thus, cellular metabolism is a rhythmic modifier for the circadian gene expression at the epigenetic level and subsequently drives circadian changes in physiology, mood, and behavior.

3.6 Metabolic and Epigenetic Reprogramming Under Depression

Dynamic alterations in epigenetic modifications have been linked to depression and depression-like behaviors (Deussing and Jakovcevski 2017;

Mahgoub and Monteggia 2013; Nagy et al. 2018). Activation of Ca^{2+}-dependent cAMP response element-binding protein (CREB) pathway appears to play a central role. Phosphorylation and activation of CREB initiate downstream transcriptional activation in response to stress stimuli-induced calcium influx. The histone acetyltransferase (HAT) CREB-binding protein (CBP) is recruited once CREB becomes activated. This pathway is central in regulating target gene expression such as brain-derived neurotrophic factor (*Bdnf*), leading to dysfunction of neuronal activity (Tsankova et al. 2006). CBP is a so-called "writer" of acetyl groups, whereas HDACs, so-called "erasers" are also critically important to remodel chromatin structures under neuronal activity (Borrelli et al. 2008; Gallegos et al. 2018). Indeed, spontaneous electrical activity of neurons results in nuclear export of HDAC4, whereas translocation of HDAC5 to the cytoplasm is stimulated by the activation of NMDA receptors (Chawla et al. 2003). Thus, the coordinated regulation of histone acetylation by HATs and HDACs underscores the central role played by chromatin remodeling in neuronal gene expression. Different types of stress exposures result in dynamic alterations of histone acetylation in several brain regions (Covington et al. 2009; Fuchikami et al. 2009; Hinwood et al. 2011; Kenworthy et al. 2014; Mahan et al. 2012; Montagud-Romero et al. 2016).

Important findings have linked components of the circadian clock machinery to epigenetic control. Specifically, CLOCK, a core component of the circadian pacemaker, has been shown to have HAT activity (Doi et al. 2006). Notably, chronic unpredictable stress has been shown to decrease the expression of CLOCK and BMAL1 proteins in the SCN, with the peak of CLOCK protein oscillation shifted from dark phase into light phase (Jiang et al. 2013). Moreover, depressive-like behaviors have been observed in mice upon knockdown of the *Clock* gene in the hippocampus. Finally, the NAD^+-dependent HDAC SIRT1 is also involved in the response to chronic stress stimuli (Abe-Higuchi et al. 2016). Pharmacological and genetic inhibitions of hippocampal SIRT1 exacerbate depressive-like behaviors. While the link between the circadian HAT/HDAC components and remodeling of specific histone marks at neural gene promoters is still undefined, these findings place the circadian clock in a pivotal position with respect to neuronal gene expression and related epigenetic modifications.

3.7 Antidepressive Therapy: Epigenetic Treatment

Development of depression and depressive-like behaviors appears to be directly linked to histone deacetylation via the activation and nuclear localization of HDACs. Indeed, a number of studies have focused on the effect of HDAC inhibitors on depressive behaviors (Table 3.1). In addition to well-established HDAC inhibitors, novel synthetic and natural compounds have been shown to inhibit HDACs activity and reduce depression-like behaviors. In cognitively-impaired mice, the CA1 region of the hippocampus display decreased acetylation of histone H4 at Lys12, an event tightly associated with genes linked to synaptic plasticity (Benito et al. 2015). Notably, the HDAC inhibitor suberoylanilide hydroxamic acid (SAHA) reinstates the expression of genes involved in synaptic plasticity, which is associated to increase Lys12 acetylation of histone H4. Also, the synthetic compound 2-{3-(3-flurophenthyloxy)phenyl-amino}benzamide (33i)that operates as a selective inhibitor of the deacetylase SIRT2, increases histone H3 and H4 acetylation, leading to the up-regulation of several synaptic plasticity markers linked to glutamate neurotransmission such as NMDA receptor subunits (Erburu et al. 2017). Also, chronic SIRT2 inhibition reverts anhedonia and social avoidance. Another study in the mouse identified two phytochemicals, dihydrocaffeic acid (DHCA) and malvidin-3'-O-glucoside (Mal-gluc), involved in reduction of depression-like behaviors (Wang et al. 2018). Remarkably, Mal-gluc modulates synaptic plasticity by inhibiting HDAC2 and increasing histone acetylation at the promoter of the gene *Rac1*

Table 3.1 Antidepressant action of HDAC inhibitors

HDAC inhibitor	Epigenetic action	Antidepressant action	References
Vorinostat (suberoylanilide hydroxamic acid, SAHA)	Normalized global H4 Lys12 acetylation	Improved memory function Reduced depression-like behaviors Increased Gdnf expression in stressed animals	Benito et al. (2015), Uchida et al. (2011)
2-{3-(3-fluorophenethyloxy) phenylamino} benzamide (33i)	Increased global H3/H4 acetylation	Reduced depression-like behaviors Increased GluN2A and GluN2B expression	Erburu et al. (2017)
MS-275	Increased H4 Lys 12 acetylation at Bdnf promoter Increased global histone H3 acetylation	Reduced depression-like behaviors Increased CREB and BDNF expression	Covington et al. (2009, 2011), Lin et al. (2012), Schmauss (2015)
Sodium butyrate	Normalized global histone H3 acetylation level Normalized HDAC2 expression Increased global H3 and H4 acetylation Increased H4 Lys 12 acetylation at Bdnf promoter Increased H4 acetylation at Ttr promoter	Reduced depression-like behaviors	Han et al. (2014), Schmauss (2015), Schroeder et al. (2007), Yamawaki et al. (2012)
TSA	Increased H4 Lys 12 acetylation at Bdnf promoter	Reduced depression-like behaviors	Schmauss (2015)
Malvidin-3'-O-glucoside (mal-gluc)	Decreased HDAC2 expression Increased H3 acetylation at Rac1 promoter	Reduced depression-like behaviors	Wang et al. (2018)
Sirtinol, EX537	Inhibition of SIRT1	Led to depression-like behavior	Abe-Higuchi et al. (2016)
SRT2104, resveratrol	Activation of SIRT1	Led to stress resistance	Abe-Higuchi et al. (2016)

(RAS-related C3 botulinum toxin substrate 1), which encodes a regulator of dendric spines and excitatory synapses.

In addition to the classical HDAC inhibitors, rapid antidepressant action of ketamine is shown to be associated with the phosphorylation and nuclear export of HDAC5 (Choi et al. 2015). Intriguingly, L-acetylcarnitine (LAC), a regulator of mitochondrial metabolism, exerts rapid antidepressant effects (Bigio et al. 2016). The effects are mediated through enhancing acetylation of the nonhistone protein NFkB-p65. This event results in increased expression of type 2 metabotropic glutamate (mGlu2) receptor and in reverting the reduction of acetylated histone H3 levels at the promoter of *Grm2* gene, encoding mGlu2 receptor, in hippocampus and prefrontal cortex from depressed animals (Nasca et al. 2013). Thus, effective treatment of depression may include both inhibition of HDACs and increased availability of acetyl group.

3.8 Is Chronobiological Treatment a Potential Antidepressive Therapy?

An intimate connection exists between circadian rhythms and behaviors, as well as circadian molecular regulation of neuronal functions and behaviors, which could contribute to developing a variety of antidepressant therapies based on chronobiology. A variety of chronobiological strategies are designed to control light/dark environments and sleep/wake cycles, which in turn can directly influence circadian rhythms (Dallaspezia and Benedetti 2011; Dallaspezia et al. 2015; Germain and Kupfer 2008; Khalifeh 2017; Ohdo 2010). We list some regimes that appear to be effective in the treatment of depression as well as mood and sleep disorders, in both animal and human models (Dallaspezia et al. 2015; Khalifeh 2017). While the mechanisms behind the rapid antidepressant action of chronotherapies are yet to be satisfactorily explored, these various classes of chronotherapies provide important leads to be further studied:

- *Sleep deprivation (SD).*

SD targets endocrine systems such as serotonin and dopamine secretions (Benedetti and Smeraldi 2009), which can also influence the circadian clock (Mongrain et al. 2011; Orozco-Solis et al. 2017; Sato et al. 2020).

- *Sleep phase advance.*

Advancing the timing of the sleep/wake cycles also aims at normalizing biological rhythms in depressed patients as well as operating on the circadian clock.

- *Light therapy.*

Light is a strong *zeitgeber* for the central clock and light therapy is utilized to rewire misaligned biological clock functions. Light therapy is developed to extend daytime photoperiod and counteract winter darkness and thereby exhibits powerful antidepressant effects especially for seasonal affective disorders (Rosenthal et al. 1984).

- *Dark therapy.*

Photoreceptors in the mammalian retina transmit the light signal to the master clock in the SCN (Berson et al. 2002). These photoreceptors, melanopsin-containing intrinsically photosensitive retina ganglion cells (ipRGCs), are the most sensitive to blue light with a wavelength of 460 nm (Lockley et al. 2003). Exposure to blue light from computer screens or smartphones at night can influence sleep quality, a common risk factor for depression. Dark therapy can improve sleep quality in patients by protecting from blue light at night.

3.9 Conclusion: The Promise of Epigenetics in Pharmacological Approaches of Depression

Genome-wide association meta-analysis has revealed a number of genomic risk variants for MDD (Wray et al. 2018). Critical genetic risk factors for MDD appear to be related to body mass, educational attainment, and schizophrenia and associated with the development of depression-like behaviors. Importantly, genetic variations of the circadian clock are associated with behavioral dysfunctions, including sleep disorders. This is the case for the human gene *Cry1*, which is linked to familial delayed sleep phase disorder (Patke et al. 2017).

In addition to the genetic insights for depression, basic and clinical research has highlighted the role of epigenetic regulation in stress and

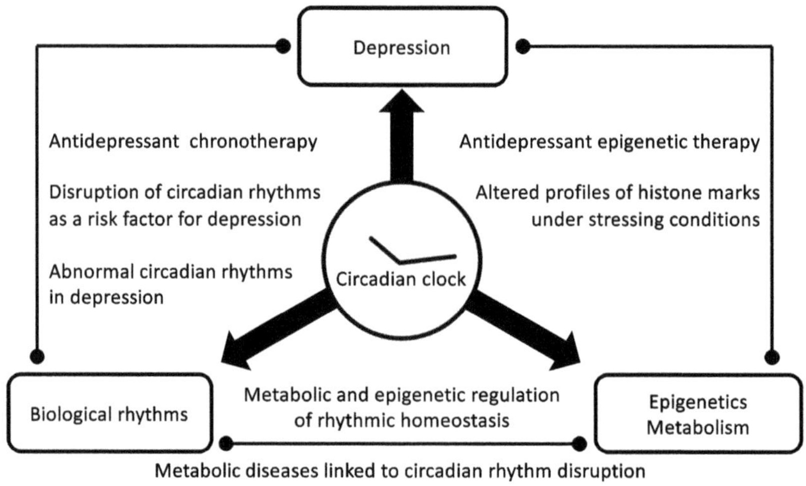

Fig. 3.2 The circadian clock mediates the connection among biological rhythms, epigenetics/metabolism, and depression. In this chapter we highlight (1) the evidence linking depression to circadian rhythms, (2) epigenetic gene regulation under depression, and (3) the crosstalk between the circadian clock and epigenetic gene regulation. The additional study will be required to develop beneficial chronobiology-based epigenetic therapies for depression-like behaviors

depression. Here we have summarized some important links between circadian rhythm and depression as well as epigenetic regulation of depression (Fig. 3.2). We have proposed that the circadian clock plays a key role in depression-like behaviors through the remodeling of epigenetic modifications. However, how the circadian clock regulates the transcription of genes involved in neuronal and synaptic activity by editing chromatin remodeling remains virtually unexplored. It is also unclear whether and how irregular light environment or sleep/wake cycles could lead to alterations in histone marks. These questions need to be addressed as they may help in developing novel antidepressant strategies which would integrate chronotherapy and epigenetics. For instance, there might be a time-of-the-day preference to exert more beneficial effects of LAC treatment to reduce depression-like behaviors.

Recent interventions for depression have shown that exercise promotes the levels of BDNF through the reduction of HDACs expression and binding to the hippocampal *Bdnf* promoter (Sleiman et al. 2016). These actions are mediated by increased levels of the ketone body β-hydroxybutyrate, known as an endogenous HDAC inhibitor (Shimazu et al. 2013). Importantly, a ketogenic diet induces β-hydroxybutyrate levels to robustly oscillate in a circadian manner, coupled to rhythmic HDAC activity and histone acetylation (Tognini et al. 2017), suggesting the efficacy of exercise intervention for the antidepressant treatment can depend on a time-of-the-day strategy. This notion may be supported by our recent findings demonstrating when to exercise is a critical determinant for its impact on skeletal muscle metabolism and endurance performance (Ezagouri et al. 2019; Sato et al. 2019). Also, both basic and clinical investigations indicate that fasting and short-term caloric restrictions improve depression-like behaviors (Zhang et al. 2015). Since not only does fasting induce ketogenesis, but also caloric restriction strongly enhances acetylation of histone and nonhistone proteins by rewiring clock-controlled metabolic pathways involved in protein acetylation (Sato et al. 2017), intervention of fasting or caloric restriction could exert antidepressant efficacy through dynamic reprogramming of cyclic epigenetic modifications. These studies pave the way to new avenues of antidepressant treatments and possibly identify the most effective treatments for patients suffering from depression.

Acknowledgments We are grieved by the unexpected loss of Prof. Paolo Sassone-Corsi. It has been an honor and pleasure to have had the opportunity to work with him on this book chapter. We thank all the members of the Sassone-Corsi lab for useful discussions and support. S.S. is supported by the Brain & Behavioral Research Foundation (NARSAD Young Investigator Grant 28681).

Conflict of Interest The authors declare no conflict of interest.

References

Abe-Higuchi N et al (2016) Hippocampal Sirtuin 1 signaling mediates depression-like behavior. Biol Psychiatry 80:815–826. https://doi.org/10.1016/j.biopsych.2016.01.009

Aguilar-Arnal L, Katada S, Orozco-Solis R, Sassone-Corsi P (2015) NAD(+)-SIRT1 control of H3K4 trimethylation through circadian deacetylation of MLL1. Nat Struct Mol Biol 22:312–318. https://doi.org/10.1038/nsmb.2990

Asher G, Sassone-Corsi P (2015) Time for food: the intimate interplay between nutrition, metabolism, and the circadian clock. Cell 161:84–92. https://doi.org/10.1016/j.cell.2015.03.015

Bellet MM, Vawter MP, Bunney BG, Bunney WE, Sassone-Corsi P (2011) Ketamine influences CLOCK:BMAL1 function leading to altered circadian gene expression. PLoS One 6:e23982. https://doi.org/10.1371/journal.pone.0023982

Benedetti F, Smeraldi E (2009) Neuroimaging and genetics of antidepressant response to sleep deprivation: implications for drug development. Curr Pharm Des 15:2637–2649

Benito E et al (2015) HDAC inhibitor-dependent transcriptome and memory reinstatement in cognitive decline models. J Clin Invest 125:3572–3584. https://doi.org/10.1172/jci79942

Berger SL, Sassone-Corsi P (2016) Metabolic signaling to chromatin. Cold Spring Harb Perspect Biol 8(11): a019463. https://doi.org/10.1101/cshperspect.a019463

Berson DM, Dunn FA, Takao M (2002) Phototransduction by retinal ganglion cells that set the circadian clock. Science 295:1070–1073. https://doi.org/10.1126/science.1067262

Bigio B, Mathe AA, Sousa VC, Zelli D, Svenningsson P, McEwen BS, Nasca C (2016) Epigenetics and energetics in ventral hippocampus mediate rapid antidepressant action: implications for treatment resistance. Proc Natl Acad Sci U S A 113:7906–7911. https://doi.org/10.1073/pnas.1603111113

Borrelli E, Nestler EJ, Allis CD, Sassone-Corsi P (2008) Decoding the epigenetic language of neuronal plasticity. Neuron 60:961–974. https://doi.org/10.1016/j.neuron.2008.10.012

Bunney BG, Bunney WE (2013) Mechanisms of rapid antidepressant effects of sleep deprivation therapy: clock genes and circadian rhythms. Biol Psychiatry 73:1164–1171. https://doi.org/10.1016/j.biopsych.2012.07.020

Chawla S, Vanhoutte P, Arnold FJ, Huang CL, Bading H (2003) Neuronal activity-dependent nucleocytoplasmic shuttling of HDAC4 and HDAC5. J Neurochem 85:151–159

Choi M et al (2015) Ketamine produces antidepressant-like effects through phosphorylation-dependent nuclear export of histone deacetylase 5 (HDAC5) in rats. Proc Natl Acad Sci U S A 112:15755–15760. https://doi.org/10.1073/pnas.1513913112

Covington HE 3rd et al (2009) Antidepressant actions of histone deacetylase inhibitors. J Neurosci 29:11451–11460. https://doi.org/10.1523/jneurosci.1758-09.2009

Covington HE 3rd, Vialou VF, LaPlant Q, Ohnishi YN, Nestler EJ (2011) Hippocampal-dependent antidepressant-like activity of histone deacetylase inhibition. Neurosci Lett 493:122–126

Dallaspezia S, Benedetti F (2011) Chronobiological therapy for mood disorders. Expert Rev Neurother 11:961–970. https://doi.org/10.1586/ern.11.61

Dallaspezia S, Suzuki M, Benedetti F (2015) Chronobiological therapy for mood disorders. Curr Psychiatry Rep 17:95. https://doi.org/10.1007/s11920-015-0633-6

Deussing JM, Jakovcevski M (2017) Histone modifications in major depressive disorder and related rodent models. Adv Exp Med Biol 978:169–183. https://doi.org/10.1007/978-3-319-53889-1_9

Doi M, Hirayama J, Sassone-Corsi P (2006) Circadian regulator CLOCK is a histone acetyltransferase. Cell 125:497–508. https://doi.org/10.1016/j.cell.2006.03.033

Erburu M, Munoz-Cobo I, Diaz-Perdigon T, Mellini P, Suzuki T, Puerta E, Tordera RM (2017) SIRT2 inhibition modulate glutamate and serotonin systems in the prefrontal cortex and induces antidepressant-like action. Neuropharmacology 117:195–208. https://doi.org/10.1016/j.neuropharm.2017.01.033

Ezagouri S et al (2019) Physiological and molecular dissection of daily variance in exercise capacity. Cell Metab 30:78–91.e74. https://doi.org/10.1016/j.cmet.2019.03.012

Fuchikami M, Morinobu S, Kurata A, Yamamoto S, Yamawaki S (2009) Single immobilization stress differentially alters the expression profile of transcripts of the brain-derived neurotrophic factor (BDNF) gene and histone acetylation at its promoters in the rat

hippocampus. Int J Neuropsychopharmacol 12:73–82. https://doi.org/10.1017/s1461145708008997
Gallegos DA, Chan U, Chen LF, West AE (2018) Chromatin regulation of neuronal maturation and plasticity. Trends Neurosci 41:311–324. https://doi.org/10.1016/j.tins.2018.02.009
Germain A, Kupfer DJ (2008) Circadian rhythm disturbances in depression human. Psychopharmacology 23:571–585. https://doi.org/10.1002/hup.964
Greco CM, Sassone-Corsi P (2018) Circadian blueprint of metabolic pathways in the brain. Nature Rev Neurosci 20:71–82. https://doi.org/10.1038/s41583-018-0096-y
Green CB, Takahashi JS, Bass J (2008) The meter of metabolism. Cell 134:728–742. https://doi.org/10.1016/j.cell.2008.08.022
Han A, Sung Y-B, Chung S-Y, Kwon M-S (2014) Possible additional antidepressant-like mechanism of sodium butyrate: Targeting the hippocampus. Neuropharmacology 81:292–302
Hinwood M, Tynan RJ, Day TA, Walker FR (2011) Repeated social defeat selectively increases deltaFosB expression and histone H3 acetylation in the infralimbic medial prefrontal cortex. Cerebral Cortex (New York, NY: 1991) 21:262–271. https://doi.org/10.1093/cercor/bhq080
Jiang WG et al (2013) Hippocampal CLOCK protein participates in the persistence of depressive-like behavior induced by chronic unpredictable stress. Psychopharmacology 227:79–92. https://doi.org/10.1007/s00213-012-2941-4
Katada S, Sassone-Corsi P (2010) The histone methyltransferase MLL1 permits the oscillation of circadian gene expression. Nat Struct Mol Biol 17:1414–1421. https://doi.org/10.1038/nsmb.1961
Kenworthy CA, Sengupta A, Luz SM, Ver Hoeve ES, Meda K, Bhatnagar S, Abel T (2014) Social defeat induces changes in histone acetylation and expression of histone modifying enzymes in the ventral hippocampus, prefrontal cortex, and dorsal raphe nucleus. Neuroscience 264:88–98. https://doi.org/10.1016/j.neuroscience.2013.01.024
Khalifeh AH (2017) The effect of chronotherapy on depressive symptoms. Evidence-based practice. Saudi Med J 38:457–464. https://doi.org/10.15537/smj.2017.5.18062
Koike N, Yoo SH, Huang HC, Kumar V, Lee C, Kim TK, Takahashi JS (2012) Transcriptional architecture and chromatin landscape of the core circadian clock in mammals. Science 338:349–354. https://doi.org/10.1126/science.1226339
Landgraf D, Long JE, Proulx CD, Barandas R, Malinow R, Welsh DK (2016) Genetic disruption of circadian rhythms in the suprachiasmatic nucleus causes helplessness, behavioral despair, and anxiety-like. Behavior in Mice Biol Psychiatry 80:827–835. https://doi.org/10.1016/j.biopsych.2016.03.1050
Lee IT et al (2015) Neuromedin s-producing neurons act as essential pacemakers in the suprachiasmatic nucleus to couple clock neurons and dictate circadian rhythms. Neuron 85:1086–1102. https://doi.org/10.1016/j.neuron.2015.02.006
LeGates TA, Fernandez DC, Hattar S (2014) Light as a central modulator of circadian rhythms, sleep and affect nature reviews. Neuroscience 15:443–454. https://doi.org/10.1038/nrn3743
Li JZ et al (2013) Circadian patterns of gene expression in the human brain and disruption in major depressive disorder. Proc Natl Acad Sci U S A 110:9950–9955. https://doi.org/10.1073/pnas.1305814110
Lin H, Geng X, Dang W, Wu B, Dai Z, Li Y, Yang Y, Zhang H, Shi J (2012) Molecular mechanisms associated with the antidepressant effects of the class I histone deacetylase inhibitor MS-275 in the rat ventrolateral orbital cortex. Brain Res 1447:119–125
Lockley SW, Brainard GC, Czeisler CA (2003) High sensitivity of the human circadian melatonin rhythm to resetting by short wavelength light. J Clin Endocrinol Metab 88:4502–4505. https://doi.org/10.1210/jc.2003-030570
Lyall LM et al (2018) Association of disrupted circadian rhythmicity with mood disorders, subjective wellbeing, and cognitive function: a cross-sectional study of 91 105 participants from the UK biobank the lancet psychiatry. Lancet Psychiatry 5(6):507–514. https://doi.org/10.1016/s2215-0366(18)30139-1
Mahan AL, Mou L, Shah N, Hu JH, Worley PF, Ressler KJ (2012) Epigenetic modulation of Homer1a transcription regulation in amygdala and hippocampus with pavlovian fear conditioning. J Neurosci 32: 4651–4659. https://doi.org/10.1523/jneurosci.3308-11.2012
Mahgoub M, Monteggia LM (2013) Epigenetics and psychiatry. Neurotherapeutics : The Journal of the American Society for Experimental Neurotherapeutics 10: 734–741. https://doi.org/10.1007/s13311-013-0213-6
Masri S, Sassone-Corsi P (2014) Sirtuins and the circadian clock: bridging chromatin and metabolism. Sci Signal 7:re6
Masri S, Sassone-Corsi P (2018) The emerging link between cancer, metabolism, and circadian rhythms. Nat Med 24:1795–1803. https://doi.org/10.1038/s41591-018-0271-8
Mongrain V, La Spada F, Curie T, Franken P (2011) Sleep loss reduces the DNA-binding of BMAL1, CLOCK, and NPAS2 to specific clock genes in the mouse cerebral cortex. PLoS One 6:e26622. https://doi.org/10.1371/journal.pone.0026622
Montagud-Romero S et al (2016) Up-regulation of histone acetylation induced by social defeat mediates the conditioned rewarding effects of cocaine. Prog Neuro-Psychopharmacol Biol Psychiatry 70:39–48. https://doi.org/10.1016/j.pnpbp.2016.04.016
Nagy C, Vaillancourt K, Turecki G (2018) A role for activity-dependent epigenetics in the development and treatment of major depressive disorder. Genes Brain Behav 17:e12446. https://doi.org/10.1111/gbb.12446

Nakahata Y et al (2008) The NAD+-dependent deacetylase SIRT1 modulates CLOCK-mediated chromatin remodeling and circadian control. Cell 134:329–340. https://doi.org/10.1016/j.cell.2008.07.002
Nakahata Y, Sahar S, Astarita G, Kaluzova M, Sassone-Corsi P (2009) Circadian control of the NAD+ salvage pathway by CLOCK-SIRT1. Science 324:654–657
Nasca C et al (2013) L-acetylcarnitine causes rapid antidepressant effects through the epigenetic induction of mGlu2 receptors. Proc Natl Acad Sci U S A 110: 4804–4809. https://doi.org/10.1073/pnas.1216100110
Ohdo S (2010) Chronotherapeutic strategy: rhythm monitoring, manipulation and disruption. Adv Drug Deliv Rev 62:859–875. https://doi.org/10.1016/j.addr.2010.01.006
Orozco-Solis R et al (2017) A circadian genomic signature common to ketamine and sleep deprivation in the anterior cingulate cortex. Biol Psychiatry 82:351–360. https://doi.org/10.1016/j.biopsych.2017.02.1176
Panda S (2016) Circadian physiology of metabolism. Science 354:1008–1015. https://doi.org/10.1126/science.aah4967
Patke A, Murphy PJ, Onat OE, Krieger AC, Ozcelik T, Campbell SS, Young MW (2017) Mutation of the human circadian clock gene CRY1 in familial delayed sleep phase disorder. Cell 169:203–215.e213. https://doi.org/10.1016/j.cell.2017.03.027
Ramsey KM et al (2009) Circadian clock feedback cycle through NAMPT-mediated NAD+ biosynthesis. Science 324:651–654. https://doi.org/10.1126/science.1171641
Rosenthal NE et al (1984) Seasonal affective disorder. A description of the syndrome and preliminary findings with light therapy Archives of general psychiatry 41: 72–80
Sato S, Sassone-Corsi P (2019) Circadian and epigenetic control of depression-like behaviors. Curr Opin Behav Sci 25:15–22. https://doi.org/10.1016/j.cobeha.2018.05.010
Sato S et al (2017) Circadian reprogramming in the liver identifies metabolic pathways of aging. Cell 170:664–677. e611. https://doi.org/10.1016/j.cell.2017.07.042
Sato S et al (2019) Time of exercise specifies the impact on muscle metabolic pathways and systemic energy homeostasis. Cell Metab 30(1):92–110.e4. https://doi.org/10.1016/j.cmet.2019.03.013
Sato S, Bunney BG, Vawter MP, Bunney WE, Sassone-Corsi P (2020) Homer1a undergoes bimodal transcriptional regulation by CREB and the circadian clock. Neuroscience 434:161–170. https://doi.org/10.1016/j.neuroscience.2020.03.031
Schmauss C (2015) An HDAC-dependent epigenetic mechanism that enhances the efficacy of the antidepressant drug fluoxetine. Sci Rep 5(1)
Schroeder FA, Lin CL, Crusio WE, Akbarian S (2007) Antidepressant-like effects of the histone deacetylase inhibitor, sodium butyrate, in the mouse. Biol Psychiatry 62(1):55–64
Shen EY et al (2016) Neuronal deletion of Kmt2a/Mll1 histone methyltransferase in ventral striatum is associated with defective spike-timing-dependent striatal synaptic plasticity, altered response to dopaminergic drugs, and increased anxiety. Neuropsychopharmacology 41:3103–3113. https://doi.org/10.1038/npp.2016.144
Shimazu T et al (2013) Suppression of oxidative stress by beta-hydroxybutyrate, an endogenous histone deacetylase inhibitor. Science 339:211–214. https://doi.org/10.1126/science.1227166
Sleiman SF et al (2016) Exercise promotes the expression of brain derived neurotrophic factor (BDNF) through the action of the ketone body beta-hydroxybutyrate. eLife 5:e15092. https://doi.org/10.7554/eLife.15092
Takahashi JS (2017) Transcriptional architecture of the mammalian circadian clock nature reviews. Genetics 18:164–179. https://doi.org/10.1038/nrg.2016.150
Tognini P et al (2017) Distinct circadian signatures in liver and gut clocks revealed by ketogenic. Diet Cell Metab 26(523-538):e525. https://doi.org/10.1016/j.cmet.2017.08.015
Top D, Young MW (2017) Coordination between differentially regulated circadian clocks generates rhythmic behavior. Cold Spring Harb Perspect Biol 10(7): a033589. https://doi.org/10.1101/cshperspect.a033589
Tsankova NM, Berton O, Renthal W, Kumar A, Neve RL, Nestler EJ (2006) Sustained hippocampal chromatin regulation in a mouse model of depression and antidepressant action. Nat Neurosci 9:519–525. https://doi.org/10.1038/nn1659
Uchida S, Hara K, Kobayashi A, Otsuki K, Yamagata H, Hobara T, Suzuki T, Miyata N, Watanabe Y (2011) Epigenetic status of Gdnf in the ventral striatum determines susceptibility and adaptation to daily stressful events. Neuron 69(2):359–372
Wang J et al (2018) Epigenetic modulation of inflammation and synaptic plasticity promotes resilience against stress in mice. Nat Commun 9:477. https://doi.org/10.1038/s41467-017-02794-5
Wray NR et al (2018) Genome-wide association analyses identify 44 risk variants and refine the genetic architecture of major depression. Nat Genet 50:668–681. https://doi.org/10.1038/s41588-018-0090-3
Yamawaki Y, Fuchikami M, Morinobu S, Segawa M, Matsumoto T, Yamawaki S (2012) Antidepressant-like effect of sodium butyrate (HDAC inhibitor) and its molecular mechanism of action in the rat hippocampus. World J Biol Psychiatry 13(6):458–467
Zaki NFW, Spence DW, BaHammam AS, Pandi-Perumal SR, Cardinali DP, Brown GM (2018) Chronobiological theories of mood disorder. Eur Arch Psychiatry Clin Neurosci 268:107–118. https://doi.org/10.1007/s00406-017-0835-5
Zhang Y, Liu C, Zhao Y, Zhang X, Li B, Cui R (2015) The effects of calorie restriction in depression and potential mechanisms. Curr Neuropharmacol 13:536–542

Part II

Brain Regions Implicated in Circadian Rhythms

The Brain's Reward System in Health and Disease

4

Robert G. Lewis, Ermanno Florio, Daniela Punzo, and Emiliana Borrelli

Abstract

Rhythmic gene expression is found throughout the central nervous system. This harmonized regulation can be dependent on- and independent of- the master regulator of biological clocks, the suprachiasmatic nucleus (SCN). Substantial oscillatory activity in the brain's reward system is regulated by dopamine. While light serves as a primary time-giver (*zeitgeber*) of physiological clocks and synchronizes biological rhythms in 24-h cycles, nonphotic stimuli have a profound influence over circadian biology. Indeed, reward-related activities (e.g., feeding, exercise, sex, substance use, and social interactions), which lead to an elevated level of dopamine, alters rhythms in the SCN and the brain's reward system. In this chapter, we will discuss the influence of the dopaminergic reward pathways on circadian system and the implication of this interplay on human health.

Keywords

Dopamine · Ventral tegmental area · Mesolimbic system · Striatum · Reward

4.1 The Dopaminergic Mesolimbic System and Reward

The mesolimbic system, also known as the reward system, is composed of brain structures that are responsible for mediating the physiological and cognitive processing of reward. Reward is a natural process during which the brain associates diverse stimuli (substances, situations, events, or activities) with a positive or desirable outcome. This results in adjustments of an individual's behavior, ultimately leading them to search for that particular positive stimulus. Reward requires the coordinated release of heterogenous neurotransmitters. However, of the brain substrates implicated in reward, dopamine has a central position. Dopamine plays a critical role in mediating the reward value of food, drink, sex, social interaction, and substance abuse (Hernandez and Hoebel 1988; Everitt 1990; Robbins and Everitt 1996; Bardo 1998; Beninger and Miller 1998).

The dopaminergic pathway mostly involved in reward is the so-called mesolimbic system, which is formed by projections of midbrain dopamine

R. G. Lewis · E. Florio · D. Punzo
School of Medicine, Department of Microbiology and Molecular Genetics, INSERMU1233, Center for Epigenetics and Metabolism, University of California - Irvine, Irvine, CA, USA
e-mail: rglewis@uci.edu; eflorio@hs.uci.edu; dpunzo@uci.edu

E. Borrelli (✉)
School of Medicine, Department of Microbiology and Molecular Genetics, INSERMU1233, Center for Epigenetics and Metabolism, University of California - Irvine, Irvine, CA, USA

University of California - Irvine, Irvine, CA, USA
e-mail: Borrelli@hs.uci.edu

O. Engmann, M. Brancaccio (eds.), *Circadian Clock in Brain Health and Disease*, Advances in Experimental Medicine and Biology 1344, https://doi.org/10.1007/978-3-030-81147-1_4

neurons of the ventral tegmental area (VTA) to the striatum, prefrontal cortex, amygdala, hippocampus, and many other structures of the limbic system. When rewarding stimuli are experienced, the dopaminergic mesolimbic system is activated which causes the release of dopamine to the targeted nuclei (Small et al. 2003; Cameron et al. 2014). The ventral striatum, including the nucleus accumbens (NAcc), is a major substrate involved in reward (Marche et al. 2017). The dorsal striatum is critically involved in action selection and initiation components of decision making and also seems to mediate feedback properties such as valiance and magnitude in addition to controlling habitual behavior (Balleine et al. 2007; Burton et al. 2015; Lipton et al. 2019). Therefore, both dorsal and ventral regions have collaborative roles in mediating reward. Nevertheless, the NAcc is most appreciated for its involvement in reward processing and its role in evaluation and incentive-based learning (Schultz et al. 1992; Daniel and Pollmann 2014).

The most prominent striatal neurons are the γ-aminobutyric acid (GABA) producing medium spiny neurons (MSNs). These cells make up to 90–95% of the neuronal population and serve as the sole output from the striatum (Kemp and Powell 1971; Graveland and DiFiglia 1985). MSNs outputs generate two pathways: the direct pathway formed by dopamine D1 receptor (D1R) expressing medium spiny neurons (dMSNs) and the indirect pathway by dopamine D2 receptor (D2R) expressing medium spiny neurons (iMSNs). Coordinated dopamine signaling to dMSNs and iMSNs within the striatum is critical for integrating and responding to rewarding stimuli.

The other 5–10% of striatal neurons are interneurons, which serve as intrastriatal regulators of MSNs activity (Oorschot 2013). The majority of interneurons are inhibitory GABAergic interneurons which modulate reward through their signaling to MSNs and expression of a variety of modulatory peptides (Gittis et al. 2010). About 1–2% are formed by the tonically active cholinergic interneurons which, despite their low abundance, critically regulate MSNs (Kharkwal et al. 2016a; Lewis et al. 2020). Indeed, activation of cholinergic interneurons has been linked to the salience of events (Gittis and Kreitzer 2012). Thus, inter- and intra-striatal connections modulate striatal circuits and play a critical role in reward processing.

Natural rewards, such as eating, drinking, and mating are necessary for survival and maintenance of a species. At its core, the reward system determines the valence of a stimulus and signals whether it is to be avoided or approached, as well as assigning the priority of one stimulus over another. Substances of abuse, whether illicit (e.g. cocaine, heroin, etc.) or licit (e.g. alcohol, nicotine, etc.), hijack the mesolimbic system by offering a reward without an obvious biological function. However, the pleasure and reward linked to initial substance use are then lost by their abuse, which leads to a vicious circle of addiction (Volkow et al. 2016).

Recent studies have shown that reward is subjective and is highly influenced by the chemistry of the individual, homeostatic state (Paulus 2007; Keramati and Gutkin 2014) and genetics (Comings and Blum 2000; Jia et al. 2016), as well as by the environment and epigenetics (Xu et al. 2007; Solinas et al. 2009; De Decker et al. 2017). Indeed, how, when, and where rewarding stimuli are experienced can have a profound influence on reward-related behaviors, as a result of activation of several circuits located in the striatum as well as in other brain regions responsible for encoding and storing memory of events. Importantly, the mesolimbic system is connected to the suprachiasmatic nucleus (SCN)—the master regulator of circadian rhythms (Grippo et al. 2017). The SCN is known to influence reward-related behavior and reciprocally, rewarding stimuli can serve as time-givers (zeitgebers) to entrain the SCN as well as peripheral clocks through the release of dopamine (Honma and Honma 1995; Davidson et al. 2005; Baba et al. 2017).

4.2 Rhythmic Variation in Dopamine-Related Activity

Rhythmic control of an organism's behavior is a critical part of adapting and anticipating environmental changes in light, temperature, and resources. Though time-keeping mechanisms are

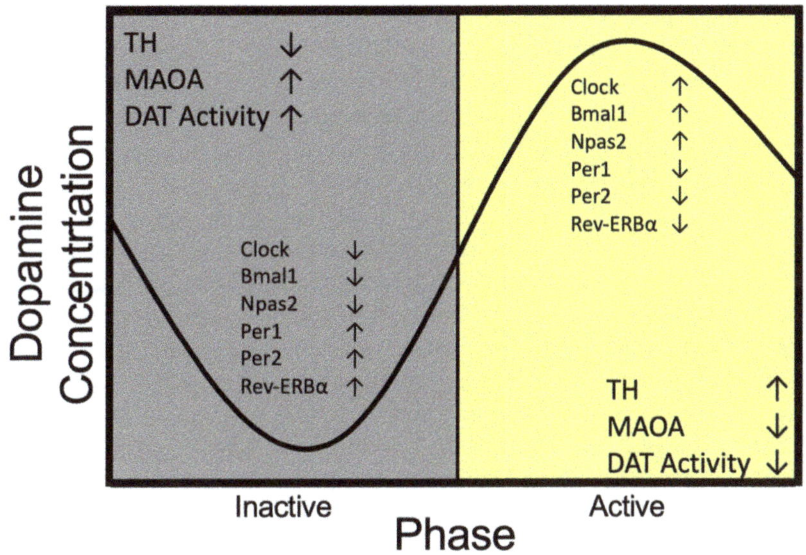

Fig. 4.1 Overview of dopamine-related activity in the reward system. Dopamine levels peak in the active phase when tyrosine hydroxylase (TH) levels are high while monoamine oxidase A (MAOA) levels are low and dopamine transporter (DAT) activity is decreased. This corresponds to core clock gene expression in the striatum, which regulates the expression of these dopamine metabolism-related gene expressions

more complex and developed in mammals, diurnal control is conserved throughout nature (Edgar et al. 2012). In mammals, the SCN organizes behavior and its correlated cellular activity through hormone and neurotransmitter release, in a 24-hour cycle based on daily light and dark phases (Dunlap 1999).

Support of a circadian regulation of reward was initially highlighted by admittance of patients experiencing substance overdose into the emergency room predominantly in the evening (Morris 1987; Raymond et al. 1992). Thus, the night spikes in overdose are likely related to differences in the metabolism of drugs of abuse during different times of day (Baird and Gauvin 2000; Abarca et al. 2002). Importantly, a variety of medications have been shown to have better clinical efficacy at precise times during day (Musiek and FitzGerald 2013; Nobis et al. 2019; Samir et al. 2020). Timing effects of rewarding stimuli also extend to natural rewards where time of day influences physiological responses as well as anticipatory rhythms (Castro 2004; Landry et al. 2012; Johnston 2014).

Dopamine levels in SN and VTA follow circadian oscillations, rising in the active phase and falling in the resting phase of the day (Smith et al. 1992; Hood et al. 2010; Ferris et al. 2014), as does its precursor and metabolites (Paulson and Robinson 1996; Castañeda et al. 2004) (Fig. 4.1). Rhythmic expressions of clock genes including *Clock*, *Rev-ERBα, Per, Npas2,* and *Bmal1* are involved in dopamine metabolism (McClung et al. 2005; Chung et al. 2014). Indeed, *Clock* and *Rev-ERBα* negatively regulate the expression of tyrosine hydroxylase (TH), the rate limiting enzyme in dopamine synthesis (Musacchio 1975). Levels of TH increase during the active phase, which is opposite to that of *Clock* and *Rev-ERBα;* loss of either circadian gene results in disrupted rhythmic TH expression (McClung et al. 2005; Chung et al. 2014). Transcription of monoamine oxidase A (MAOA), the enzyme responsible for dopamine breakdown, is regulated by the expression of NPAS2, BMAL1, and PER2 (Hampp et al. 2008). Deletion of *Per2* causes a lack of MAOA expression during the resting phase, which leads to elevated basal levels of dopamine in the NAcc (Hampp et al. 2008).

Psychostimulants increase extracellular dopamine levels and alter the expression of clock genes in the striatum (Nikaido et al. 2001; Uz et al. 2003; Lynch et al. 2008). Though drugs like cocaine and methamphetamine simultaneously alter levels of other neurotransmitters such as serotonin (Haughey et al. 2000; Andrews and Lucki 2001), their impact on clock gene expression is largely dependent on dopamine signaling. Indeed, administration of the D1R agonist, SKF-38393, increases mRNA levels of *Per1*, *Clock*, *Bmal1*, and *Npas2* while the D2R agonist, quinpirole, decreases *Clock* and *Per1* expression (Imbesi et al. 2009). D1R signaling plays a critical role in *Per2* expression, as D1R-null mice have reduced *Per2* in the striatum (Gallardo et al. 2014). Interestingly, depletion of dopamine by 6-hydroxydopamine lesions of dopaminergic neurons results into suppression of PER2 oscillations which can be rescued by D2R agonists (Hood et al. 2010). These results imply that the simultaneous activation of both D1R and D2R is necessary for the normal *Per2* oscillations in the striatum. The effect of D2R on *Per2* expression might not be direct, but mediated by the inhibitory regulation of iMSNs on dMSNs through collaterals (Lemos et al. 2016; Kharkwal et al. 2016b).

Dopamine's influence on the SCN was inferred by expression of both D1R and D5R on its neurons (Weiner et al. 1990; Rivkees and Lachowicz 1997; Doyle et al. 2002). In neonatal hamsters, light pulses mirror the effects of D1R agonists suggesting that the maternal levels of DA correspond to the active phase in the fetal SCN (Viswanathan and Davis 1997). Dopamine has been reported to play a critical role in entraining fetal development through the SCN and that after this period the SCN's responsiveness to dopamine declines (Weaver and Reppert 1995; Mendoza and Challet 2014). Nevertheless, D1R activation in the SCN shifts the phase of circadian rhythms and a direct connection between the VTA and the SCN has been described (Grippo and Güler 2019). Furthermore, D2R seems to be absolutely required for the light-induced suppression of locomotor activity (masking), whereas other visual or nonvisual photic responses seem to be D2R independent (Doi et al. 2006). These results showed a yet unappreciated function of D2R-mediated signaling in regulating the proper organization of daily locomotor activity in light-dark cycles.

Thus, the daily fluctuation in VTA dopamine neuron activity has a substantial role in SCN entrainment and other circadian activities.

4.3 Food and its Relationship to the Circadian Control of the Mesolimbic System

It is a complex process that both the type of food we consume and how much is consumed integrates the homeostatic and reward systems. Controlled food intake relies on balanced responses between orexigenic and anorexigenic neurons of the hypothalamus, which respond to circulating hormones and nutrients (Kalra et al. 1999; Meister 2007). The hypothalamus regulates the production of neuropeptides like ghrelin, leptin, and neuropeptide Y (NPY) in a diurnal manner, which contributes to appetite regulation (Kalra et al. 1999). Genetically engineered mice with deletions of genes encoding either ghrelin, leptin, or NPY have aberrant feeding behaviors or metabolic fuel preference (Bannon et al. 2000; Wortley et al. 2004; Cristino et al. 2013; Schéle et al. 2016). In a simplistic model, low levels of nutrients such as glucose, fats, and amino acids increase levels of ghrelin and decrease leptin (Weigle et al. 1997; Tschöp et al. 2000; Klok et al. 2007). Ghrelin acts on NPY-producing neurons in the hypothalamus which cause the release of NPY (Kohno et al. 2003). Food intake restores deficits in nutrients, decreasing ghrelin and causes the release of leptin from adipose tissue (Izadi et al. 2014). Leptin acts on NPY-producing neurons in the hypothalamus, reducing the amount of NPY released (Baver et al. 2014). An intact control of homeostatic regulation through integration of these signals and the subsequent response is necessary for the maintenance of a stable body weight. Dysregulation of this system leads to obesity and its associated comorbidities including heart disease and diabetes (Turek et al. 2005; Depner et al. 2014; Reutrakul and Knutson 2015).

Taste, smell, texture, and temperature all contribute to the subjective pleasantness of food and rely on the mesolimbic dopamine system. The taste of saccharin sweetened water, for example, is chosen over intravenous cocaine administration in mice (Lenoir et al. 2007). Food that is more palatable, and as a result more rewarding, is expected to cause increased release of dopamine in the NAcc (Volkow et al. 2010, 2012). Indeed, palatable foods containing high levels of sugars (Rada et al. 2005) and fats (Rada et al. 2012; Cone et al. 2013) are known to stimulate the release of dopamine into the NAcc. Dopamine has an essential role in mediating appetite which goes above the homeostatic system. Dopamine-deficient mice with inactive TH in VTA neurons (Szczypka et al. 2001) as well as mice with constitutive deletions of both D1R and D2R (Kobayashi et al. 2004) develop early fatal hypophagia. Dopaminergic pathways have been found to be altered in obese subjects. Striatal D2R expression is reduced by a palatable food diet in mice (Johnson and Kenny 2010) and in humans striatal D2R availability is significantly lower in obese patients compared to control individuals (Wang et al. 2001).

Repeated exposure to food with high fat and sugar content results in compulsive food consumption, poor control of food intake, and food stimulus conditioning (Jauch-Chara and Oltmanns 2014). These results suggest that palatable food can disrupt endogenous homeostatic regulation of food intake through activation of the reward system. Interestingly, leptin receptors have been found in the VTA and SNpc, and a putative role in regulating dopamine release has been proposed (Figlewicz et al. 2003). Moreover, ghrelin is known to stimulate VTA dopamine neurons to release dopamine into the NAcc (Abizaid et al. 2006). Thus, endogenous and exogenous signals control appetite through important interactions between the physiological need for food and the reward system.

Food consumption follows circadian rhythms. Through regulation of complex networks involving the homeostatic and reward systems, food intake sets time. One hypothesis posits that orexigenic pathways, which increase feeding behavior, become gradually activated during fasting while sleeping. However, this hypothesis contrasts evidence in humans showing that hunger has an endogenous circadian rhythm with lowest levels in the morning (8am) and greatest in the evening (8pm) regardless of the type of food intake (Scheer et al. 2013). Moreover, in the absence of external time cues individuals seek 2–3 meals during their active phase; however, the timing when these meals occur shows massive subject variability and is influenced by differences in circadian period and wakefulness (Aschoff et al. 1986).

A number of clocks in the brain can be reset by peripheral metabolic signals, which may contribute to food anticipation. Palatable foods can trigger anticipatory bouts of locomotor activity and arousal indicating an activation of the mesolimbic dopamine system (Mistlberger 1994). Despite this insight, the anatomical locations and molecular mechanisms for the food clock remain elusive. Mice with genetic deletions of *Bmal1*, *Per1*, and *Per2* have normal food anticipatory behavior as do SCN-lesioned mice (Storch and Weitz 2009). This information indicates that the central clock is not required for food anticipation. However, mutations in *Per1* have been shown to shift food intake to the resting phase, which leads to obesity in mice (Liu et al. 2014). Additionally, mice carrying deletions of *Bmal1* and *Per2* become obese from eating food equally during day and night as do ClockΔ19 mutant mice (Turek et al. 2005; Guo et al. 2012). The most likely candidates for the food clock lie in other regions of hypothalamus as well as in the striatum (Gallardo et al. 2014).

Nutrition, metabolism, and circadian rhythms are intricately linked with each other (Fig. 4.2). Timing of food intake can alter the circadian system positively or negatively. Indeed, meal timing can affect sleep/wake cycles, body temperature, performance, and alertness (Hotz et al. 1987; Hawley and Burke 1997; GRANT et al. 2017; Hou et al. 2019). These effects are enhanced by calorie restriction, high-fat and high-sugar, among others. Rhythmic dopamine levels from the VTA to the NAcc underlie motivation, food craving, and anticipation (Parekh et al. 2015). The SCN indirectly projects to the VTA through the medial preoptic nucleus of the hypothalamus (Luo and Aston-Jones 2009); this connection might conceivably allow for food-seeking directed movement through modulation of dopamine signaling in the striatum. This connection may also affect reinforcement and

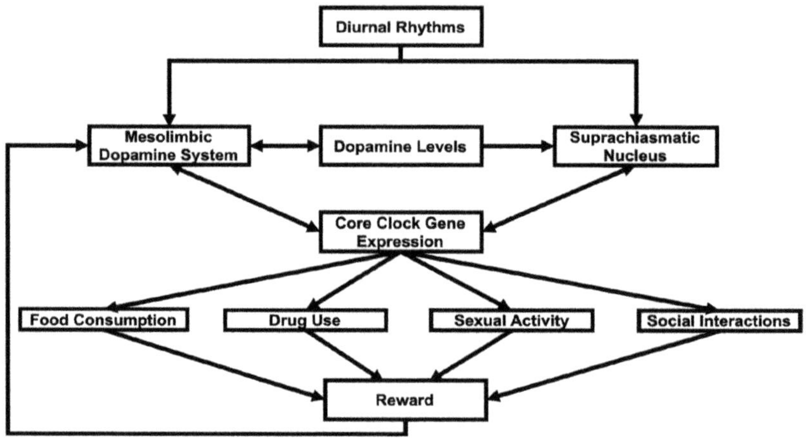

Fig. 4.2 A schematic representation of the brain's reward system in relation to circadian rhythms. Diurnal rhythms of the mesolimbic dopamine system and suprachiasmatic nucleus directly influence the activity of these brain regions. Rhythmic dopamine levels influence the activity of the mesolimbic dopamine system and suprachiasmatic nucleus. The activation of dopamine receptors in these brain regions alters core clock gene expression. The expression of core clock gene affects rewarding behaviors including food consumption, drug use, sexual activity, and social interactions, which activate the mesolimbic dopamine system

conditioned learning associated with food intake. Thus, the control of food intake is dependent on a balanced interaction between metabolic and hedonic circadian brain circuits.

4.4 Rhythmicity of Mating Behavior and Sex-Driven Reward

To facilitate necessity for species survival, mating activity highly engages the reward system and principally involves dopamine (Balfour et al. 2004). Dopamine release critically affects mating at the motor, arousal, motivation, and reward levels. In rats, systemic pharmacological treatments, which increase or decrease dopamine signaling, improve or worsen parameters of copulatory activity, respectively (Melis and Argiolas 1995). Dopamine signaling in the striatum has been postulated to mediate the reinforcing properties of sexual reproductive activity (Becker et al. 2001; Sanna et al. 2020). Regardless, dopamine signaling appears to play a critical role in sex-driven reward.

Like almost all physiological parameters in animals and humans, mating also shows some rhythmicity. In humans, most sexual encounters occur around midnight (Refinetti 2005). Environmental factors, namely partner availability, is the predominant factor important for human sexual activity. Peripheral tissues in the reproductive axis have been shown to have rhythmic clock gene expression, which might influence or synchronize with sexual behavior (Sen and Hoffmann 2020).

In animals, mating rhythmicity is important for avoiding predation and is also important for finding the right mating partner. Strong seasonal rhythms, which are linked to the amount of light, are apparent in males of many species including rodents and sheep, which are better suited models in this area of chronobiology (Reiter et al. 1980). As an example, rams are sensitive to daily changes in light across the year which induce hormonal variations and modulate gonadal function as well as libido without changes in hormone secretion (Lincoln et al. 2003). This provides evidence that, unlike other natural rewards like food, reproductive behavior is not under homeostatic regulation.

Anticipation rhythms have been observed in rodents in response to schedules; thus, suggesting that anticipatory rhythms may be located within the reward system or could be entrained by stimuli, which also engage the reward system. Indeed, circadian clock genes in the dopaminergic pathways can be shifted by natural rewards as well as dopaminergic compounds. These findings imply that copulation could also induce robust circadian anticipatory rhythms. Male rodents can anticipate daily opportunities to mate (Landry et al. 2012). Interestingly, rats can anticipate scheduled mating toward the end of their daily active phase and in the middle of their resting phase. Reproductive behavior also shows diurnal variation as does sex-related reward, which peaks in the daily active phase and corresponds with dopamine levels in the striatum (Webb et al. 2009). These results suggest that sexual anticipation and reward are linked with diurnal rhythms in the dopaminergic mesolimbic system (Melis and Argiolas 1995) though the molecular mechanisms remain elusive.

4.5 Drugs of Abuse

Although our understanding of the specific actions of drugs on the reward system has been growing, the complexity of the fundamental mechanisms underlying drug abuse and dependence increases. Drugs of abuse share one common mechanism: they raise dopamine levels in the brain, which elicits reward, driving vulnerable (Swendsen and Moal 2011) substance users to seek for more drugs leading to addiction. At the cellular level, the drug-induced dopamine increase alters neuronal plasticity at the molecular level leading to alterations of gene expression and the consequent modification of neuronal circuits.

A growing body of evidence connects perturbations of circadian rhythms and clock genes to the development and progression of addictive disorders (Webb 2017). People with addiction have highly disrupted rhythms which could be a result of genetic and/or epigenetic factors like sleep deprivation (Logan et al. 2018). Indeed, those with an evening chronotype (night owls) have been linked to disorders of the mesolimbic dopamine system including depression, insomnia, and substance abuse (Merikanto et al. 2013; Kivelä et al. 2018). Many behaviors that depend on the mesolimbic system, such as psychomotor sensitization and drug-seeking, show rhythmic patterns and are under the control of circadian genes (Abarca et al. 2002). Surprisingly, substance abuse leads to lasting changes in circadian rhythms, which can persist even after cessation of the drug intake (Jones et al. 2003).

Like for natural rewards, there are diurnal variations in the behavioral response to substances of abuse. Addictive drugs are known to influence behavioral rhythms, through modifications of the expression of clock genes such as *Clock, Per1,* and *Per2*. *Clock* is expressed in the VTA and NAcc and has been implicated in modulating reward processing. Mice with Clock null mutations show enhanced sensitivity to cocaine which has been demonstrated by conditioned place preference (CPP) (McClung et al. 2005) and self-administration (Ozburn et al. 2012) models of substance abuse. Similarly, *Per1* and *Per2* seem to have roles in cocaine sensitization (Uz et al. 2003), which is thought to be a critical component of drug craving that leads to dependence (Robinson and Berridge 2008). *Per1* and *Per2* expressions appear modulated by D1R and D2R. Interestingly, *Per1* and *Per2* mutants show increased alcohol CPP compared to WT controls (Gamsby et al. 2013). *Per1* null mice show decreased morphine CPP (Perreau-Lenz et al. 2017) and absence of cocaine CPP (Abarca et al. 2002). In contrasts *Per2* mutants show no difference from WT littermates when tested for cocaine CPP (Abarca et al. 2002).

4.6 Social Reward, Electronics, and the Clock

The developed world has a long-held fascination for technologies with entertainment purposes, which continues to grow. Adults in the United States spend 2–4 h per day using electronic devices, making technology a deeply engrained part of our lives (Dyck et al. 2011). The aberrant and persistent usage of these devices has called into question whether one could become addicted to them. Indeed, research focusing on television

(Horvath 2004), internet (Caplan 2010), and smartphone (van Deursen et al. 2015) use has sought to understand these behaviors in terms of addiction. As previously discussed, natural rewards release dopamine through activation of the mesolimbic system to promote survival and maintenance of the species. Like for substance use disorders, could technology equally hijack the reward system? Social media platforms leverage the reward system in ways similar to what gambling does to promote usage as much as possible through activation of the dopaminergic pathways (Izuma et al. 2008). Evidence has recently been presented which connects successful social interactions and the dopaminergic mesolimbic system (Torquet et al. 2018).

Growing evidence suggests that electronic devices can negatively impact circadian rhythms. Studies recently emerged have linked smartphone usage to increased anxiety and depression as well as poor sleep quality (Demirci et al. 2015). Indeed, lights from backlit screens can delay and advance circadian timing causing asynchronization (Blume et al. 2019). Associations between loss of sleep and electronic media exposure have been extensively reported in adolescents and adults (Suganuma et al. 2007; Fossum et al. 2014; Lemola et al. 2015). The alerting effects of night time use of electronics could be due to the suppression of melatonin by blue light exposure from the device in the retina (West et al. 2010). This emphasizes the utility of using "night shifting" modes, which switch displays to decreased blue light. Beyond the effects light has on sleep disruption, nighttime use of electronics and media engagement likely has profound effects on circadian rhythms. This effect can in part be ascribed to the release of dopamine in the striatum from the rewarding nature of social interactions. It is clear that more studies are required to trace the links between dopamine-mediated reward and circadian rhythms. While the connection is warranted, the molecular mechanisms that define dopamine-circadian interactions and their consequences on our health are still in their infancy.

References

Abarca C, Albrecht U, Spanagel R (2002) Cocaine sensitization and reward are under the influence of circadian genes and rhythm. Proc Natl Acad Sci U S A 99:9026–9030. https://doi.org/10.1073/pnas.142039099

Abizaid A, Liu Z-W, Andrews ZB et al (2006) Ghrelin modulates the activity and synaptic input organization of midbrain dopamine neurons while promoting appetite. J Clin Invest 116:3229–3239. https://doi.org/10.1172/JCI29867

Andrews CM, Lucki I (2001) Effects of cocaine on extracellular dopamine and serotonin levels in the nucleus accumbens. Psychopharmacology 155:221–229. https://doi.org/10.1007/s002130100704

Aschoff J, Von Goetz C, Wildgruber C, Wever RA (1986) Meal timing in humans during isolation without time cues. J Biol Rhythm 1:151–162. https://doi.org/10.1177/074873048600100206

Baba K, DeBruyne JP, Tosini G (2017) Dopamine 2 receptor activation entrains circadian clocks in mouse retinal pigment epithelium. Sci Rep 7:5103. https://doi.org/10.1038/s41598-017-05394-x

Baird TJ, Gauvin D (2000) Characterization of cocaine self-administration and pharmacokinetics as a function of time of day in the rat. Pharmacol Biochem Behav 65:289–299. https://doi.org/10.1016/s0091-3057(99)00207-5

Balfour ME, Yu L, Coolen LM (2004) Sexual behavior and sex-associated environmental cues activate the mesolimbic system in male rats. Neuropsychopharmacology 29:718–730. https://doi.org/10.1038/sj.npp.1300350

Balleine BW, Delgado MR, Hikosaka O (2007) The role of the dorsal striatum in reward and decision-making. J Neurosci 27:8161–8165. https://doi.org/10.1523/JNEUROSCI.1554-07.2007

Bannon AW, Seda J, Carmouche M et al (2000) Behavioral characterization of neuropeptide Y knockout mice. Brain Res 868:79–87. https://doi.org/10.1016/S0006-8993(00)02285-X

Bardo MT (1998) Neuropharmacological mechanisms of drug reward: beyond dopamine in the nucleus Accumbens. CRN 12:37–67. https://doi.org/10.1615/CritRevNeurobiol.v12.i1-2.30

Baver SB, Hope K, Guyot S et al (2014) Leptin modulates the intrinsic excitability of AgRP/NPY neurons in the arcuate nucleus of the hypothalamus. J Neurosci 34:5486–5496. https://doi.org/10.1523/JNEUROSCI.4861-12.2014

Becker JB, Rudick CN, Jenkins WJ (2001) The role of dopamine in the nucleus Accumbens and striatum during sexual behavior in the female rat. J Neurosci 21:3236–3241. https://doi.org/10.1523/JNEUROSCI.21-09-03236.2001

Beninger RJ, Miller R (1998) Dopamine D1-like receptors and reward-related incentive learning. Neurosci

Biobehav Rev 22:335–345. https://doi.org/10.1016/S0149-7634(97)00019-5
Blume C, Garbazza C, Spitschan M (2019) Effects of light on human circadian rhythms, sleep and mood. Somnologie (Berl) 23:147–156. https://doi.org/10.1007/s11818-019-00215-x
Burton AC, Nakamura K, Roesch MR (2015) From ventral-medial to dorsal-lateral striatum: neural correlates of reward-guided decision-making. Neurobiol Learn Mem 0:51–59. https://doi.org/10.1016/j.nlm.2014.05.003
Cameron CM, Wightman RM, Carelli RM (2014) Dynamics of rapid dopamine release in the nucleus accumbens during goal-directed behaviors for cocaine versus natural rewards. Neuropharmacology 86:319–328. https://doi.org/10.1016/j.neuropharm.2014.08.006
Caplan SE (2010) Theory and measurement of generalized problematic internet use: a two-step approach. Comput Hum Behav 26:1089–1097. https://doi.org/10.1016/j.chb.2010.03.012
Castañeda TR, de Prado BM, Prieto D, Mora F (2004) Circadian rhythms of dopamine, glutamate and GABA in the striatum and nucleus accumbens of the awake rat: modulation by light. J Pineal Res 36:177–185. https://doi.org/10.1046/j.1600-079x.2003.00114.x
de MJ C (2004) The time of day of food intake influences overall intake in humans. J Nutr 134:104–111. https://doi.org/10.1093/jn/134.1.104
Chung S, Lee EJ, Yun S et al (2014) Impact of circadian nuclear receptor REV-ERBα on midbrain dopamine production and mood regulation. Cell 157:858–868. https://doi.org/10.1016/j.cell.2014.03.039
Comings DE, Blum K (2000) Reward deficiency syndrome: genetic aspects of behavioral disorders. Prog Brain Res 126:325–341. https://doi.org/10.1016/S0079-6123(00)26022-6
Cone JJ, Chartoff EH, Potter DN et al (2013) Prolonged high fat diet reduces dopamine reuptake without altering DAT gene expression. PLoS One 8:e58251. https://doi.org/10.1371/journal.pone.0058251
Cristino L, Busetto G, Imperatore R et al (2013) Obesity-driven synaptic remodeling affects endocannabinoid control of orexinergic neurons. Proc Natl Acad Sci U S A 110(24):E2229–E2238. https://doi.org/10.1073/pnas.1219485110
Daniel R, Pollmann S (2014) A universal role of the ventral striatum in reward-based learning: evidence from human studies. Neurobiol Learn Mem 0:90–100. https://doi.org/10.1016/j.nlm.2014.05.002
Davidson AJ, Tataroglu Ö, Menaker M (2005) Circadian effects of timed meals (and other rewards). In: Young MW (ed) Methods in enzymology. Academic Press, San Diego, CA, pp 509–523
De Decker A, Verbeken S, Sioen I et al (2017) Palatable food consumption in children: interplay between (food) reward motivation and the home food environment. Eur J Pediatr 176:465–474. https://doi.org/10.1007/s00431-017-2857-4
Demirci K, Akgönül M, Akpinar A (2015) Relationship of smartphone use severity with sleep quality, depression, and anxiety in university students. J Behav Addict 4:85–92. https://doi.org/10.1556/2006.4.2015.010
Depner CM, Stothard ER, Wright KP (2014) Metabolic consequences of sleep and circadian disorders. Curr Diab Rep 14:507. https://doi.org/10.1007/s11892-014-0507-z
Doi M, Yujnovsky I, Hirayama J et al (2006) Impaired light masking in dopamine D2 receptor-null mice. Nat Neurosci 9:732–734. https://doi.org/10.1038/nn1711
Doyle SE, Grace MS, McIvor W, Menaker M (2002) Circadian rhythms of dopamine in mouse retina: the role of melatonin. Vis Neurosci 19:593–601. https://doi.org/10.1017/s0952523802195058
Dunlap JC (1999) Molecular bases for circadian clocks. Cell 96:271–290. https://doi.org/10.1016/S0092-8674(00)80566-8
Dyck DV, Cardon G, Deforche B et al (2011) Socio-demographic, psychosocial and home-environmental attributes associated with adults' domestic screen time. BMC Public Health 11:1–10. https://doi.org/10.1186/1471-2458-11-668
Edgar RS, Green EW, Zhao Y et al (2012) Peroxiredoxins are conserved markers of circadian rhythms. Nature 485:459–464. https://doi.org/10.1038/nature11088
Everitt BJ (1990) Sexual motivation: a neural and behavioural analysis of the mechanisms underlying appetitive and copulatory responses of male rats. Neurosci Biobehav Rev 14:217–232. https://doi.org/10.1016/S0149-7634(05)80222-2
Ferris MJ, España RA, Locke JL et al (2014) Dopamine transporters govern diurnal variation in extracellular dopamine tone. PNAS 111:E2751–E2759. https://doi.org/10.1073/pnas.1407935111
Figlewicz DP, Evans SB, Murphy J et al (2003) Expression of receptors for insulin and leptin in the ventral tegmental area/substantia nigra (VTA/SN) of the rat. Brain Res 964:107–115. https://doi.org/10.1016/S0006-8993(02)04087-8
Fossum IN, Nordnes LT, Storemark SS et al (2014) The association between use of electronic Media in bed before Going to sleep and insomnia symptoms, daytime sleepiness, Morningness, and Chronotype. Behav Sleep Med 12:343–357. https://doi.org/10.1080/15402002.2013.819468
Gallardo CM, Darvas M, Oviatt M et al (2014) Dopamine receptor 1 neurons in the dorsal striatum regulate food anticipatory circadian activity rhythms in mice. eLife 3:e03781. https://doi.org/10.7554/eLife.03781
Gamsby J, Templeton E, Bonvini L et al (2013) The circadian Per1 and Per2 genes influence alcohol intake, reinforcement, and blood alcohol levels. Behav Brain Res 249:15–21. https://doi.org/10.1016/j.bbr.2013.04.016
Gittis AH, Kreitzer AC (2012) Striatal microcircuitry and movement disorders. Trends Neurosci 35:557–564. https://doi.org/10.1016/j.tins.2012.06.008

Gittis AH, Nelson AB, Thwin MT et al (2010) Distinct roles of GABAergic interneurons in the regulation of striatal output pathways. J Neurosci 30:2223–2234. https://doi.org/10.1523/JNEUROSCI.4870-09.2010
Grant CL, Dorrian J, Coates AM et al (2017) The impact of meal timing on performance, sleepiness, gastric upset, and hunger during simulated night shift. Ind Health 55:423–436. https://doi.org/10.2486/indhealth.2017-0047
Graveland GA, DiFiglia M (1985) The frequency and distribution of medium-sized neurons with indented nuclei in the primate and rodent neostriatum. Brain Res 327:307–311. https://doi.org/10.1016/0006-8993(85)91524-0
Grippo RM, Güler AD (2019) Dopamine signaling in circadian Photoentrainment: consequences of Desynchrony. Yale J Biol Med 92:271–281
Grippo RM, Purohit AM, Zhang Q et al (2017) Direct midbrain dopamine input to the suprachiasmatic nucleus accelerates circadian entrainment. Curr Biol 27:2465–2475.e3. https://doi.org/10.1016/j.cub.2017.06.084
Guo B, Chatterjee S, Li L et al (2012) The clock gene, brain and muscle Arnt-like 1, regulates adipogenesis via Wnt signaling pathway. FASEB J 26:3453–3463. https://doi.org/10.1096/fj.12-205781
Hampp G, Ripperger JA, Houben T et al (2008) Regulation of monoamine oxidase A by circadian-clock components implies clock influence on mood. Curr Biol 18:678–683. https://doi.org/10.1016/j.cub.2008.04.012
Haughey HM, Fleckenstein AE, Metzger RR, Hanson GR (2000) The effects of methamphetamine on serotonin transporter activity: role of dopamine and hyperthermia. J Neurochem 75:1608–1617. https://doi.org/10.1046/j.1471-4159.2000.0751608.x
Hawley JA, Burke LM (1997) Effect of meal frequency and timing on physical performance. Br J Nutr 77:S91–S103. https://doi.org/10.1079/BJN19970107
Hernandez L, Hoebel BG (1988) Food reward and cocaine increase extracellular dopamine in the nucleus accumbens as measured by microdialysis. Life Sci 42:1705–1712. https://doi.org/10.1016/0024-3205(88)90036-7
Honma S, Honma K (1995) Phase-dependent phase shift of methamphetamine-induced circadian rhythm by haloperidol in SCN-lesioned rats. Brain Res 674:283–290. https://doi.org/10.1016/0006-8993(95)00027-N
Hood S, Cassidy P, Cossette M-P et al (2010) Endogenous dopamine regulates the rhythm of expression of the clock protein PER2 in the rat dorsal striatum via daily activation of D2 dopamine receptors. J Neurosci 30:14046–14058. https://doi.org/10.1523/JNEUROSCI.2128-10.2010
Horvath CW (2004) Measuring television addiction. J Broadcast Electron Media 48:378–398. https://doi.org/10.1207/s15506878jobem4803_3
Hotz MM, Connolly MS, Lynch CB (1987) Adaptation to daily meal-timing and its effect on circadian temperature rhythms in two inbred strains of mice. Behav Genet 17:37–51. https://doi.org/10.1007/BF01066009
Hou T, Wang C, Joshi S et al (2019) Active time-restricted feeding improved sleep-wake cycle in db/db Mice. Front Neurosci 13:969. https://doi.org/10.3389/fnins.2019.00969
Imbesi M, Yildiz S, Arslan AD et al (2009) Dopamine receptor-mediated regulation of neuronal "clock" gene expression. Neuroscience 158:537–544. https://doi.org/10.1016/j.neuroscience.2008.10.044
Izadi V, Saraf-Bank S, Azadbakht L (2014) Dietary intakes and leptin concentrations. ARYA Atheroscler 10:266–272
Izuma K, Saito DN, Sadato N (2008) Processing of social and monetary rewards in the human striatum. Neuron 58:284–294. https://doi.org/10.1016/j.neuron.2008.03.020
Jauch-Chara K, Oltmanns KM (2014) Obesity – a neuropsychological disease? Systematic review and neuropsychological model. Prog Neurobiol 114:84–101. https://doi.org/10.1016/j.pneurobio.2013.12.001
Jia T, Macare C, Desrivières S et al (2016) Neural basis of reward anticipation and its genetic determinants. Proc Natl Acad Sci U S A 113:3879–3884. https://doi.org/10.1073/pnas.1503252113
Johnson PM, Kenny PJ (2010) Addiction-like reward dysfunction and compulsive eating in obese rats: role for dopamine D2 receptors. Nat Neurosci 13:635–641. https://doi.org/10.1038/nn.2519
Johnston JD (2014) Physiological responses to food intake throughout the day. Nutr Res Rev 27:107–118. https://doi.org/10.1017/S0954422414000055
Jones EM, Knutson D, Haines D (2003) Common problems in patients recovering from chemical dependency. Am Fam Physician 68:1971–1978
Kalra SP, Dube MG, Pu S et al (1999) Interacting appetite-regulating pathways in the hypothalamic regulation of body weight. Endocr Rev 20:68–100. https://doi.org/10.1210/edrv.20.1.0357
Kemp JM, Powell TP (1971) The structure of the caudate nucleus of the cat: light and electron microscopy. Philos Trans R Soc Lond Ser B Biol Sci 262:383–401. https://doi.org/10.1098/rstb.1971.0102
Keramati M, Gutkin B (2014) Homeostatic reinforcement learning for integrating reward collection and physiological stability. eLife 3:e04811. https://doi.org/10.7554/eLife.04811
Kharkwal G, Brami-Cherrier K, Lizardi-Ortiz JE et al (2016a) Parkinsonism driven by antipsychotics originates from dopaminergic control of striatal cholinergic interneurons. Neuron 91:67–78. https://doi.org/10.1016/j.neuron.2016.06.014
Kharkwal G, Radl D, Lewis RG, Borrelli E (2016b) Dopamine D2 receptors in striatal output neurons enable the psychomotor effects of cocaine. PNAS 113:11609–11614. https://doi.org/10.1073/pnas.1608362113

Kivelä L, Papadopoulos MR, Antypa N (2018) Chronotype and psychiatric disorders. Curr Sleep Med Rep 4:94–103. https://doi.org/10.1007/s40675-018-0113-8

Klok MD, Jakobsdottir S, Drent ML (2007) The role of leptin and ghrelin in the regulation of food intake and body weight in humans: a review. Obes Rev 8:21–34. https://doi.org/10.1111/j.1467-789X.2006.00270.x

Kobayashi M, Iaccarino C, Saiardi A et al (2004) Simultaneous absence of dopamine D1 and D2 receptor-mediated signaling is lethal in mice. Proc Natl Acad Sci U S A 101:11465–11470. https://doi.org/10.1073/pnas.0402028101

Kohno D, Gao H-Z, Muroya S et al (2003) Ghrelin directly interacts with neuropeptide-Y-containing neurons in the rat arcuate nucleus: Ca2+ signaling via protein kinase A and N-type channel-dependent mechanisms and cross-talk with leptin and orexin. Diabetes 52:948–956. https://doi.org/10.2337/diabetes.52.4.948

Landry GJ, Opiol H, Marchant EG et al (2012) Scheduled daily mating induces circadian anticipatory activity rhythms in the Male Rat. PLoS One 7(7):e40895. https://doi.org/10.1371/journal.pone.0040895

Lemola S, Perkinson-Gloor N, Brand S et al (2015) Adolescents' electronic media use at night, sleep disturbance, and depressive symptoms in the smartphone age. J Youth Adolescence 44:405–418. https://doi.org/10.1007/s10964-014-0176-x

Lemos JC, Friend DM, Kaplan AR et al (2016) Enhanced GABA transmission drives bradykinesia following loss of dopamine D2 receptor signaling. Neuron 90:824–838. https://doi.org/10.1016/j.neuron.2016.04.040

Lenoir M, Serre F, Cantin L, Ahmed SH (2007) Intense sweetness surpasses cocaine reward. PLoS One 2:e698. https://doi.org/10.1371/journal.pone.0000698

Lewis RG, Serra M, Radl D et al (2020) Dopaminergic control of striatal cholinergic interneurons underlies cocaine-induced psychostimulation. Cell Rep 31(3):107527. https://doi.org/10.1016/j.celrep.2020.107527

Lincoln GA, Andersson H, Hazlerigg D (2003) Clock genes and the long-term regulation of prolactin secretion: evidence for a photoperiod/circannual timer in the pars Tuberalis. J Neuroendocrinol 15:390–397. https://doi.org/10.1046/j.1365-2826.2003.00990.x

Lipton DM, Gonzales BJ, Citri A (2019) Dorsal striatal circuits for habits, compulsions and addictions. Front Syst Neurosci 13:28. https://doi.org/10.3389/fnsys.2019.00028

Liu Z, Huang M, Wu X et al (2014) PER1 phosphorylation specifies feeding rhythm in mice. Cell Rep 7:1509–1520. https://doi.org/10.1016/j.celrep.2014.04.032

Logan RW, Hasler BP, Forbes EE et al (2018) Impact of sleep and circadian rhythms on addiction vulnerability in adolescents. Biol Psychiatry 83:987–996. https://doi.org/10.1016/j.biopsych.2017.11.035

Luo AH, Aston-Jones G (2009) Circuit projection from suprachiasmatic nucleus to ventral tegmental area: a novel circadian output pathway. Eur J Neurosci 29:748–760. https://doi.org/10.1111/j.1460-9568.2008.06606.x

Lynch WJ, Girgenti MJ, Breslin FJ et al (2008) Gene profiling the response to repeated cocaine self-administration in dorsal striatum: a focus on circadian genes. Brain Res 1213:166–177. https://doi.org/10.1016/j.brainres.2008.02.106

Marche K, Martel A-C, Apicella P (2017) Differences between dorsal and ventral striatum in the sensitivity of Tonically active neurons to rewarding events. Front Syst Neurosci 11:52. https://doi.org/10.3389/fnsys.2017.00052

McClung CA, Sidiropoulou K, Vitaterna M et al (2005) Regulation of dopaminergic transmission and cocaine reward by the clock gene. Proc Natl Acad Sci U S A 102:9377–9381

Meister B (2007) Neurotransmitters in key neurons of the hypothalamus that regulate feeding behavior and body weight. Physiol Behav 92:263–271. https://doi.org/10.1016/j.physbeh.2007.05.021

Melis MR, Argiolas A (1995) Dopamine and sexual behavior. Neurosci Biobehav Rev 19:19–38. https://doi.org/10.1016/0149-7634(94)00020-2

Mendoza J, Challet E (2014) Circadian insights into dopamine mechanisms. Neuroscience 282:230–242. https://doi.org/10.1016/j.neuroscience.2014.07.081

Merikanto I, Lahti T, Kronholm E et al (2013) Evening types are prone to depression. Chronobiol Int 30:719–725. https://doi.org/10.3109/07420528.2013.784770

Mistlberger RE (1994) Circadian food-anticipatory activity: formal models and physiological mechanisms. Neurosci Biobehav Rev 18:171–195. https://doi.org/10.1016/0149-7634(94)90023-X

Morris RW (1987) Circadian and circannual rhythms of emergency room drug-overdose admissions. Prog Clin Biol Res 227B:451–457

Musacchio JM (1975) Enzymes involved in the biosynthesis and degradation of Catecholamines. In: Iversen LL, Iversen SD, Snyder SH (eds) Biochemistry of biogenic amines. Springer US, Boston, MA, pp 1–35

Musiek ES, FitzGerald GA (2013) Molecular clocks in pharmacology. Handb Exp Pharmacol 217:243–260. https://doi.org/10.1007/978-3-642-25950-0_10

Nikaido T, Akiyama M, Moriya T, Shibata S (2001) Sensitized increase of period gene expression in the mouse caudate/putamen caused by repeated injection of methamphetamine. Mol Pharmacol 59:894–900. https://doi.org/10.1124/mol.59.4.894

Nobis CC, Laramée GD, Kervezee L et al (2019) The circadian clock of CD8 T cells modulates their early response to vaccination and the rhythmicity of related signaling pathways. PNAS 116:20077–20086. https://doi.org/10.1073/pnas.1905080116

Oorschot DE (2013) The percentage of interneurons in the dorsal striatum of the rat, cat, monkey and human: a critique of the evidence. Basal Ganglia 3:19–24. https://doi.org/10.1016/j.baga.2012.11.001

Ozburn AR, Larson EB, Self DW, McClung CA (2012) Cocaine self-administration behaviors in ClockΔ19 mice. Psychopharmacology 223:169–177. https://doi.org/10.1007/s00213-012-2704-2

Parekh PK, Ozburn AR, McClung CA (2015) Circadian clock genes: effects on dopamine, reward and addiction. Alcohol 49:341–349. https://doi.org/10.1016/j.alcohol.2014.09.034

Paulson PE, Robinson TE (1996) Regional differences in the effects of amphetamine withdrawal on dopamine dynamics in the striatum. Neuropsychopharmacology 14:325–337. https://doi.org/10.1016/0893-133X(95)00141-Y

Paulus MP (2007) Neural basis of reward and craving--a homeostatic point of view. Dialogues Clin Neurosci 9:379–387

Perreau-Lenz S, Hoelters L-S, Leixner S et al (2017) mPer1 promotes morphine-induced locomotor sensitization and conditioned place preference via histone deacetylase activity. Psychopharmacology 234:1713–1724. https://doi.org/10.1007/s00213-017-4574-0

Rada P, Avena NM, Hoebel BG (2005) Daily bingeing on sugar repeatedly releases dopamine in the accumbens shell. Neuroscience 134:737–744. https://doi.org/10.1016/j.neuroscience.2005.04.043

Rada P, Avena NM, Barson JR et al (2012) A high-fat meal, or intraperitoneal Administration of a fat Emulsion, increases extracellular dopamine in the nucleus Accumbens. Brain Sci 2:242–253. https://doi.org/10.3390/brainsci2020242

Raymond RC, Warren M, Morris RW, Leikin JB (1992) Periodicity of presentations of drugs of abuse and overdose in an emergency department. J Toxicol Clin Toxicol 30:467–478. https://doi.org/10.3109/15563659209021561

Refinetti R (2005) Time for sex: nycthemeral distribution of human sexual behavior. J Circad Rhythms 3:art.4. https://doi.org/10.1186/1740-3391-3-4

Reiter RJ, Richardson BA, Johnson LY et al (1980) Pineal melatonin rhythm: reduction in aging Syrian hamsters. Science 210:1372–1373. https://doi.org/10.1126/science.7434032

Reutrakul S, Knutson KL (2015) Consequences of circadian disruption on Cardiometabolic health. Sleep Med Clin 10:455–468. https://doi.org/10.1016/j.jsmc.2015.07.005

Rivkees SA, Lachowicz JE (1997) Functional D1 and D5 dopamine receptors are expressed in the suprachiasmatic, supraoptic, and paraventricular nuclei of primates. Synapse 26:1–10. https://doi.org/10.1002/(SICI)1098-2396(199705)26:1<1::AID-SYN1>3.0.CO;2-D

Robbins TW, Everitt BJ (1996) Neurobehavioural mechanisms of reward and motivation. Curr Opin Neurobiol 6:228–236. https://doi.org/10.1016/S0959-4388(96)80077-8

Robinson TE, Berridge KC (2008) The incentive sensitization theory of addiction: some current issues. Philos Trans R Soc Lond Ser B Biol Sci 363:3137–3146. https://doi.org/10.1098/rstb.2008.0093

Samir R, Prabhu SD, Young ME (2020) Chronobiological influence over cardiovascular function. Circ Res 126:258–279. https://doi.org/10.1161/CIRCRESAHA.119.313349

Sanna F, Bratzu J, Serra MP et al (2020) Altered sexual behavior in dopamine transporter (DAT) knockout male rats: a behavioral, neurochemical and intracerebral microdialysis study. Front Behav Neurosci 14:58. https://doi.org/10.3389/fnbeh.2020.00058

Scheer FAJL, Morris CJ, Shea SA (2013) The internal circadian clock increases hunger and appetite in the evening independent of food intake and other behaviors. Obesity 21:421–423. https://doi.org/10.1002/oby.20351

Schéle E, Bake T, Rabasa C, Dickson SL (2016) Centrally administered ghrelin acutely influences food choice in rodents. PLoS One 11(2):e0149456. https://doi.org/10.1371/journal.pone.0149456

Schultz W, Apicella P, Scarnati E, Ljungberg T (1992) Neuronal activity in monkey ventral striatum related to the expectation of reward. J Neurosci 12:4595–4610. https://doi.org/10.1523/JNEUROSCI.12-12-04595.1992

Sen A, Hoffmann HM (2020) Role of core circadian clock genes in hormone release and target tissue sensitivity in the reproductive axis. Mol Cell Endocrinol 501:110655. https://doi.org/10.1016/j.mce.2019.110655

Small DM, Jones-Gotman M, Dagher A (2003) Feeding-induced dopamine release in dorsal striatum correlates with meal pleasantness ratings in healthy human volunteers. NeuroImage 19:1709–1715. https://doi.org/10.1016/S1053-8119(03)00253-2

Smith AD, Olson RJ, Justice JB (1992) Quantitative microdialysis of dopamine in the striatum: effect of circadian variation. J Neurosci Methods 44:33–41. https://doi.org/10.1016/0165-0270(92)90111-P

Solinas M, Thiriet N, Rawas RE et al (2009) Environmental enrichment during early stages of life reduces the behavioral, neurochemical, and molecular effects of cocaine. Neuropsychopharmacology 34:1102–1111. https://doi.org/10.1038/npp.2008.51

Storch K-F, Weitz CJ (2009) Daily rhythms of food-anticipatory behavioral activity do not require the known circadian clock. PNAS 106:6808–6813. https://doi.org/10.1073/pnas.0902063106

Suganuma N, Kikuchi T, Yanagi K et al (2007) Using electronic media before sleep can curtail sleep time and result in self-perceived insufficient sleep. Sleep Biol Rhythms 5:204–214. https://doi.org/10.1111/j.1479-8425.2007.00276.x

Swendsen J, Moal ML (2011) Individual vulnerability to addiction. Ann N Y Acad Sci 1216:73–85. https://doi.org/10.1111/j.1749-6632.2010.05894.x

Szczypka MS, Kwok K, Brot MD et al (2001) Dopamine production in the caudate putamen restores feeding in dopamine-deficient mice. Neuron 30:819–828. https://doi.org/10.1016/s0896-6273(01)00319-1

Torquet N, Marti F, Campart C et al (2018) Social interactions impact on the dopaminergic system and drive individuality. Nat Commun 9(1):3081. https://doi.org/10.1038/s41467-018-05526-5

Tschöp M, Smiley DL, Heiman ML (2000) Ghrelin induces adiposity in rodents. Nature 407:908–913. https://doi.org/10.1038/35038090

Turek FW, Joshu C, Kohsaka A et al (2005) Obesity and metabolic syndrome in circadian clock mutant mice. Science 308:1043–1045. https://doi.org/10.1126/science.1108750

Uz T, Akhisaroglu M, Ahmed R, Manev H (2003) The pineal gland is critical for circadian Period1 expression in the striatum and for circadian cocaine sensitization in mice. Neuropsychopharmacology 28:2117–2123. https://doi.org/10.1038/sj.npp.1300254

van Deursen AJAM, Bolle CL, Hegner SM, Kommers PAM (2015) Modeling habitual and addictive smartphone behavior: the role of smartphone usage types, emotional intelligence, social stress, self-regulation, age, and gender. Comput Hum Behav 45:411–420. https://doi.org/10.1016/j.chb.2014.12.039

Viswanathan N, Davis FC (1997) Single prenatal injections of melatonin or the D1-dopamine receptor agonist SKF 38393 to pregnant hamsters sets the offsprings' circadian rhythms to phases 180° apart. J Comp Physiol A 180:339–346. https://doi.org/10.1007/s003590050053

Volkow ND, Wang G-J, Fowler JS et al (2010) Addiction: decreased reward sensitivity and increased expectation sensitivity conspire to overwhelm the brain's control circuit. BioEssays 32:748–755. https://doi.org/10.1002/bies.201000042

Volkow ND, Wang GJ, Fowler JS et al (2012) Food and drug reward: overlapping circuits in human obesity and addiction. In: Carter CS, Dalley JW (eds) Brain imaging in behavioral neuroscience. Springer, Berlin, Heidelberg, pp 1–24

Volkow ND, Koob GF, McLellan AT (2016) Neurobiologic advances from the brain disease model of addiction. N Engl J Med 374:363–371. https://doi.org/10.1056/NEJMra1511480

Wang G-J, Volkow ND, Logan J et al (2001) Brain dopamine and obesity. Lancet 357:354–357. https://doi.org/10.1016/S0140-6736(00)03643-6

Weaver DR, Reppert SM (1995) Definition of the developmental transition from dopaminergic to photic regulation of c-fos gene expression in the rat suprachiasmatic nucleus. Mol Brain Res 33:136–148. https://doi.org/10.1016/0169-328X(95)00117-B

Webb IC (2017) Circadian rhythms and substance abuse: Chronobiological considerations for the treatment of addiction. Curr Psychiatry Rep 19:12. https://doi.org/10.1007/s11920-017-0764-z

Webb IC, Baltazar RM, Wang X et al (2009) Diurnal variations in natural and drug reward, mesolimbic tyrosine hydroxylase, and clock gene expression in the male rat. J Biol Rhythm 24:465–476. https://doi.org/10.1177/0748730409346657

Weigle DS, Duell PB, Connor WE et al (1997) Effect of fasting, refeeding, and dietary fat restriction on plasma leptin levels. J Clin Endocrinol Metab 82:561–565. https://doi.org/10.1210/jcem.82.2.3757

Weiner DM, Levey AI, Brann MR (1990) Expression of muscarinic acetylcholine and dopamine receptor mRNAs in rat basal ganglia. Proc Natl Acad Sci U S A 87:7050–7054

West KE, Jablonski MR, Warfield B et al (2010) Blue light from light-emitting diodes elicits a dose-dependent suppression of melatonin in humans. J Appl Physiol 110:619–626. https://doi.org/10.1152/japplphysiol.01413.2009

Wortley KE, Anderson KD, Garcia K et al (2004) Genetic deletion of ghrelin does not decrease food intake but influences metabolic fuel preference. PNAS 101:8227–8232. https://doi.org/10.1073/pnas.0402763101

Xu Z, Hou B, Gao Y et al (2007) Effects of enriched environment on morphine-induced reward in mice. Exp Neurol 204:714–719. https://doi.org/10.1016/j.expneurol.2006.12.027

5 Brain Clocks, Sleep, and Mood

Xiao Yu, Nicholas P. Franks, and William Wisden

Abstract

The suprachiasmatic nucleus houses the master clock, but the genes which encode the circadian clock components are also expressed throughout the brain. Here, we review how circadian clock transcription factors regulate neuromodulator systems such as histamine, dopamine, and orexin that promote arousal. These circadian transcription factors all lead to repression of the histamine, dopamine, and orexin systems during the sleep period, so ensuring integration with the ecology of the animal. If these transcription factors are deleted or mutated, in addition to the global disturbances in circadian rhythms, this causes a chronic up-regulation of neuromodulators leading to hyperactivity, elevated mood, and reduced sleep, which have been suggested to be states resembling mania.

Keywords

NREM sleep · Histamine · Histidine decarboxylase · Orexin · Dopamine · Tyrosine dehydroxylase · Hypothalamus · Ventral tegmental area · Lateral habenula · Mania

X. Yu
Division of Endocrinology, Boston Children's Hospital, Harvard Medical School, Boston, USA

N. P. Franks · W. Wisden (✉)
Department of Life Sciences, Imperial College London, London, UK

UK Dementia Research Institute, Imperial College London, London, UK
e-mail: w.wisden@imperial.ac.uk

5.1 Introduction: The Universal Nature of Sleep

Sleep seems a universal phenomenon in animals and is a feature that varies with time of day. Most animals have a distributed sleep pattern throughout the 24 h, but tend to be more active at a particular time. For example, mice and rats are more active during the night (although they also sleep then), and are more likely to be asleep during the day. On the other hand, we primates have a consolidated block of sleep, and we are active mainly during the day. On top of this basic rhythm, there are further "harmonics"—for example, many people tend to feel transiently sleepy during mid-afternoon (the siesta phase). Multiple factors determine these patterns, but classic experiments, for example studying human volunteers in free-running conditions in underground chambers, showed that the basic rest-activity rhythm is under circadian control (Aschoff 1965). In free-running conditions, such as in continuous light or darkness, with no external cues, these activities continue with a fixed rhythm.

Within the microarchitecture of sleep itself, there are also circadian changes—for example, delta power (0.5–4 Hz) frequencies in NREM sleep show a low amplitude circadian modulation

O. Engmann, M. Brancaccio (eds.), *Circadian Clock in Brain Health and Disease*, Advances in Experimental Medicine and Biology 1344, https://doi.org/10.1007/978-3-030-81147-1_5

(Dijk and Czeisler 1995). Similarly, REM sleep shows circadian variations (Wurts and Edgar 2000; Dijk and Czeisler 1995). Sleep timing, however, is just one aspect of the circadian process, albeit the most apparent to us in our daily lives. The circadian process determines not just onset and duration of sleep, but also numerous other parameters such as body temperature, blood pressure, cortisol levels, appetite, thirst, mood, and aggression; in fact, nearly every process in the body is under circadian control (Rijo-Ferreira and Takahashi 2019; Takahashi 2017; Cedernaes et al. 2019; Gamble and Silver 2020; Patke et al. 2020; Gizowski and Bourque 2018; Blum et al. 2018; Mistlberger 2020; Todd et al. 2018; Walker et al. 2020; Wulff et al. 2010; Logan and McClung 2019).

5.2 Process C and Process S: The Circadian and Homeostatic Drives to Sleep

The most influential theory of mammalian sleep is that the timing of when we sleep and wake up is determined by two independent drives, the circadian and homeostatic, termed Process C and Process S, respectively (Borbely et al. 2016; Deboer 2018; Tobler et al. 1983; Borbely 1982). Experimental evidence for this model is strong, and we discuss this in detail in the sections below. The model is that the circadian process waxes and wanes over the 24 h, and adds and subtracts from the homeostatic process, which tracks the time spent awake. The interaction of the two drives determines in this model when we fall asleep and when we wake. The two drives can be uncoupled experimentally in forced desynchrony experiments (Dijk and Czeisler 1995).

The circadian drive determines when we sleep, so it is the ecological aspect of sleep, influencing, when, for example, an animal is hunting or avoiding predators. The homeostatic drive tracks time spent awake and is thought to reflect the (still unknown) function of sleep. Unlike the circadian drive, which waxes and wanes over the 24 h, the homeostatic drive increases if sleep has not been permitted until the urge to enter sleep is overwhelming. After a period of sleep deprivation, the immediate NREM sleep (known as recovery sleep) is deeper (represented by increased delta power in the EEG) and there is some attempt to catch up on lost sleep. The mechanisms for this homeostatic sleep drive recovery sleep are not fully understood, but are likely to involve a distributed system, requiring in part activation of GABA/galanin neurons in the preoptic hypothalamus and GABA neurons in the ventral tegmental area in the midbrain (Ma et al. 2019; Reichert et al. 2019; Yu et al. 2020), as well as signals from skeletal muscle (Ehlen et al. 2017). As time awake continues, the process is also tracked by increasing phosphorylation of numerous synaptic proteins (Wang et al. 2018).

5.3 The SCN: Does It Directly Promote Sleep?

In mammals, the suprachiasmatic nucleus (SCN) in the hypothalamus is the center of the body's control of most circadian rhythms (Patton and Hastings 2018; Michel and Meijer 2020; Stephan and Zucker 1972; Moore and Eichler 1972; Eastman et al. 1984). In the SCN, neurons and astrocytes, each with a circadian clock, interact to govern the overall circadian output (Brancaccio et al. 2019; Brancaccio et al. 2017). The transcriptional-translation feedback loops that generate the circadian rhythm in plants and animals are understood in detail (Hastings et al. 2018; Cox and Takahashi 2019), and were the topic of the Nobel Prize awarded to J. Hall, M. Rosbash, and M. Young in 2017 (Siwicki et al. 2018). In mammals, transcriptional oscillations in the *Per1* & 2, *Cry1* & 2, *Bmal1* and *Clock,* and Rev-erbα genes set up a sequential alternation, via negative feedback, in protein products. Any one of the review articles cited above all have excellent diagrams of this transcriptional translational feedback clocks. These changes in gene expression lead to changes in expression of numerous output genes which, in turn, determine excitability of SCN neurons, probably via changes in ion channel expression and membrane potential, that then drive the circadian rhythm (Hastings et al. 2018; Belle et al. 2009; Paul et al. 2020). For example, the BMAL1 and CLOCK proteins form dimers that bind to E-box enhancer elements in the regulatory regions of output genes to usually stimulate, and occasionally inhibit, many target genes (Koike et al. 2012). Because the levels of BMAL1 and CLOCK proteins are oscillating, the extent of stimulation or inhibition of output genes will also oscillate. However, the genes that regulate circadian rhythms are actually expressed in all cells of the body, and it is now appreciated that all cells in all organs, including the brain, seem to have transcriptional clocks that could be synchronized to the SCN master clock (Kyriacou and Hastings 2010; Takahashi 2017; Koike et al. 2012; Paul et al. 2020).

The SCN is believed to impose overall sleep-wake architecture by receiving retinal input, i.e., light synchronizes the clock. This seems to occur by retinal inputs triggering changes in gene expression in SCN neurons (Jones et al. 2018; Rusak et al. 1990; Foster et al. 2020). In this model, the circadian clock governs the principal timing of sleep. When the SCN is disrupted, either physically by lesioning or by genetic ablation of the transcriptional clock, the overall architecture of sleep is lost, but sleep still occurs in the same amount, if not more so (Mistlberger et al. 1983; Mistlberger 2005). SCN-lesioned monkeys have more NREM sleep within the 24-h natural sleep-wake cycle, although the distinction between sleep during light and dark is abolished (Qiu et al. 2019). Similarly, mice with no Bmal1 gene have no circadian clock and sleep seemingly randomly over the 24 h (Laposky et al. 2005). A similar result occurs in *Cry1/Cry2* double knock-out mice (Wisor et al. 2002). Yet if the mice lacking this transcriptional clock are sleep deprived, they still display sleep homeostasis (Fig. 5.1). This means in the recovery sleep that follows sleep deprivation, mice display an increase in the EEG delta power of NREM sleep (sleep homeostasis) and sleep longer (Wisor et al. 2002; Deboer 2018; Franken and Dijk 2009).

Sleep deprivation still causes an increase of sleep and increase in NREM sleep delta power in suprachiasmatic nuclei-lesioned rats (Tobler et al. 1983) (Fig. 5.1a). Genetic studies showed that mice lacking circadian rhythms (by deleting the cryptochrome (*Cry*) 1 and 2 genes) increase their overall 24-h NREM sleep, especially during the "dark" active phase (Fig. 5.1b) (Wisor et al. 2002). Of note, like SCN lesions, the NREM sleep cycle, has generally more NREM sleep during the "light" than the "dark" period, was flattened. Unexpectedly, the hallmark of sleep homeostasis, increased delta power in the recovery sleep, was also remarkably elevated during the natural sleep-wake cycle of *Cry1,2* knock-out mice (Fig. 5.1b). Following sleep deprivation, delta power was substantially higher than control mice (Fig. 5.1b), implicating cryptochromes as functionally involved in the homeostatic regulation of sleep (Wisor et al. 2002). Using the same approach, the deletion of the Bmal1 gene in mice increases EEG delta power in the light phase. Nevertheless, as for SCN-lesioned animals or *Cry1,2* knock-out mice, NREM sleep still appears in Bmal1 knockouts (Fig. 5.1c), although the amount of sleep and delta power during the initial recovery sleep after sleep deprivation was reduced compared with littermate controls (Laposky et al. 2005) (Fig. 5.1c). Using CRISPR/Cas9-mediated editing, Qiu et al. generated *BMAL1* knockout cynomolgus monkeys. Of note, these monkeys showed reduced sleep in males but varied effects in females. However, the rhythm of sleep still appeared in *BMAL1*-KO monkeys, implicating different roles of *BMAL1* in sleep-wake regulation between nocturnal and diurnal animals (Qiu et al. 2019).

Using chromatin immunoprecipitation, it was found that sleep deprivation alters DNA-binding of the circadian transcription factors CLOCK and BMAL1 to the cis-regulatory sequences of target clock genes in mice, which suggests that sleep-wake could drive the DNA-binding of core clock components (Mongrain et al. 2011). All these evidences demonstrated that sleep deprivation drives changes in clock gene expression, and on the other hand, alternations of circadian genes contribute to homeostatic sleep. These studies suggest that circadian rhythms regulate the sleep-wake cycle, total sleep time, and/or sleep

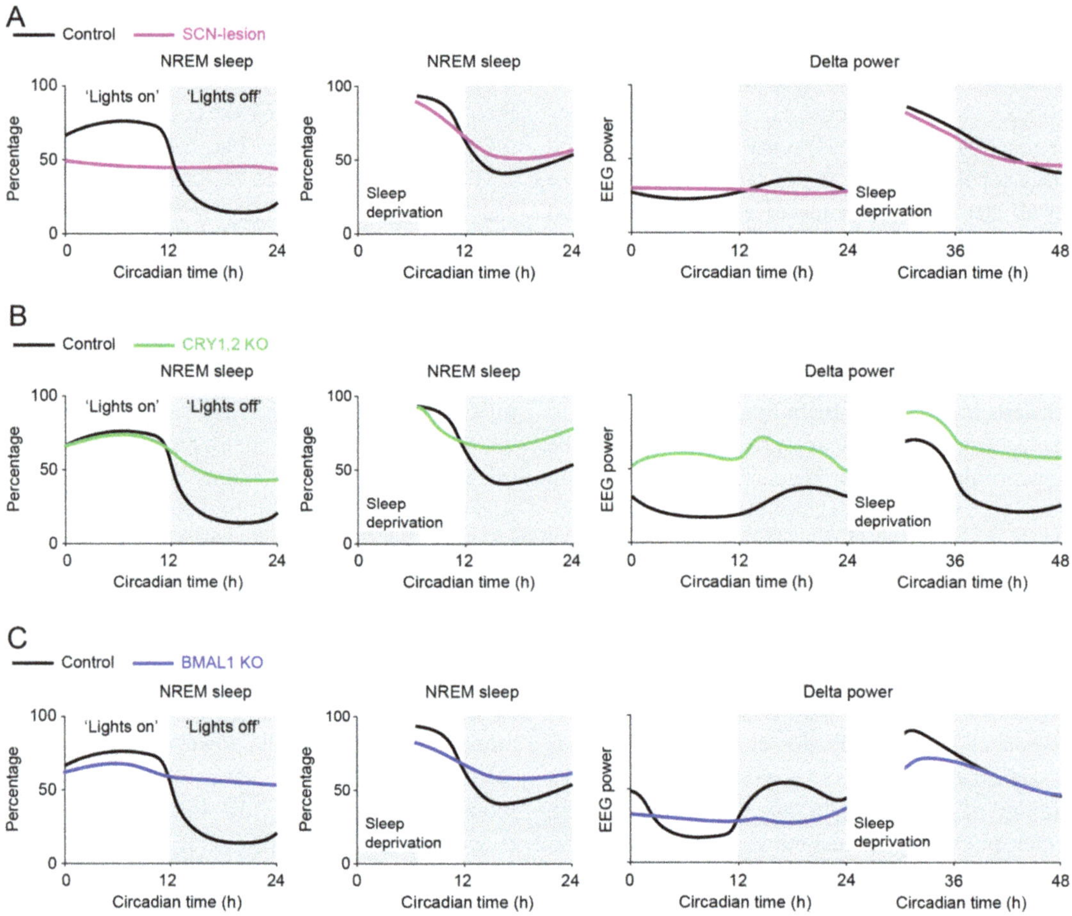

Fig. 5.1 Sleep-wake and sleep homeostasis parameters in control and SCN-lesioned or circadian gene knock-out mice. Percentage of NREM sleep during the natural 24-h sleep-wake cycle (left panel), percentage of NREM sleep (middle panel), and EEG delta power (right panel) during the natural 24-h sleep-wake cycle and after sleep deprivation in control and SCN-lesioned rats (**a**), control and *Cry1,2* knock-out mice (**b**), and control and *Bmal1*knock-out mice (**c**). Data are redrawn and presented schematically from **a**, Tobler et al., 1983; **b**, Wisor et al. 2002; and **c**, Laposky et al. 2005

bouts, but homeostatic regulation of sleep is distinct from the circadian rhythm. Overall, although deletion or mutation of core clock components differentially affects sleep-wake architecture and sleep homeostasis, circadian genes are dispensable for the homeostatic sleep drive, thus supporting the Process S and Process C models.

In humans, mutations in various of the clock control genes generate advanced and delayed sleep phase syndromes (Zhang and Fu 2020; Archer et al. 2003; Archer et al. 2018; Ashbrook et al. 2020). For example, delayed sleep phase disorder can result from a mutation in human *Cry 1* (Patke et al. 2017). Similarly, a length polymorphism in the human Per3 gene, where there are either 4 or 5 repeats of a domain that influences the stability of the protein, correlates with people who have different sleeping tendencies: people homozygous for the 5 repeat allele of *per3* have a morning diurnal preference; those homozygous for the 4 repeat allele tend to have delayed sleep phase disorder (Archer et al. 2018). Similarly, another mutation in humans that destabilizes *Per3* produces familial advanced sleep phase disorder (Zhang et al. 2016). Intriguingly, the *per3* gene is not actually required for the operation of the circadian clock (this is taken care of by *per1* and *per 2*), but the *per3* gene strongly oscillates in its expression over 24 h in many brain areas and body organs, suggesting the long and short sleep phenotypes arise from local-type clocks involving *per 3* expression (Archer et al. 2018). Although the effects of these mutations are clear, the detailed mechanisms of how they influence sleep timing, and where in the brain, and in which circuits this happens, are not understood (Ashbrook et al. 2020).

5.4 The SCN Master Clock's Effects on Timing of Sleep and Wake

The SCN governs globally most circadian rhythms for the brain and the body, with perhaps exceptions, for example, entrainment to food cues (Mistlberger 2020). It is not yet understood how the relevant SCN neurons, which are GABAergic/peptidergic (e.g., GABA/arginine vasopressin, and GABA/vasoactive intestinal polypeptide (VIP, GABA/neuromedin S, prokineticin 2, and others)) (Lee et al. 2015; Castel and Morris 2000; Okamura et al. 1989; Zhang et al. 2009; Moore and Speh 1993), communicate with the neurons that promote sleep (or arousal) in the hypothalamus or elsewhere. The precise connections have not yet been discovered. Furthermore, it is clear that the SCN has a number of subtypes of cell, not all of which might be pacemakers (Belle et al. 2009). Subsets of mouse SCN pacemakers govern different outputs, e.g., drinking before sleep to ensure that mice have adequate hydration before entering sleep (Gizowski et al. 2016; Gizowski and Bourque 2018).

Most SCN cells, regardless of mammalian species, are mostly firing when it is light (i.e., with retinal input), so they do not encode behavioral state per se, but external conditions. Nevertheless, many SCN neurons are vigilance state-dependent: they are active during wake and during REM sleep, and show less activity during NREM sleep (Deboer et al. 2003). It has been suggested that, in fact, the vigilance state, such as REM sleep, influences the circadian firing (Deboer et al. 2003). In any case, the SCN receives many afferents from neural types that promote arousal (e.g., histamine, acetylcholine) (Michelsen et al. 2005). From unit recordings, there are uniquely sleep-active neurons in the lateral parts of the SCN (Sakai 2014), although the precise identity of these cells has not yet been determined.

Photometry recordings to measure calcium activity in a subset of genetically identified cells in the SCN, those cells that express VIP, found that although the cells are activated by light (Jones et al. 2018), they are otherwise vigilance state-independent, i.e., they fire the same amount during wake, NREM, and REM sleep (Todd et al. 2020). Some VIP SCN neurons determine circadian locomotion and the circadian variation in core body temperature, but do not obviously influence sleep-wake, except when they are selectively lesioned the variation between the light and dark cycles in terms of distribution of sleep and wake is somewhat flattened out—but sleep still occurs just as much as before with no change in its microarchitecture (Todd et al. 2020). These particular SCN VIP cells project to the dorsal medial hypothalamus (Todd et al. 2020). About 20% of rodent SCN neurons are active during the dark (Deboer et al. 2003; Collins et al. 2020). In the mouse SCN, a subset of GABA/VIP cells are active in the dark and influence NREM sleep, albeit by a small amount, selectively in the dark phase (Collins et al. 2020). It is not known if the SCN neurons that are sleep-active cells in the SCN are driving sleep or are responding to sleep-promoting neurons elsewhere.

5.5 Neurons Which Control Sleep and Wake Are Dispersed Throughout the Brain: A Master Sleep Center Has Not Been Identified

The experience of jet lag shows the potent influence the SCN has on the sleep-wake circuitry.

What are the target cells in the brain that the SCN neurons might act on? The short answer to this is that we still do not know. Neurons which promote the two types of sleep, NREM and REM, are in distributed systems (Liu and Dan 2019; Peever and Fuller 2016; Adamantidis et al. 2019; Jones 2020). Unlike the SCN master clock for circadian rhythmicity, there seems to be no "master center" for sleep in the brain. There are many types of neurons in the brain, which have been identified to induce NREM sleep (Scammell et al. 2017; Yu et al. 2019a; Oishi et al. 2017; Zhang et al. 2015; Anaclet et al. 2014; Zhong et al. 2019; Harding et al. 2018; Zhang et al. 2019). These include NREM-promoting neurons in the hypothalamus, midbrain, and brainstem. It seems unclear if there is a primary NREM sleep-promoting center. Most of these cells induce NREM sleep when artificially activated by chemogenetics or optogenetics. For REM sleep, although the basis for muscle atonia in the brainstem seems reasonably clear, the neurons which promote the theta EEG rhythm seem distributed throughout the forebrain and brain stem (Peever and Fuller 2017; Adamantidis et al. 2019; Jego et al. 2013; Weber et al. 2015).

Similarly, there are a diverse range of cells which can promote wakefulness and which could be targets for SCN modulation, or which could have their own local clocks: these include the aminergic histamine, noradrenaline, and dopamine and acetylcholine cells, the cells in the lateral hypothalamus releasing orexin/hypocretin (a peptide), as well as certain GABA and glutamate cells in the midbrain and hypothalamus. A common feature is that most if not all of the arousal-promoting neurons have long-range projections. Again, it is not yet clear if one of these systems is more important than others, or if different features of arousal (e.g., motivation, attention) are produced by a mixture of different transmitters. Thus, overall, it is not yet clear if the SCN controls the timing of sleep by mainly inhibiting (or exciting) wake-promoting cells, or inhibiting (or exciting) sleep-promoting cells, or a mixture of both (Mistlberger 2005). In support of the first possibility, that the SCN controls the timing of sleep by inhibiting wake-promoting cells, it has been discovered that a subset of GABA SCN cells inhibit corticotropin-releasing factor (CRF) neurons in the paraventricular hypothalamic nucleus, preventing CRF cells in turn from exciting arousal-inducing orexin neurons in the lateral hypothalamus (Ono et al. 2020).

5.6 Local Clocks in the Brain?

Transcriptional clock genes are expressed in most brain cells of all types, i.e., neurons, glia, endothelial cells, and the transcript and protein levels of these genes oscillate (Franken 2013; Guilding and Piggins 2007; Jan et al. 2020; Paul et al. 2020). Thus, it has been suggested that local clocks exist in the brain (Kyriacou and Hastings 2010; Guilding and Piggins 2007; Jan et al. 2020) (Fig. 5.2). These local clocks, presumed to be coordinated by the SCN, may further enhance the timing of circadian processes in the brain. Indeed, most brain regions show circadian rhythmicity when assessed for the time of day at which they show maximal responsiveness to stimuli (as assessed in humans by, e.g., MRI) (Muto et al. 2016). Many circadian clock type genes such as *Per2* and *Per3* show strong oscillations in expression throughout the brain (Archer et al. 2018; Jan et al. 2020; Paul et al. 2020).

5.7 Circadian Transcription Factors Repress the Histamine, Dopamine, and Orexin Arousal Systems

The bioamines histamine and dopamine and the peptide orexin (also known as hypocretin) are examples of transmitters that are made in highly localized and small population of neurons in the midbrain and hypothalamus that project widely in the brain and promote arousal (Haas and Panula 2003; Scammell et al. 2019; Soya and Sakurai 2020; Li and de Lecea 2020; Eban-Rothschild et al. 2016). It turns out that the key genes encoding enzymes responsible for histamine and

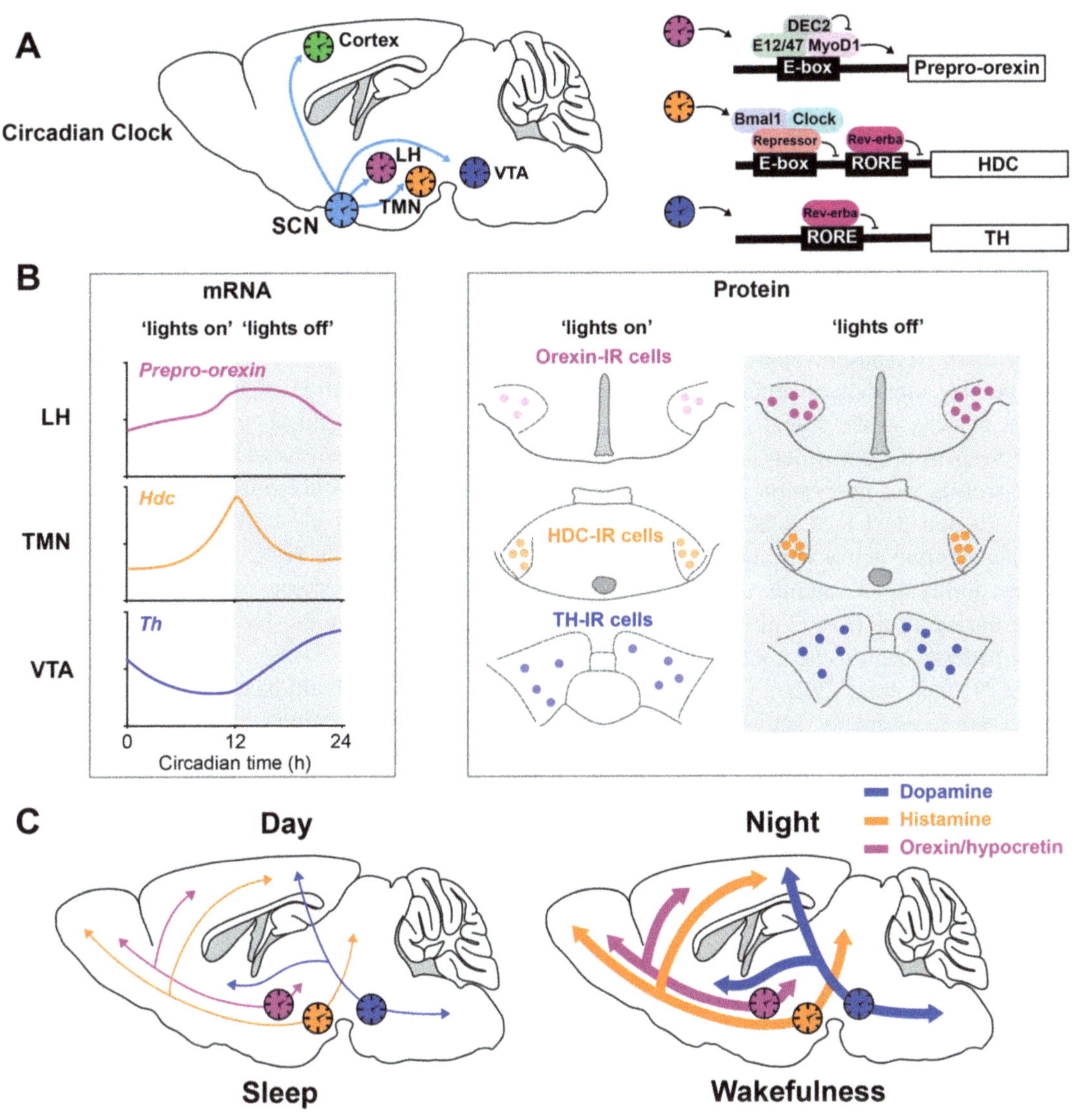

Fig. 5.2 Putative local circadian control of orexin, histamine, and dopamine synthesis in the mouse brain. (**a**) Central and local clocks in the TMN, LH, VTA or cortex in the brain, and transcriptional regulation of *prepro-orexin*, *hdc,* and *th* genes by circadian components. (**b**) Circadian rhythms tune the expression of orexin expression in the LH, *hdc* in the TMN, and *th* in the VTA. Transcripts of prepro-orexin, *hdc,* and *th* vary during the day in a diurnal manner, and an increased number of orexin/hypocretin-, HDC-, or TH-IR positive cells or intenser fluorescent staining are visualized during the night. (**c**) More dopamine, histamine, and orexin/hypocretin are released, and thereby promote wakefulness at night. *LH* lateral hypothalamus, *TMN* tuberomammillary nucleus, *VTA* ventral tegmental area

dopamine synthesis, histidine decarboxylase (*hdc*), and tyrosine hydroxylase (*th*) respectively, or in the case of orexin/hypocretin, the transcription of its gene prepro-orexin, are all repressed by circadian transcription factors that oscillate daily, which in turn cause daily oscillating levels of histamine, dopamine, and orexin (Fig. 5.2).

5.8 Orexin Synthesis Is Controlled by Local Dec 2 Repression, but the Circadian Variation of Firing of Orexin Neurons Is Influenced by SCN Input

The lateral hypothalamus contains heterogeneous neuronal populations that control many behaviors, including sleep and wakefulness (Venner et al. 2016; Jego et al. 2013; Herrera et al. 2016; Yamashita and Yamanaka 2017). The peptide orexin/hypocretin is produced in a subset of cells in the lateral hypothalamus, and these neurons project widely in the brain to promote arousal and motivation (Soya and Sakurai 2020; Li and de Lecea 2020). Orexin neurons discharge during active waking, but decrease discharge during quiet waking, and virtually cease firing during sleep (Lee et al. 2005). Part of this firing pattern seems to be under direct control of the SCN (Ono et al. 2020). SCN GABA cells, which are presumably active during the day, release GABA onto paraventricular hypothalamic CRF cells. CRF cells excite orexin neurons. So, if CRF cells are inhibited by the SCN, then the excitatory drive onto orexin neurons will be diminished, ensuring less orexin is released during the day (Ono et al. 2020).

There also seems to be local control of orexin synthesis. Orexin/hypocretin mRNA levels vary in a diurnal manner (Espana et al. 2002). In mice, there is more hypocretin/orexin expression in the dark (active) period (McGregor et al. 2017) (Fig. 5.2). Unlike orexin/hypocretin neurons in the LH, the number of NREM/REM sleep-promoting melanin-concentrating hormone (MCH) neurons in the LH did not differ between dark and light phases (McGregor et al. 2017). The orexin gene is under direct control of the transcription factor DEC2/SHARP-1, a repressor and/or co-activator of the circadian clock (Hirano et al. 2018; Rossner et al. 2008) (Fig. 5.2a). DEC2, a basic helix loop helix transcription factor, binds to E-box elements in target genes (Rossner et al. 2008). Humans with *Dec 2* mutations are short sleepers—they sleep only about 6.25 h on average (He et al. 2009), possibly because of raised orexin levels.

5.9 Bmal1 Represses Histamine Synthesis

The neurons that produce histamine are located in part of the posterior hypothalamus, the tuberomammillary nucleus (TMN), and send projections throughout the brain (Panula et al. 1989). Both deletion of the *hdc* gene or selectively lesioning histamine neurons in mice severely disrupt sleep-wake architecture, inducing strong sleep-wake fragmentation (Sakai et al. 2010; Parmentier et al. 2002; Yu et al. 2019b). Mice deficient in HDC or lacking histamine neurons fail to become aroused at the transition from dusk to night (i.e., from lights on to lights off).

Using in situ hybridization, *hdc* mRNA levels in human hypothalamic samples were significantly higher for people who had died during the daytime rather than at night (Shan et al. 2012). In mice, the number of HDC-positive cells using antibody staining were more significant in the dark than in the light (McGregor et al. 2017; Yu et al. 2014). These results suggested that both *hdc* mRNA and protein levels have a diurnal expression, about 1.5-fold in the case of the mRNA, which might be under circadian control (Fig. 5.2b, c). When Bmal1 is specifically deleted in histaminergic neurons, the diurnal expression of *hdc* transcripts and protein was flattened but elevated, and substantially caused a tonic increase of histamine levels in the brain which no longer varied diurnally (Yu et al. 2014). As a consequence, wakefulness and sleep became fragmented, the interrupted sleep probably due to increased histamine levels. At the cellular level, the resting membrane potential did not differ between day and night of HDC neurons. Knocking-out Bmal1 in HDC neurons does not affect electrophysiological properties, suggesting local clock regulates histamine synthesis but not neuronal activity per se (Yu et al. 2014). Although the precise mechanism was not worked out, suggests Bmal1 either directly represses *hdc* transcription, or this happens indirectly via diminished levels of the transcriptional repressor REV-ERBα (whose expression is reduced as a consequence of Bmal1 knockout in histaminergic neurons) (Yu et al. 2014). It would be expected that mice with tonically elevated histamine would also have some kind of psychiatric like illness, although this has not been examined.

Following sleep deprivation, mice with disruption of *Bmal1* in histaminergic neurons had less recovery sleep and lower EEG NREM delta power during the recovery sleep after sleep deprivation, which is probably due to *hdc* expression being remarkably higher in sleep deprivation than in recovery sleep in wild-type mice (Yu et al. 2014). In contrast, this difference was diminished in HDC-Bmal1 knock-out mice with an overall high level of HDC, suggesting that the local BMAL1-dependent clock mechanism suppresses histaminergic tone during sleep and thereby facilitates sleep homeostasis. This is possible because sleep deprivation alters Bmal1 expression in histaminergic neurons, therefore changing HDC expression (Yu et al. 2014).

There could also be circadian feedback from the "histamine clock" to the master clock in the SCN. Histamine can reset the SCN clock (Kim et al. 2015; Kim et al. 2016). Histamine projects to the SCN, and SCN neurons seem to take up histamine (Michelsen et al. 2005). Global knockouts of mouse histamine receptors, however, do not disturb the locomotor periodicity but flattens the amplitude (Rozov et al. 2015). On the other hand, mice with no circadian clock in histamine neurons, and so having potentially elevated histamine release onto the SCN, have no alteration in their global circadian wheel-running behavior in free-running conditions (Yu et al. 2014), suggesting that feedback from a local clock to the master clock is not that critical.

5.10 CLOCK and REV-ERBα Repress Dopamine Synthesis

Like histidine decarboxylase, the gene encoding the enzyme that is part of the pathway producing dopamine, tyrosine hydroxylase (*th*), shows diurnal expression in the rodent VTA and substantia nigra (Weber et al. 2004; Webb et al. 2009; Chung et al. 2014) (Fig. 5.2). The *th* gene is under repressive control (direct and indirect) by transcription factors involved in the circadian cycle (Kim et al. 2017). Mice with a mutation in the *clock* gene have increased (twofold) *th* mRNA and protein expression in the VTA (McClung et al. 2005). Similarly, the transcription factor nuclear receptor REV-ERBα directly represses *th* gene transcription. Deletion of the circadian gene or pharmacological inhibition of REV-ERBα activity disrupted circadian expression of *th* with an overall increase of brain dopamine levels (Chung et al. 2014). REV-ERBα antagonizes another transcription factor, NURR1, which activates the *th* gene promotor and thereby generates rhythmicity of the dopaminergic system (Kim et al. 2017; Chung et al. 2014) (Fig. 5.2).

Globally knocking-out the *REV-ERBα* gene in mice lowers delta power during baseline NREM and reduces EEG delta power during the recovery NREM sleep after sleep deprivation (Mang et al. 2016). These effects could well arise because of chronically elevated dopamine in these mice, which is acting to undermine sleep quality by promoting wake. Another prominent phenotype of the *REV-ERBα* knockout mice is "mania," i.e., hyperactivity, reduced sleep, increased activity on the tail suspension and forced swim tests, and increased sensitivity to D-amphetamine, all phenotypes probably arising as a consequence of chronically elevated dopamine (Chung et al. 2014). In the mania experienced by humans, usually as a part of bipolar disorder, patients sleep little and have elevated mood (Logan and McClung 2016; Gold and Sylvia 2016; Harrison et al. 2018). As part of this mechanism, CLOCK is another circadian transcription factor whose presence is required in mouse VTA dopamine neurons to maintain balanced levels of dopamine (see above). Mice with mutations in the *clock* gene have a mania-like phenotype (Roybal et al. 2007), which can be rescued by selective expression of wild-type clock protein in the VTA of clock-mutant mice, thus suggesting the idea of local circadian-like control (Roybal et al. 2007).

As well as transmitter synthesis, neuronal firing of dopamine neurons is also controlled by circadian factors. The VTA neurons of mice with a global *clock* gene mutation have elevated rates of firing (Roybal et al. 2007; McClung et al. 2005). Using single-unit extracellular recordings in anesthetized rats, dopamine firing rates presented a diurnal rhythm (Fifel et al. 2018). In addition, Fifel et al., found a significant circadian modulation with increased firing rates during the active phase in the VTA in mice. They combined neural activity recordings with electroencephalogram (EEG) recordings and revealed a strong vigilance state-dependent neuronal activity modulation with increased activity during wakefulness (Fifel et al. 2018).

The diurnal phenotype of raised VTA dopamine neuronal firing during waking is further emphasized and pushed to an extreme when GABA neurons in the VTA are selectively lesioned, which causes nearly 100% continuous wakefulness for the entire night (lights off period) but more normal amounts of sleep during the day (Yu et al. 2020; Yu et al. 2019a). The GABA neurons contribute to inhibition of the VTA dopamine neurons (Yu et al. 2019a); without this inhibition, the circadian-regulated increased firing of dopamine neurons reaches a maximum, and mania-like behavior (hyperactivity, reduced sleep, sensitization to D-amphetamine, increased mobility on the tail suspension, and forced swim tests) emerges during the night (Yu et al. 2020). It is likely these VTA GABA neurons will also have some kind of local circadian control, although this has not been studied, and they receive excitatory glutamate inputs from the lateral habenula (see next section). Glutamate release from the lateral habenula is itself under circadian control, being highest during the day (Sakhi et al. 2014).

5.11 The Lateral Habenula: A Collection of Glutamatergic Projection Neurons in the Midbrain Required for Sleep Consolidation in the Day

The lateral habenula, which is a mainly glutamatergic group of neurons situated in the midbrain and projecting to target areas such as GABA neurons in the VTA and the serotonergic dorsal raphe, has a strong circadian clock as assessed by *Per2* gene expression in brain slices (Bano-Otalora and Piggins 2017; Sakhi et al. 2014). The LHb not only receives light input via the SCN, but also via a direct input from the retina, which could synchronize the clock (Bano-Otalora and Piggins 2017). The LHb is required for NREM sleep (Gelegen et al. 2018). This could happen because LHb projections exciting VTA GABA neurons could induce NREM sleep (Yu et al. 2019a). In mice, blocking glutamate release from LHb neurons by genetic expression of tetanus toxin light-chain leads to a striking high sleep-wake fragmentation, but selectively during the day time ("lights on") (Gelegen et al. 2018). This suggests that these LHb neurons would be most active then, during the light phase. Indeed, mouse LHb neurons have a higher mean firing rate during the late part of the lights-on phase (Sakhi et al. 2014). This variation in firing rate is abolished in LHb slices prepared from *Cry1/Cry2* double knockout mice, confirming that this is a local clock-controlled phenomenon (Sakhi et al. 2014). Thus, as well as being possibily supervised directly from the SCN in the way that orexin neurons in the LH receive SCN input (Ono et al. 2020), the putative local clock in the LHb would serve to reinforce consolidated sleep in the day (for mice). Thus, rhythmic circadian signaling runs throughout the midbrain circuitry regulating NREM sleep.

5.12 Conclusion

Overall, there is a strong link between circadian rhythm and neuronal activity in sleep-wake regulatory centers. Nevertheless, it is still unclear if these transcription factors are really acting as genuine local clocks. In any case, together with input from the SCN that determines firing rate (Ono et al. 2020), the locally expressed circadian-type transcription factors ensure higher expression of histamine, dopamine, and orexin in the wake phase. Arousal is a multidimensional quality. The modulators that promote arousal are engaged with all aspects of mental function. If this fine tuning by circadian mechanisms goes awry, large changes in sleep and mood can result, which could sometimes be pathological, contributing for example to mania-like states.

Acknowledgments Our work was supported by the Wellcome Trust (107839/Z/15/Z, NPF and 107841/Z/15/Z, WW) and the the UK Dementia Research Institute (WW and NPF).

References

Adamantidis AR, Gutierrez Herrera C, Gent TC (2019) Oscillating circuitries in the sleeping brain. Nat Rev Neurosci 20(12):746–762. https://doi.org/10.1038/s41583-019-0223-4

Anaclet C, Ferrari L, Arrigoni E, Bass CE, Saper CB, Lu J, Fuller PM (2014) The GABAergic parafacial zone is a medullary slow wave sleep-promoting center. Nat Neurosci 17(9):1217–1224. https://doi.org/10.1038/nn.3789

Archer SN, Robilliard DL, Skene DJ, Smits M, Williams A, Arendt J, von Schantz M (2003) A length polymorphism in the circadian clock gene Per3 is linked to delayed sleep phase syndrome and extreme diurnal preference. Sleep 26(4):413–415. https://doi.org/10.1093/sleep/26.4.413

Archer SN, Schmidt C, Vandewalle G, Dijk DJ (2018) Phenotyping of PER3 variants reveals widespread effects on circadian preference, sleep regulation, and health. Sleep Med Rev 40:109–126. https://doi.org/10.1016/j.smrv.2017.10.008

Aschoff J (1965) Circadian rhythms in man. Science 148 (3676):1427–1432. https://doi.org/10.1126/science.148.3676.1427

Ashbrook LH, Krystal AD, Fu YH, Ptacek LJ (2020) Genetics of the human circadian clock and sleep homeostat. Neuropsychopharmacology 45(1):45–54. https://doi.org/10.1038/s41386-019-0476-7

Bano-Otalora B, Piggins HD (2017) Contributions of the lateral habenula to circadian timekeeping. Pharmacol Biochem Behav 162:46–54. https://doi.org/10.1016/j.pbb.2017.06.007

Belle MD, Diekman CO, Forger DB, Piggins HD (2009) Daily electrical silencing in the mammalian circadian clock. Science 326(5950):281–284. https://doi.org/10.1126/science.1169657

Blum ID, Bell B, Wu MN (2018) Time for bed: genetic mechanisms mediating the circadian regulation of sleep. Trends Genet 34(5):379–388. https://doi.org/10.1016/j.tig.2018.01.001

Borbely AA (1982) A two process model of sleep regulation. Hum Neurobiol 1(3):195–204

Borbely AA, Daan S, Wirz-Justice A, Deboer T (2016) The two-process model of sleep regulation: a reappraisal. J Sleep Res 25(2):131–143. https://doi.org/10.1111/jsr.12371

Brancaccio M, Patton AP, Chesham JE, Maywood ES, Hastings MH (2017) Astrocytes control circadian timekeeping in the suprachiasmatic nucleus via glutamatergic signaling. Neuron 93(6):1420–1435. https://doi.org/10.1016/j.neuron.2017.02.030

Brancaccio M, Edwards MD, Patton AP, Smyllie NJ, Chesham JE, Maywood ES, Hastings MH (2019) Cell-autonomous clock of astrocytes drives circadian behavior in mammals. Science 363(6423):187–192. https://doi.org/10.1126/science.aat4104

Castel M, Morris JF (2000) Morphological heterogeneity of the GABAergic network in the suprachiasmatic nucleus, the brain's circadian pacemaker. J Anat 196 (Pt 1):1–13. https://doi.org/10.1046/j.1469-7580.2000.19610001.x

Cedernaes J, Waldeck N, Bass J (2019) Neurogenetic basis for circadian regulation of metabolism by the hypothalamus. Genes Dev 33(17–18):1136–1158. https://doi.org/10.1101/gad.328633.119

Chung S, Lee EJ, Yun S, Choe HK, Park SB, Son HJ, Kim KS, Dluzen DE, Lee I, Hwang O, Son GH, Kim K (2014) Impact of circadian nuclear receptor REV-ERBalpha on midbrain dopamine production and mood regulation. Cell 157(4):858–868. https://doi.org/10.1016/j.cell.2014.03.039

Collins B, Pierre-Ferrer S, Muheim C, Lukacsovich D, Cai Y, Spinnler A, Herrera CG, Wen S, Winterer J, Belle MDC, Piggins HD, Hastings M, Loudon A, Yan J, Foldy C, Adamantidis A, Brown SA (2020) Circadian VIPergic neurons of the suprachiasmatic nuclei sculpt the sleep-wake cycle. Neuron. https://doi.org/10.1016/j.neuron.2020.08.001

Cox KH, Takahashi JS (2019) Circadian clock genes and the transcriptional architecture of the clock mechanism. J Mol Endocrinol 63(4):R93–R102. https://doi.org/10.1530/Jme-19-0153

Deboer T (2018) Sleep homeostasis and the circadian clock: do the circadian pacemaker and the sleep homeostat influence each other's functioning?

Neurobiol Sleep Circadian Rhythms 5:68–77. https://doi.org/10.1016/j.nbscr.2018.02.003
Deboer T, Vansteensel MJ, Detari L, Meijer JH (2003) Sleep states alter activity of suprachiasmatic nucleus neurons. Nat Neurosci 6(10):1086–1090. https://doi.org/10.1038/nn1122
Dijk DJ, Czeisler CA (1995) Contribution of the circadian pacemaker and the sleep homeostat to sleep propensity, sleep structure, electroencephalographic slow waves, and sleep spindle activity in humans. J Neurosci 15 (5 Pt 1):3526–3538
Eastman CI, Mistlberger RE, Rechtschaffen A (1984) Suprachiasmatic nuclei lesions eliminate circadian temperature and sleep rhythms in the rat. Physiol Behav 32(3):357–368. https://doi.org/10.1016/0031-9384(84)90248-8
Eban-Rothschild A, Rothschild G, Giardino WJ, Jones JR, de Lecea L (2016) VTA dopaminergic neurons regulate ethologically relevant sleep-wake behaviors. Nat Neurosci 19(10):1356–1366. https://doi.org/10.1038/nn.4377
Ehlen JC, Brager AJ, Baggs J, Pinckney L, Gray CL, DeBruyne JP, Esser KA, Takahashi JS, Paul KN (2017) Bmal1 function in skeletal muscle regulates sleep. elife 6:e26557. https://doi.org/10.7554/eLife.26557
Espana RA, Plahn S, Berridge CW (2002) Circadian-dependent and circadian-independent behavioral actions of hypocretin/orexin. Brain Res 943 (2):224–236. https://doi.org/10.1016/s0006-8993(02)02653-7
Fifel K, Meijer JH, Deboer T (2018) Circadian and homeostatic modulation of multi-unit activity in midbrain dopaminergic structures. Sci Rep 8(1):7765. https://doi.org/10.1038/s41598-018-25770-5
Foster RG, Hughes S, Peirson SN (2020) Circadian Photoentrainment in mice and humans. Biology (Basel) 9(7):180. https://doi.org/10.3390/biology9070180
Franken P (2013) A role for clock genes in sleep homeostasis. Curr Opin Neurobiol 23(5):864–872. https://doi.org/10.1016/j.conb.2013.05.002
Franken P, Dijk DJ (2009) Circadian clock genes and sleep homeostasis. Eur J Neurosci 29(9):1820–1829. https://doi.org/10.1111/j.1460-9568.2009.06723.x
Gamble KL, Silver R (2020) Circadian rhythmicity and the community of clockworkers. Eur J Neurosci 51 (12):2314–2328. https://doi.org/10.1111/ejn.14626
Gelegen C, Miracca G, Ran MZ, Harding EC, Ye Z, Yu X, Tossell K, Houston CM, Yustos R, Hawkins ED, Vyssotski AL, Dong HL, Wisden W, Franks NP (2018) Excitatory pathways from the lateral Habenula enable Propofol-induced sedation. Curr Biol 28 (4):580–587. e585. https://doi.org/10.1016/j.cub.2017.12.050
Gizowski C, Bourque CW (2018) The neural basis of homeostatic and anticipatory thirst. Nat Rev Nephrol 14(1):11–25. https://doi.org/10.1038/nrneph.2017.149
Gizowski C, Zaelzer C, Bourque CW (2016) Clock-driven vasopressin neurotransmission mediates anticipatory thirst prior to sleep. Nature 537(7622):685–688. https://doi.org/10.1038/nature19756
Gold AK, Sylvia LG (2016) The role of sleep in bipolar disorder. Nat Sci Sleep 8:207–214. https://doi.org/10.2147/NSS.S85754
Guilding C, Piggins HD (2007) Challenging the omnipotence of the suprachiasmatic timekeeper: are circadian oscillators present throughout the mammalian brain? Eur J Neurosci 25(11):3195–3216. https://doi.org/10.1111/j.1460-9568.2007.05581.x
Haas H, Panula P (2003) The role of histamine and the tuberomamillary nucleus in the nervous system. Nat Rev Neurosci 4(2):121–130. https://doi.org/10.1038/nrn1034
Harding EC, Yu X, Miao A, Andrews N, Ma Y, Ye Z, Lignos L, Miracca G, Ba W, Yustos R, Vyssotski AL, Wisden W, Franks NP (2018) A neuronal hub binding sleep initiation and body cooling in response to a warm external stimulus. Curr Biol 28(14):2263–2273. e2264. https://doi.org/10.1016/j.cub.2018.05.054
Harrison PJ, Geddes JR, Tunbridge EM (2018) The emerging neurobiology of bipolar disorder. Trends Neurosci 41(1):18–30. https://doi.org/10.1016/j.tins.2017.10.006
Hastings MH, Maywood ES, Brancaccio M (2018) Generation of circadian rhythms in the suprachiasmatic nucleus. Nat Rev Neurosci 19(8):453–469. https://doi.org/10.1038/s41583-018-0026-z
He Y, Jones CR, Fujiki N, Xu Y, Guo B, Holder JL Jr, Rossner MJ, Nishino S, Fu YH (2009) The transcriptional repressor DEC2 regulates sleep length in mammals. Science 325(5942):866–870. https://doi.org/10.1126/science.1174443
Herrera CG, Cadavieco MC, Jego S, Ponomarenko A, Korotkova T, Adamantidis A (2016) Hypothalamic feedforward inhibition of thalamocortical network controls arousal and consciousness. Nat Neurosci 19 (2):290–298. https://doi.org/10.1038/nn.4209
Hirano A, Hsu PK, Zhang L, Xing L, McMahon T, Yamazaki M, Ptacek LJ, Fu YH (2018) DEC2 modulates orexin expression and regulates sleep. Proc Natl Acad Sci U S A 115(13):3434–3439. https://doi.org/10.1073/pnas.1801693115
Jan M, O'Hara BF, Franken P (2020) Recent advances in understanding the genetics of sleep. F1000Res 9: F1000. https://doi.org/10.12688/f1000research.22028.1
Jego S, Glasgow SD, Herrera CG, Ekstrand M, Reed SJ, Boyce R, Friedman J, Burdakov D, Adamantidis AR (2013) Optogenetic identification of a rapid eye movement sleep modulatory circuit in the hypothalamus. Nat Neurosci 16(11):1637–1643. https://doi.org/10.1038/nn.3522
Jones BE (2020) Arousal and sleep circuits. Neuropsychopharmacology 45(1):6–20. https://doi.org/10.1038/s41386-019-0444-2

Jones JR, Simon T, Lones L, Herzog ED (2018) SCN VIP neurons are essential for Normal light-mediated resetting of the circadian system. J Neurosci 38 (37):7986–7995. https://doi.org/10.1523/JNEUROSCI.1322-18.2018

Kim YS, Kim YB, Kim WB, Yoon BE, Shen FY, Lee SW, Soong TW, Han HC, Colwell CS, Lee CJ, Kim YI (2015) Histamine resets the circadian clock in the suprachiasmatic nucleus through the H1R-CaV 1.3-RyR pathway in the mouse. Eur J Neurosci 42 (7):2467–2477. https://doi.org/10.1111/ejn.13030

Kim YS, Kim YB, Kim WB, Lee SW, Oh SB, Han HC, Lee CJ, Colwell CS, Kim YI (2016) Histamine 1 receptor-Gbetagamma-cAMP/PKA-CFTR pathway mediates the histamine-induced resetting of the suprachiasmatic circadian clock. Mol Brain 9(1):49. https://doi.org/10.1186/s13041-016-0227-1

Kim J, Jang S, Choe HK, Chung S, Son GH, Kim K (2017) Implications of circadian rhythm in dopamine and mood regulation. Mol Cells 40(7):450–456. https://doi.org/10.14348/molcells.2017.0065

Koike N, Yoo SH, Huang HC, Kumar V, Lee C, Kim TK, Takahashi JS (2012) Transcriptional architecture and chromatin landscape of the core circadian clock in mammals. Science 338(6105):349–354. https://doi.org/10.1126/science.1226339

Kyriacou CP, Hastings MH (2010) Circadian clocks: genes, sleep, and cognition. Trends Cogn Sci 14 (6):259–267. https://doi.org/10.1016/j.tics.2010.03.007

Laposky A, Easton A, Dugovic C, Walisser J, Bradfield C, Turek F (2005) Deletion of the mammalian circadian clock gene BMAL1/Mop3 alters baseline sleep architecture and the response to sleep deprivation. Sleep 28 (4):395–409. https://doi.org/10.1093/sleep/28.4.395

Lee MG, Hassani OK, Jones BE (2005) Discharge of identified orexin/hypocretin neurons across the sleep-waking cycle. J Neurosci 25(28):6716–6720. https://doi.org/10.1523/JNEUROSCI.1887-05.2005

Lee IT, Chang AS, Manandhar M, Shan Y, Fan J, Izumo M, Ikeda Y, Motoike T, Dixon S, Seinfeld JE, Takahashi JS, Yanagisawa M (2015) Neuromedin s-producing neurons act as essential pacemakers in the suprachiasmatic nucleus to couple clock neurons and dictate circadian rhythms. Neuron 85 (5):1086–1102. https://doi.org/10.1016/j.neuron.2015.02.006

Li SB, de Lecea L (2020) The hypocretin (orexin) system: from a neural circuitry perspective. Neuropharmacology 167:107993. https://doi.org/10.1016/j.neuropharm.2020.107993

Liu D, Dan Y (2019) A motor theory of sleep-wake control: arousal-action circuit. Annu Rev Neurosci 42:27–46. https://doi.org/10.1146/annurev-neuro-080317-061813

Logan RW, McClung CA (2016) Animal models of bipolar mania: the past, present and future. Neuroscience 321:163–188. https://doi.org/10.1016/j.neuroscience.2015.08.041

Logan RW, McClung CA (2019) Rhythms of life: circadian disruption and brain disorders across the lifespan. Nat Rev Neurosci 20(1):49–65. https://doi.org/10.1038/s41583-018-0088-y

Ma Y, Miracca G, Yu X, Harding EC, Miao A, Yustos R, Vyssotski AL, Franks NP, Wisden W (2019) Galanin neurons unite sleep homeostasis and alpha2-adrenergic sedation. Curr Biol 29(19):3315–3322. e3313. https://doi.org/10.1016/j.cub.2019.07.087

Mang GM, La Spada F, Emmenegger Y, Chappuis S, Ripperger JA, Albrecht U, Franken P (2016) Altered sleep homeostasis in rev-erbalpha knockout mice. Sleep 39(3):589–601. https://doi.org/10.5665/sleep.5534

McClung CA, Sidiropoulou K, Vitaterna M, Takahashi JS, White FJ, Cooper DC, Nestler EJ (2005) Regulation of dopaminergic transmission and cocaine reward by the clock gene. Proc Natl Acad Sci U S A 102 (26):9377–9381. https://doi.org/10.1073/pnas.0503584102

McGregor R, Shan L, Wu MF, Siegel JM (2017) Diurnal fluctuation in the number of hypocretin/orexin and histamine producing: implication for understanding and treating neuronal loss. PLoS One 12(6): e0178573. https://doi.org/10.1371/journal.pone.0178573

Michel S, Meijer JH (2020) From clock to functional pacemaker. Eur J Neurosci 51(1):482–493. https://doi.org/10.1111/ejn.14388

Michelsen KA, Lozada A, Kaslin J, Karlstedt K, Kukko-Lukjanov TK, Holopainen I, Ohtsu H, Panula P (2005) Histamine-immunoreactive neurons in the mouse and rat suprachiasmatic nucleus. Eur J Neurosci 22 (8):1997–2004. https://doi.org/10.1111/j.1460-9568.2005.04387.x

Mistlberger RE (2005) Circadian regulation of sleep in mammals: role of the suprachiasmatic nucleus. Brain Res Brain Res Rev 49(3):429–454. https://doi.org/10.1016/j.brainresrev.2005.01.005

Mistlberger RE (2020) Food as circadian time cue for appetitive behavior. F1000Res 9:F1000. https://doi.org/10.12688/f1000research.20829.1

Mistlberger RE, Bergmann BM, Waldenar W, Rechtschaffen A (1983) Recovery sleep following sleep deprivation in intact and suprachiasmatic nuclei-lesioned rats. Sleep 6(3):217–233. https://doi.org/10.1093/sleep/6.3.217

Mongrain V, La Spada F, Curie T, Franken P (2011) Sleep loss reduces the DNA-binding of BMAL1, CLOCK, and NPAS2 to specific clock genes in the mouse cerebral cortex. PLoS One 6(10):e26622. https://doi.org/10.1371/journal.pone.0026622

Moore RY, Eichler VB (1972) Loss of a circadian adrenal corticosterone rhythm following suprachiasmatic lesions in the rat. Brain Res 42(1):201–206. https://doi.org/10.1016/0006-8993(72)90054-6

Moore RY, Speh JC (1993) GABA is the principal neurotransmitter of the circadian system. Neurosci Lett 150

(1):112–116. https://doi.org/10.1016/0304-3940(93)90120-a
Muto V, Jaspar M, Meyer C, Kusse C, Chellappa SL, Degueldre C, Balteau E, Shaffii-Le Bourdiec A, Luxen A, Middleton B, Archer SN, Phillips C, Collette F, Vandewalle G, Dijk DJ, Maquet P (2016) Local modulation of human brain responses by circadian rhythmicity and sleep debt. Science 353 (6300):687–690. https://doi.org/10.1126/science.aad2993
Oishi Y, Xu Q, Wang L, Zhang BJ, Takahashi K, Takata Y, Luo YJ, Cherasse Y, Schiffmann SN, de Kerchove d'Exaerde A, Urade Y, Qu WM, Huang ZL, Lazarus M (2017) Slow-wave sleep is controlled by a subset of nucleus accumbens core neurons in mice. Nat Commun 8(1):734. https://doi.org/10.1038/s41467-017-00781-4
Okamura H, Berod A, Julien JF, Geffard M, Kitahama K, Mallet J, Bobillier P (1989) Demonstration of GABAergic cell bodies in the suprachiasmatic nucleus: in situ hybridization of glutamic acid decarboxylase (GAD) mRNA and immunocytochemistry of GAD and GABA. Neurosci Lett 102(2–3):131–136. https://doi.org/10.1016/0304-3940(89)90067-0
Ono D, Mukai Y, Hung CJ, Chowdhury S, Sugiyama T, Yamanaka A (2020) The mammalian circadian pacemaker regulates wakefulness via CRF neurons in the paraventricular nucleus of the hypothalamus. Sci Adv 6(45):eabd0384. https://doi.org/10.1126/sciadv.abd0384
Panula P, Pirvola U, Auvinen S, Airaksinen MS (1989) Histamine-immunoreactive nerve fibers in the rat brain. Neuroscience 28(3):585–610. https://doi.org/10.1016/0306-4522(89)90007-9
Parmentier R, Ohtsu H, Djebbara-Hannas Z, Valatx JL, Watanabe T, Lin JS (2002) Anatomical, physiological, and pharmacological characteristics of histidine decarboxylase knock-out mice: evidence for the role of brain histamine in behavioral and sleep-wake control. J Neurosci 22(17):7695–7711
Patke A, Murphy PJ, Onat OE, Krieger AC, Ozcelik T, Campbell SS, Young MW (2017) Mutation of the human circadian clock gene CRY1 in familial delayed sleep phase disorder. Cell 169(2):203–215. e213. https://doi.org/10.1016/j.cell.2017.03.027
Patke A, Young MW, Axelrod S (2020) Molecular mechanisms and physiological importance of circadian rhythms. Nat Rev Mol Cell Biol 21(2):67–84. https://doi.org/10.1038/s41580-019-0179-2
Patton AP, Hastings MH (2018) The suprachiasmatic nucleus. Curr Biol 28(15):R816–R822. https://doi.org/10.1016/j.cub.2018.06.052
Paul JR, Davis JA, Goode LK, Becker BK, Fusilier A, Meador-Woodruff A, Gamble KL (2020) Circadian regulation of membrane physiology in neural oscillators throughout the brain. Eur J Neurosci 51 (1):109–138. https://doi.org/10.1111/ejn.14343
Peever J, Fuller PM (2016) Neuroscience: a distributed neural network controls REM sleep. Curr Biol 26(1): R34–R35. https://doi.org/10.1016/j.cub.2015.11.011
Peever J, Fuller PM (2017) The biology of REM sleep. Curr Biol 27(22):R1237–R1248. https://doi.org/10.1016/j.cub.2017.10.026
Qiu P, Jiang J, Liu Z, Cai Y, Huang T, Wang Y, Liu Q, Nie Y, Liu F, Cheng J, Li Q, Tang Y-C, Poo M, Sun Q, Chang H-C (2019) BMAL1 knockout macque monkeys display reduced sleep and psychiatric disorders. Natl Sci Rev 6:87–100. https://doi.org/10.1093/nsr/nwz002
Reichert S, Pavon Arocas O, Rihel J (2019) The neuropeptide Galanin is required for homeostatic rebound sleep following increased neuronal activity. Neuron 104 (2):370–384. e375. https://doi.org/10.1016/j.neuron.2019.08.010
Rijo-Ferreira F, Takahashi JS (2019) Genomics of circadian rhythms in health and disease. Genome Med 11 (1):82. https://doi.org/10.1186/s13073-019-0704-0
Rossner MJ, Oster H, Wichert SP, Reinecke L, Wehr MC, Reinecke J, Eichele G, Taneja R, Nave KA (2008) Disturbed clockwork resetting in Sharp-1 and Sharp-2 single and double mutant mice. PLoS One 3(7): e2762. https://doi.org/10.1371/journal.pone.0002762
Roybal K, Theobold D, Graham A, DiNieri JA, Russo SJ, Krishnan V, Chakravarty S, Peevey J, Oehrlein N, Birnbaum S, Vitaterna MH, Orsulak P, Takahashi JS, Nestler EJ, Carlezon WA Jr, McClung CA (2007) Mania-like behavior induced by disruption of CLOCK. Proc Natl Acad Sci U S A 104 (15):6406–6411. https://doi.org/10.1073/pnas.0609625104
Rozov SV, Porkka-Heiskanen T, Panula P (2015) On the role of histamine receptors in the regulation of circadian rhythms. PLoS One 10(12):e0144694. https://doi.org/10.1371/journal.pone.0144694
Rusak B, Robertson HA, Wisden W, Hunt SP (1990) Light pulses that shift rhythms induce gene expression in the suprachiasmatic nucleus. Science 248 (4960):1237–1240. https://doi.org/10.1126/science.2112267
Sakai K (2014) Single unit activity of the suprachiasmatic nucleus and surrounding neurons during the wake-sleep cycle in mice. Neuroscience 260:249–264. https://doi.org/10.1016/j.neuroscience.2013.12.020
Sakai K, Takahashi K, Anaclet C, Lin JS (2010) Sleep-waking discharge of ventral tuberomammillary neurons in wild-type and histidine decarboxylase knock-out mice. Front Behav Neurosci 4:53. https://doi.org/10.3389/fnbeh.2010.00053
Sakhi K, Wegner S, Belle MD, Howarth M, Delagrange P, Brown TM, Piggins HD (2014) Intrinsic and extrinsic cues regulate the daily profile of mouse lateral habenula neuronal activity. J Physiol 592 (22):5025–5045. https://doi.org/10.1113/jphysiol.2014.280065
Scammell TE, Arrigoni E, Lipton JO (2017) Neural circuitry of wakefulness and sleep. Neuron 93

(4):747–765. https://doi.org/10.1016/j.neuron.2017.01.014
Scammell TE, Jackson AC, Franks NP, Wisden W, Dauvilliers Y (2019) Histamine: neural circuits and new medications. Sleep 42(1):zsy183. https://doi.org/10.1093/sleep/zsy183
Shan L, Hofman MA, van Wamelen DJ, Van Someren EJ, Bao AM, Swaab Dick F (2012) Diurnal fluctuation in histidine decarboxylase expression, the rate limiting enzyme for histamine production, and its disorder in neurodegenerative diseases. Sleep 35(5):713–715. https://doi.org/10.5665/sleep.1838
Siwicki KK, Hardin PE, Price JL (2018) Reflections on contributing to "big discoveries" about the fly clock: our fortunate paths as post-docs with 2017 Nobel laureates Jeff hall, Michael Rosbash, and Mike Young. Neurobiol Sleep Circadian Rhythms 5:58–67. https://doi.org/10.1016/j.nbscr.2018.02.004
Soya S, Sakurai T (2020) Evolution of orexin neuropeptide system: structure and function. Front Neurosci 14:691. https://doi.org/10.3389/fnins.2020.00691
Stephan FK, Zucker I (1972) Circadian rhythms in drinking behavior and locomotor activity of rats are eliminated by hypothalamic lesions. Proc Natl Acad Sci U S A 69(6):1583–1586. https://doi.org/10.1073/pnas.69.6.1583
Takahashi JS (2017) Transcriptional architecture of the mammalian circadian clock. Nat Rev Genet 18 (3):164–179. https://doi.org/10.1038/nrg.2016.150
Tobler I, Borbely AA, Groos G (1983) The effect of sleep deprivation on sleep in rats with suprachiasmatic lesions. Neurosci Lett 42(1):49–54. https://doi.org/10.1016/0304-3940(83)90420-2
Todd WD, Fenselau H, Wang JL, Zhang R, Machado NL, Venner A, Broadhurst RY, Kaur S, Lynagh T, Olson DP, Lowell BB, Fuller PM, Saper CB (2018) A hypothalamic circuit for the circadian control of aggression. Nat Neurosci 21(5):717–724. https://doi.org/10.1038/s41593-018-0126-0
Todd WD, Venner A, Anaclet C, Broadhurst RY, De Luca R, Bandaru SS, Issokson L, Hablitz LM, Cravetchi O, Arrigoni E, Campbell JN, Allen CN, Olson DP, Fuller PM (2020) Suprachiasmatic VIP neurons are required for normal circadian rhythmicity and comprised of molecularly distinct subpopulations. Nat Commun 11(1):4410. https://doi.org/10.1038/s41467-020-17197-2
Venner A, Anaclet C, Broadhurst RY, Saper CB, Fuller PM (2016) A novel population of wake-promoting GABAergic neurons in the ventral lateral hypothalamus. Curr Biol 26(16):2137–2143. https://doi.org/10.1016/j.cub.2016.05.078
Walker WH 2nd, Walton JC, DeVries AC, Nelson RJ (2020) Circadian rhythm disruption and mental health. Transl Psychiatry 10(1):28. https://doi.org/10.1038/s41398-020-0694-0
Wang Z, Ma J, Miyoshi C, Li Y, Sato M, Ogawa Y, Lou T, Ma C, Gao X, Lee C, Fujiyama T, Yang X, Zhou S, Hotta-Hirashima N, Klewe-Nebenius D, Ikkyu A, Kakizaki M, Kanno S, Cao L, Takahashi S, Peng J, Yu Y, Funato H, Yanagisawa M, Liu Q (2018) Quantitative phosphoproteomic analysis of the molecular substrates of sleep need. Nature 558(7710):435–439. https://doi.org/10.1038/s41586-018-0218-8
Webb IC, Baltazar RM, Wang X, Pitchers KK, Coolen LM, Lehman MN (2009) Diurnal variations in natural and drug reward, mesolimbic tyrosine hydroxylase, and clock gene expression in the male rat. J Biol Rhythm 24(6):465–476. https://doi.org/10.1177/0748730409346657
Weber M, Lauterburg T, Tobler I, Burgunder JM (2004) Circadian patterns of neurotransmitter related gene expression in motor regions of the rat brain. Neurosci Lett 358(1):17–20. https://doi.org/10.1016/j.neulet.2003.12.053
Weber F, Chung S, Beier KT, Xu M, Luo L, Dan Y (2015) Control of REM sleep by ventral medulla GABAergic neurons. Nature 526(7573):435–438. https://doi.org/10.1038/nature14979
Wisor JP, O'Hara BF, Terao A, Selby CP, Kilduff TS, Sancar A, Edgar DM, Franken P (2002) A role for cryptochromes in sleep regulation. BMC Neurosci 3:20. https://doi.org/10.1186/1471-2202-3-20
Wulff K, Gatti S, Wettstein JG, Foster RG (2010) Sleep and circadian rhythm disruption in psychiatric and neurodegenerative disease. Nat Rev Neurosci 11 (8):589–599. https://doi.org/10.1038/nrn2868
Wurts SW, Edgar DM (2000) Circadian and homeostatic control of rapid eye movement (REM) sleep: promotion of REM tendency by the suprachiasmatic nucleus. J Neurosci 20(11):4300–4310
Yamashita T, Yamanaka A (2017) Lateral hypothalamic circuits for sleep-wake control. Curr Opin Neurobiol 44:94–100. https://doi.org/10.1016/j.conb.2017.03.020
Yu X, Zecharia A, Zhang Z, Yang Q, Yustos R, Jager P, Vyssotski AL, Maywood ES, Chesham JE, Ma Y, Brickley SG, Hastings MH, Franks NP, Wisden W (2014) Circadian factor BMAL1 in histaminergic neurons regulates sleep architecture. Curr Biol 24 (23):2838–2844. https://doi.org/10.1016/j.cub.2014.10.019
Yu X, Li W, Ma Y, Tossell K, Harris JJ, Harding EC, Ba W, Miracca G, Wang D, Li L, Guo J, Chen M, Li Y, Yustos R, Vyssotski AL, Burdakov D, Yang Q, Dong H, Franks NP, Wisden W (2019a) GABA and glutamate neurons in the VTA regulate sleep and wakefulness. Nat Neurosci 22(1):106–119. https://doi.org/10.1038/s41593-018-0288-9
Yu X, Ma Y, Harding EC, Yustos R, Vyssotski AL, Franks NP, Wisden W (2019b) Genetic lesioning of histamine neurons increases sleep-wake fragmentation and reveals their contribution to modafinil-induced wakefulness. Sleep 42(5):zsz031. https://doi.org/10.1093/sleep/zsz031
Yu X, Ba W, Zhao G, Ma Y, Harding EC, Yin L, Wang D, Li H, Zhang P, Shi Y, Yustos R, Vyssotski AL, Dong H, Franks NP, Wisden W (2020) Dysfunction

of ventral tegmental area GABA neurons causes mania-like behavior. Mol Psychiatry. https://doi.org/10.1038/s41380-020-0810-9
Zhang L, Fu YH (2020) The molecular genetics of human sleep. Eur J Neurosci 51(1):422–428. https://doi.org/10.1111/ejn.14132
Zhang C, Truong KK, Zhou QY (2009) Efferent projections of prokineticin 2 expressing neurons in the mouse suprachiasmatic nucleus. PLoS One 4(9): e7151. https://doi.org/10.1371/journal.pone.0007151
Zhang Z, Ferretti V, Guntan I, Moro A, Steinberg EA, Ye Z, Zecharia AY, Yu X, Vyssotski AL, Brickley SG, Yustos R, Pillidge ZE, Harding EC, Wisden W, Franks NP (2015) Neuronal ensembles sufficient for recovery sleep and the sedative actions of alpha2 adrenergic agonists. Nat Neurosci 18(4):553–561. https://doi.org/10.1038/nn.3957
Zhang L, Hirano A, Hsu PK, Jones CR, Sakai N, Okuro M, McMahon T, Yamazaki M, Xu Y, Saigoh N, Saigoh K, Lin ST, Kaasik K, Nishino S, Ptacek LJ, Fu YH (2016) A PERIOD3 variant causes a circadian phenotype and is associated with a seasonal mood trait. Proc Natl Acad Sci U S A 113(11):E1536–E1544. https://doi.org/10.1073/pnas.1600039113
Zhang Z, Zhong P, Hu F, Barger Z, Ren Y, Ding X, Li S, Weber F, Chung S, Palmiter RD, Dan Y (2019) An excitatory circuit in the Perioculomotor midbrain for non-REM sleep control. Cell 177(5):1293–1307.e1216. https://doi.org/10.1016/j.cell.2019.03.041
Zhong P, Zhang Z, Barger Z, Ma C, Liu D, Ding X, Dan Y (2019) Control of non-REM sleep by midbrain Neurotensinergic neurons. Neuron 104(4):795–809.e796. https://doi.org/10.1016/j.neuron.2019.08.026

Astrocyte Circadian Timekeeping in Brain Health and Neurodegeneration

6

Marco Brancaccio, Anne C. Wolfes, and Natalie Ness

Abstract

Almost three decades ago, astrocytes neighbouring clock neurons of the suprachiasmatic nucleus, the hypothalamic tissue responsible for synchronising circadian timekeeping in mammals, were found to undergo morphological and protein expression changes in a cyclic 24-h pattern, suggesting that glia could harbour circadian timekeeping mechanisms and that neuron–glia interactions could play a part in the daily organisation of rhythms of physiology and behaviour. Recently, it has become clear that astrocytes are circadian timekeepers, capable of initiating daily patterns of behaviour and imposing their intrinsic circadian tempo in mammals. In this chapter, we will describe properties of intracellular timekeeping of astrocytes and the mechanisms by which astrocytes functionally integrate in brain circuits underlying circadian, sleep, and cognitive behaviours in mammals. We will then discuss how altered astrocyte timekeeping may be involved in early brain vulnerability underpinning neurodegeneration. We will focus on Alzheimer's disease as a template of how altered astrocyte timekeeping may be involved in neurodegeneration, both directly via unbalancing of inflammatory and oxidative stress cellular pathways, and indirectly, by altering sleep and cognitive functions.

Keywords

Circadian clocks · Astrocyte timekeeping · Sleep · Neurodegeneration · Alzheimer's disease

6.1 Introduction

Circadian timekeeping anticipates predictable variations of the environment following light/dark cycles generated by the earth's rotation (Woelfle et al. 2004; Bell-Pedersen et al. 2005). In their simplest form, circadian rhythms are generated by intracellular negative feedback loops of timed clock gene transcription and translation, with a periodicity of roughly 24 hours (*circa*—about; *diem*—a day) (Takahashi 2017). In mammals, a central master oscillator in the anterior hypothalamus, the suprachiasmatic nucleus (SCN), synchronises circadian cellular oscillators to orchestrate physiology and behaviour and coordinate their activities to the solar cycle (Hastings et al. 2018). An additional level of complexity arises from the vast array of neuronal and glial cell types that play an important but still poorly characterised role in

M. Brancaccio (✉) · A. C. Wolfes · N. Ness
Department of Brain Sciences, Division of Neuroscience, Imperial College London, London, UK

UK Dementia Research Institute at Imperial College London, London, UK
e-mail: m.brancaccio@imperial.ac.uk

O. Engmann, M. Brancaccio (eds.), *Circadian Clock in Brain Health and Disease*, Advances in Experimental Medicine and Biology 1344, https://doi.org/10.1007/978-3-030-81147-1_6

integrating central and local timekeeping cues fine-tuning daily regulation of brain functions and behaviours. Astrocytes are amongst the most diverse cells in the brain and are emerging as key glial players in the modulation of neuronal activities underpinning a variety of behaviours (Artiushin and Sehgal 2020; Nagai et al. 2021; Broadhead and Miles 2021). Far from playing a merely structural and supportive role, they have recently been shown to possess strong circadian timekeeping properties and to be able to initiate and support cell-autonomous circadian patterns of behaviour in adult animals (Hastings et al. 2018; Brancaccio et al. 2019).

In this chapter, we will present evidence supporting the role of astrocytes as key players in circadian timekeeping and discuss the intracellular and intercellular mechanisms by which they modulate neuronal circuit activities and impose circadian patterns of behaviour. While the available literature strongly supports a role for astrocytes in the specification of internal timekeeping, most work has been conducted in the SCN, given its central role in synchronising circadian clocks across the brain and body, and generally less is known about the specific relevance of astrocyte timekeeping in regulating other aspects of behaviour, such as sleep and cognition. Nevertheless, a growing literature implicates astrocytes in the daily organisation of brain function and behaviour, albeit the underlying timekeeping mechanisms have not yet been fully addressed. We will review these findings and highlight possible points of circadian regulation of these behaviours, operated by astrocyte timekeepers. Circadian and sleep disruption are emerging as a shared feature and cause of neurodegenerative conditions (Leng et al. 2019; Nedergaard and Goldman 2020). In the second part of the chapter, We will focus on Alzheimer's disease (AD), the most prevalent form of dementia as a template for the investigation of the role of disrupted astrocytic timekeeping in neurodegenerative conditions. We will discuss findings suggesting that astrocyte timekeeping may play a key role in preserving brain health and that its disruption may be involved in neurodegeneration both directly, via imbalance of inflammatory and oxidative stress cellular responses, and indirectly, by altering sleep and cognitive functions and accelerating disease progression.

6.2 Intracellular Circadian Timekeeping in Astrocytes

6.2.1 The Molecular Clockwork of Astrocytes

Molecular circadian timekeeping relies on cell-autonomous transcription-translation feedback loops (TTFLs). In mammals, the core clock genes include Circadian locomotor output cycles protein kaput (*Clock*) (and its paralog *Npas2*) and Brain and muscle ARNT-like 1 (*Bmal1*), positive regulators of the TTFL, as well as Period 1-2 (*Per1-2*) and Cryptochrome 1-2 (*Cry1-2*), the negative arm of the TTFL. CLOCK and BMAL1 proteins form heterodimers, activating transcription of *Per* and *Cry* genes at circadian time (CT) via enhancer box (E-box) sequences (Takahashi 2017; Hastings et al. 2018). PER and CRY subsequently accumulate in the nucleus, form repressor complexes and begin to repress their own transcription at around CT12. This transcriptional repression and the degradation of existing PER-CRY complexes lead to a decline in the repressor complexes that ultimately allows for a new cycle to be initiated. This core clock gene TTFL is stabilised by an accessory feedback loop involving nuclear receptors ROR-α/β and REV-ERBα/β, which are transcriptionally activated via E-box binding of the CLOCK: BMAL1 heterodimer (Takahashi 2017). The core clock gene TTFL and its interlocking accessory transcriptional feedback loops generate transcriptional rhythms in clock-controlled genes via interactions with regulatory enhancer elements, including E-boxes, RREs, and D-boxes. Like neurons, astrocytes in the cortex and in the SCN have been shown to possess functional TTFLs, underlying rhythmic morphological and protein expression changes (Lavialle and Serviere 1993) which can be phase-shifted in response to physiological signals (Prolo et al. 2005; Tso et al. 2017; Brancaccio et al. 2019). AAV-mediated expression of genetically encoded fluorescent calcium indicators in astrocytes has enabled closer dissection of cellular circadian timekeeping in astrocytes. Simultaneous recordings of fluorescent markers for astrocytic and neuronal $[Ca^{2+}]_i$ in SCN slices revealed that astrocytes show robust $[Ca^{2+}]_i$ rhythms that differ from neuronal $[Ca^{2+}]_i$ rhythms in their waveform and phase, with astrocytic $[Ca^{2+}]_i$ presenting a broader peak that is phased to the dark period (peaking at CT18), in contrast to neurons, which are most active during daytime, with neuronal calcium and voltage rhythms peaking at CT6 (Brancaccio et al. 2013, 2017). Rhythms of clock gene expression are also differentially phased between these two cell types, with expression of the clock gene Cry1 peaking at CT11 in neurons and CT17 in astrocytes (Brancaccio et al. 2019). Double knockout of *Cry1* and *Cry2* (Cry1/2-null) renders the SCN arrhythmic and can be rescued by cell-type specific AAV-mediated expression of Cry1 in neurons of organotypic SCN slices. Importantly, not only did expression of Cry1 in astrocytes alone rescue molecular rhythmicity, but also the periodicity of the restored behaviour was different when compared to neurons: astrocyte-initiated rhythms had a shorter free-running period, and required more time to initiate stable period oscillations of Per2 expression (Brancaccio et al. 2019). These findings suggest that the anti-phasic molecular rhythms of SCN astrocytes and neurons likely arise as a result of cell-type specific regulation of the TTFL and that astrocytes can have an active, instructive role in driving circadian oscillations of neuronal activity and behaviour (see below).

6.2.2 Behaviourally Consequential Manipulations of the Astrocyte Clockwork

In recent years, several protocols of astrocytic clock manipulations have demonstrated a role for astrocyte timekeeping mechanisms in specifying circadian patterns of behaviour. Ablating *Bmal1* clock gene expression specifically in astrocytes by somatic, virally mediated CRISPR/Cas9 lengthened the period of clock gene expression in the SCN and rest-activity behaviour in animals left in constant darkness (Tso et al. 2017). When EAAT1/GLAST-positive astrocytes lacked *Bmal1* (Bmal1cKO), the SCN showed increased vasoactive intestinal peptide (VIP) levels at the onset of the dark phase, reduced levels of *Bmal1* and *Per2* expression, and dampening of their fluctuations (Barca-Mayo et al. 2017). This did not alter locomotor activity under light-dark (LD) conditions, but resulted in a bimodal pattern of locomotor activity and a delayed active phase in constant darkness (DD) (Barca-Mayo et al. 2017). Overall, these loss-of-function experiments show that TTFL expression in astrocytes is important to support accurate circadian timekeeping in mammals, but not strictly required for circadian rhythm generation.

Another key question is whether astrocytes possess cell-autonomous circadian timekeeping properties, which may instruct circadian behavioural function. To answer this two main protocols have been used: (1) temporal mismatch or (2) rescue of TTFL in astrocytes. The first approach takes advantage of the Tau mutation of the casein kinase 1ε gene, which reduces circadian periodicity by ~4 h in homozygosis, due to accelerated degradation of PERs (Meng et al. 2008). Cre recombinase-mediated deletion of the floxed $Ck1\varepsilon^{Tau/Tau}$ locus reverts behavioural period to 24 h, thus allowing to create an internal temporal mismatch within the animal when Cre expression is restricted by cell-type specific promoters. Expression of Cre by the GFAP (virally mediated in the SCN) or Aldh1L1 (by germline breeding) astrocyte-specific promoters were both effective in reverting patterns of rest-activity behaviour to a 24 h periodicity in DD, thus showing that astrocytes can impose their intrinsic circadian tempo to behavioural function (Brancaccio et al. 2017; Tso et al. 2017). When Cre was expressed in neurons of $Ck1\varepsilon^{Tau/Tau}$ animals by the human Synapsin1 promoter, the restored rhythms were indistinguishable from the ones elicited by astrocytic manipulations, thus not allowing to demonstrate differences in the cell-autonomous properties of astrocytic vs neuronal circadian timekeepers. In the second approach of astrocytic TTFL rescue, expression of Cry1 was restored in SCN astrocytes by co-injecting AAVs expressing Cre under the GFAP promoter, together with AAVs expressing Cry1 within a flexed, Cre-activatable cassette in genetically arrhythmic Cry1/2-null mice. Remarkably, this treatment was effective in establishing newly generated patterns of circadian behaviour in DD. The period of the rescued free-running behaviour was different from 24 hours in both cases (as expected because of the *Cry2*-null status of the rescued animals) (van der Horst et al. 1999) and therefore reflective of an intrinsically generated rhythm. Importantly, rhythms expressed upon expression of Cry1 in astrocytes were different in periodicity when compared to the ones produced upon neuron-specific Cry1 rescue in littermates. Moreover, the period of the restored neuronal vs astrocytic Cry1-rescued behaviour diverged more with increasing number of respectively targeted cells. These results were confirmed in isolated SCN slices thus showing that specific rhythms of clock gene expression and behaviour, driven by astrocytic or neuronal clockwork can be distinguished in Cry1-rescued mice. Therefore, while both astrocytes and neurons of the SCN can exert central timekeeping functions, the underlying cell-autonomous mechanisms specifying the temporal details of such oscillations may differ (Brancaccio et al. 2019).

6.3 Astrocytes in Circadian Intercellular Timekeeping

6.3.1 Astrocyte–Neuron Communication

Given data of astrocyte clock rescue, a central question is whether specific signalling pathways deliver temporal signals from astrocytes to neurons and, vice versa the existence of neuronal signals that may feed back to astrocytes to maintain accurate timekeeping. Extracellular glutamate has been proposed as a potential mediator of astrocyte-neuronal interactions that could underpin the astrocytic rescue of neuronal function observed in Brancaccio et al. (2017, 2019). Extracellular glutamate measured by the genetically encoded iGluSnFR reporter (Marvin et al. 2013) revealed strong oscillations in the SCN, which were anti-phasic to neuronal calcium and voltage and elevated during night-time. Importantly, the vast majority of SCN neurons (>95%) are GABAergic and therefore the reported widespread oscillations observed in isolated SCN brain tissues are unlikely to be due to any glutamatergic neuronal release. The largest contributors to glutamate concentrations are excitatory amino acid transporters (EAATs), including EAAT1/GLAST and EEAT2/GLT-1, two predominantly astrocytic glutamate transporters, which follow circadian expression patterns and peak during the light phase (Spanagel et al. 2005), similar to the mRNA expression of EAAT3/EAAC1, a neuronal glutamate transporter (Cagampang et al. 1996). Blocking EAAT activity by DL-TBOA elevated extracellular glutamate detected by iGluSnFR, caused internal desynchronisation of the SCN, decreased amplitude, and lengthened the period of clock gene expression and neuronal calcium (Brancaccio et al. 2017). In *Per2* mutant mice, EAAT1 becomes arrhythmic, while EAAT2 expression is shifted (Spanagel et al. 2005). In vitro, EAAT1 expression levels are also significantly reduced in the absence of CLOCK, PER2, and NPAS2 (Beaulé et al. 2009), and glutamate uptake is only half as efficient in *Per2* mutant as in wild-type mice (Spanagel et al. 2005). Together, these data suggest that EAAT expression may be regulated by clock genes and is essential to maintain rhythmic extracellular glutamate concentrations, required for internal synchronisation of the SCN circuit. Extracellular glutamate in the SCN has been proposed to activate presynaptic NMDA receptors containing the NR2C subunit and to promote GABA release during night-time. Consistent with this view, inhibition of the NMDA-NR2C subunit during night-time depolarises SCN neurons and de-synchronises SCN neuronal activities and clock gene expression within the SCN (Brancaccio et al. 2017); see Hastings et al.

(2019) for an updated model. Moreover, NR2C inhibition irreversibly impairs rescue of circadian oscillations of clock gene expression in the SCN in Cry1/2 null mice in which Cry1 is restored in astrocytes (Brancaccio et al. 2019), thus showing that astrocytically released glutamate and NMDA-NR2C are an essential signalling axis mediating astrocyte–neuronal circadian interactions in the SCN.

While more data exist for glutamate, other molecules have been proposed as putative mediators of astrocyte-neuronal communication, such as ATP. Rat SCNs feature over four times the ATP concentrations of the anterior hypothalamus (Yamazaki et al. 1994), suggesting that ATP may represent a prominent signalling cue in the SCN. SCN astrocytes release ATP in a circadian fashion, which appear to be clock gene-dependent in cell cultures and SCN slices (Womac et al. 2009; Burkeen et al. 2011; Marpegan et al. 2011; Svobodova et al. 2018). ATP rhythms can be suppressed or potentiated by pharmacological blocking/agonists of purinergic and pannexin-1 receptors, suggesting that these proteins may be involved in ATP release (Svobodova et al. 2018). As extracellular ATP increases (peaking in the dark phase) in vitro and in vivo, SCN neuronal firing rate declines (Yamazaki et al. 1994; Womac et al. 2009). This anti-phasic pattern between astrocytic ATP and neuronal activity persists in constant darkness (Womac et al. 2009). In rat SCN slices, GABAergic neurons respond to extracellular ATP with increased frequency (but not amplitude) of spontaneous inhibitory postsynaptic currents, in a dose-dependent manner (Bhattacharya et al. 2013).

6.3.2 Neuron–Astrocyte Communication

As previously mentioned, a key question is whether and how neurons may feed back circadian temporal information to astrocytes, thus "closing" a neuron/astrocyte intercellular circadian feedback loop. While few studies exist that address the neuron-to-astrocyte arm of the loop, endocannabinoids and VIP signalling have been proposed as possible mediators. In rats living in 12:12 LD cycles, GFAP expression peaks during the dark phase specifically in the ventrolateral, but not the dorsomedial SCN; constant darkness or removing the adrenal glands abolishes this pattern (Becquet et al. 2008). GFAP expression changes represent some of the SCN's daily structural rearrangements: at night, glial coverage of VIP neuron dendrites increases by 29%, but decreases by 19% for AVP neurons (Becquet et al. 2008). In turn, VIP released by SCN neurons can entrain circadian rhythms of astrocytes in a dose-dependent manner in vitro (Marpegan et al. 2009).

Additional key molecules of neuron–astrocyte interaction are endocannabinoids, neuronal retrograde messengers that modulate neurotransmitter release by binding to presynaptic cannabinoid type 1 receptors (CB_1Rs). Astrocytes express CB_1Rs in vitro and in vivo, and when neuronal endocannabinoids bind to these, astrocytes additionally express the CB_1R ligand arachidonoylglycerol (2-AG) (Walter and Stella 2003; Marsicano et al. 2003; Walter et al. 2004; Navarrete and Araque 2008; Hegyi et al. 2018). Two major endocannabinoids follow circadian expression patterns in the nucleus accumbens, prefrontal cortex, striatum, and hippocampus: 2-AG reaches significantly higher levels during the light phase, while anandamide/*N*-arachidonylethanolamine reaches significantly higher levels during the dark phase (Valenti et al. 2004). Other endocannabinoids and CB_1R agonists also display circadian expression changes, depending on the brain region (Liedhegner et al. 2014). One study focusing on the SCN reported strong CB_1R immunoreactivity and found that cannabinoids diminished the ability to entrain to light zeitgebers (Acuna-Goycolea et al. 2010). However, cannabinoids left glutamate release from the retinohypothalamic tract unaffected, and free-running mice showed normal circadian rhythms (Acuna-Goycolea et al. 2010). The authors suggested endocannabinoids as non-photic cues for neuronal entrainment, but did not address astrocytic CB_1R contributions. Intriguingly, recent work introduced astrocytic retrograde cannabinoid signalling as a mediator of clock timing in the SCN (Hablitz et al. 2020a). In organotypic slices, neuronal endocannabinoids bind to astrocytic CB_1Rs and trigger Ca^{2+} signalling in SCN astrocytes, which in turn inhibits presynaptic GABA release via adenosine signalling from astrocytes (Hablitz et al. 2020a; Carney 2020). The authors suggested that endocannabinoid signalling in astrocytes serves to fine-tune neuronal responses to light.

A representative schematic of the astrocyte-neuronal mediators that have been implicated in the interplay between neurons and astrocytes is provided in Fig. 6.1.

6.3.3 Astrocyte–Astrocyte Communication

Autocrine and paracrine signalling between astrocytes, for instance via glutamate and ATP release, is enabled by a number of transmembrane channels, including connexins and pannexins (Giaume et al. 2020). Astrocytes form interconnected networks via gap junctions, which allow diffusion of molecules, such as metabolites, second messengers, glutamate, small peptides, and RNA (Giaume et al. 2010). Large gap junction-coupled networks of hundreds of astrocytes can be found in various brain regions (Houades et al. 2008; Rouach et al. 2008). Gap junctions between astrocytes are primarily formed by coupling of connexin (CX) 43 and CX30 hemichannels (Nagy et al. 2004), whose expression levels vary regionally and with age (Nagy et al. 1999; Cotrina et al. 2001) and determine the permeability of gap junctions (Harris 2007; Yum et al. 2007). Interestingly, the expression of CX30 and CX43 oscillates throughout the day in the mouse SCN, with fluctuations of CX43 persisting in constant darkness (Ali et al. 2019), suggesting that the astrocytic clock may modulate astrocytic networks through regulation of gap junction proteins. Gap junctions are key to astrocyte function, for instance by allowing them to indirectly associate with all the synapses within their network. The contribution of astrocyte–astrocyte communication is difficult to determine, as connexins are also important for astrocyte–neuron communication, where they mediate metabolite transport, glutamate recycling, and K^+ buffering (Rouach et al. 2008; Pannasch et al. 2011). Astrocytes and neurons can connect directly through heterotypic connexin gap junctions, where the prevalent astrocytic CX43 can couple with CX36 in neurons (Condorelli et al. 2000; Kay et al. 2016). In these instances, astrocytes can provide neurons with metabolites like glucose, lactate, or glutamine, and exert spatial K^+ buffering as well as Ca^{2+} signalling. Moreover, connexins in their hemichannel forms can release glutamate and ATP, shaping neuronal function in a paracrine fashion, which can directly modulate neuronal activity (Ye et al. 2003; Wang et al. 2013). Indeed, Brancaccio et al. (2019) showed that drugs selectively blocking the hemichannel form of CX43 (TAT-Gap19) disrupt clock gene expression (detected by a PER2::LUC reporter) in the SCN and impair oscillations of extracellular glutamate elicited by astrocytic Cry1 rescue in arrhythmic dCry KO mice. These data support a model in which paracrine glutamate release by astrocytes is promoted by CX43 hemichannels and is required for daily rhythms of extracellular glutamate and synchronisation of neuronal activity within the SCN.

6.3.4 Astrocyte–Glia Communication

Paracrine signalling also enables astrocyte communication with other glial cells; for instance, microglia-derived ATP can bind astrocytic P2Y1 receptors, inducing the production and release of interleukin 6 in culture (Shinozaki et al. 2014). It has also been observed that connexins can form heterotypic gap junctions between astrocytes and oligodendrocytes for K^+ buffering (Kamasawa et al. 2005). Whether such glial interplay contributes to circadian function is currently unknown.

Microglia and astrocytes closely coordinate their functions, both in normal physiology and in response to injury and neurodegeneration (Vainchtein and Molofsky 2020). For instance, astrocytes and microglia have essential roles in synapse formation and remodelling during

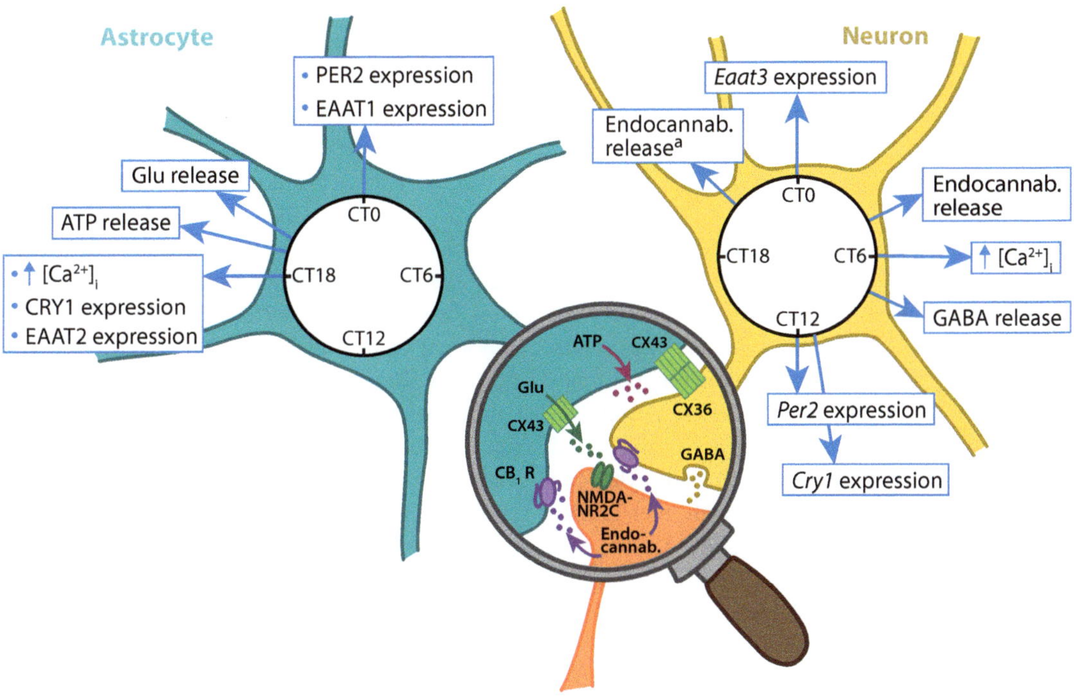

Fig. 6.1 Summary of astrocytic and neuronal intracellular circadian markers and intercellular mediators involved in circadian timekeeping in the brain. Several molecular mechanisms exhibit circadian rhythms in astrocytes and neurons, often in anti-phasic patterns. Timepoints indicated on the circadian time (CT) clock within each cell type describes peak activity. Except where described below, all mechanisms refer to the SCN. For astrocytes, arrows highlight peaks of: EAAT1 expression in the whole brain at CT0 (Spanagel et al. 2005), EAAT2 expression in the whole brain at CT18 (Spanagel et al. 2005), PER2 expression at CT0 in cultured cortical astrocytes (Beaulé et al. 2009), CRY1 expression at CT18 (Brancaccio et al. 2019), intracellular Ca^{2+} rise (Brancaccio et al. 2017), ATP release (Womac et al. 2009; Marpegan et al. 2011), and glutamate (Glu) release (Brancaccio et al. 2017). For neurons, arrows highlight peaks of: *Eaat3* expression at CT0 (Cagampang et al. 1996), intracellular Ca^{2+} rise at CT6 (Brancaccio et al. 2017), GABA release (Wagner et al. 1997; Itri et al. 2004), *Per2* expression at CT12 (Amir et al. 2004), and *Cry1* expression at CT11 (Brancaccio et al. 2019). For neuronal endocannabinoid release, note that 2-AG in the nucleus accumbens, hippocampus, prefrontal cortex, and striatum is released at CT3 (Valenti et al. 2004) and anandamide is released in the hypothalamus at CT5 (Murillo-Rodriguez et al. 2006); [a]anandamide is released at CT21 in the nucleus accumbens, hippocampus, prefrontal cortex, and striatum (Valenti et al. 2004; Murillo-Rodriguez et al. 2006). The magnifying glass shows a tripartite synapse, highlighting the interplay between a presynaptic neuron, its postsynaptic counterpart, and an astrocyte in the context of circadian mechanisms

development and their coordination is essential to balance synapse numbers. Astrocytes have a primary role in promoting synapse formation during development, and synapse maturation has been shown to increase astrocytic expression of interleukin 33, which directly increases microglial phagocytic activity, suggesting a homeostatic communication loop in which increases in synapse formation by astrocytes trigger synaptic pruning by microglia (Vainchtein et al. 2018). While it is not known whether and how glial clocks may interact, it is well established that the release of several cytokines and chemokines (including TNF-α and complement components C3, CCL2), key mediators of astrocyte-microglia crosstalk, is under direct control of the circadian clock (Wolff et al. 2020; Griffin et al. 2020). Recent evidence shows that some of these factors can induce phase shifts in glial clock rhythms (Duhart et al. 2013, 2016), suggesting a potential mechanism by which glial clocks could coordinate their circadian functions. New insights into how glial clocks communicate may also allow us to gain a deeper understanding of the functional impact of glial clocks at the circuit level, especially given the emerging evidence for a key role of the circadian clock in microglia and its potential implications in neurodegenerative conditions (Nakanishi et al. 2021).

The presence of circadian timekeeping mechanisms in oligodendrocytes is much less defined and their role in oligodendrocytic function is still debated (Colwell and Ghiani 2020). Interestingly, as with astrocyte–neuron and astrocyte–astrocyte gap junctions, astrocytes can also supply oligodendrocytes with metabolites (lactate, glucose, glutamate) and mediate ion homeostasis through hemichannels made up of astrocytic CX43 and oligodendrocytic CX47 (Nagy et al. 2003; Rinholm et al. 2011, 2016; Fasciani et al. 2018; Basu and Sarma 2018). Recent studies have revealed that astrocyte–oligodendrocyte gap junctions are essential for CNS myelination and homeostasis (Papaneophytou et al. 2019). CX43/CX47 gap junctions also modulate K^+ buffering and bi-directional Ca^{2+} signalling between astrocytes and oligodendrocytes (Parys et al. 2010), and when abolished, cause myelin vacuole build-up, leading to axonal degeneration (Menichella et al. 2006). Given circadian expression of CX43 (Ali et al. 2019), this may also affect the function of myelinating cells of the CNS.

6.4 Astrocyte Timekeepers in the Regulation of Brain Function

6.4.1 Astrocyte Timekeeping and Cognitive Function

Astrocytes have been shown to play a role in daily regulation of synaptic and network function in the hippocampus and cortex. In the hippocampus, about 10% of genes are rhythmically transcribed (Debski et al. 2020) and synaptic excitability of granule cells fluctuates in a circadian manner (Barnes et al. 1977). In fact, human lifestyle factors that mimic sleep restriction (such as jet-lag and shift work) directly correlate with reduced temporal lobe volume and poorer cognitive performance (Cho 2001; Ariznavarreta et al. 2002). Similarly, in mice, chronic circadian disruption leads to shorter dendrites and reduced neuronal complexity in the prefrontal cortex, as well as to cognitive changes such as higher error rates in learning tasks (Morris water maze) and abnormal behaviour in open field and light/dark box tests (Karatsoreos et al. 2011). Under normal conditions, the formation, recall, and extinction of fear conditioning memory varies in a circadian manner in mice, with improved recall and slower extinction of memories during the light phase, a pattern which persists in constant darkness (Chaudhury and Colwell 2002).

Bmal1-null mice are arrhythmic in constant darkness and exhibit impairments in spatial and contextual fear memory. Organotypic hippocampal slices from Bmal1-null mice also show significantly reduced long-term potentiation (LTP), which may result from impaired Bmal1-dependent signalling events, likely including circadian regulation of MAPK activity (Wardlaw et al. 2014). Circadian control of hippocampal function and plasticity is also reflected in acute slice preparations, as slices prepared during the day show more robust LTP at Schaffer collateral synapses than those prepared during the night (Raghavan et al. 1999). But what causes daily fluctuations in LTP in the hippocampus? McCauley et al. (2020) recently found that while decreased NMDA receptor expression in neurons is a contributing factor, structural and functional circadian changes in astrocytes are also involved in mediating the loss of LTP in the dark phase. Summation of AMPA excitatory postsynaptic currents (EPSCs) was lower in the dark phase with no significant effect on NMDA EPSCs, an effect explained by an increase in glutamate lifetime in the extracellular space, which prolongs AMPA receptor recovery time. Experimentally, synaptically-activated glutamate transporter currents in astrocytes were slower in the dark phase, suggesting that glutamate uptake by astrocytes indeed varies in a circadian manner.

Accordingly, the number of astrocytic processes was reduced during the dark phase, decreasing the proximity to surrounding synapses, and normalised entropy was lower in the dark phase, indicating structural changes in astrocytes throughout the circadian cycle. In line with this finding, Bmal1 deletion has also been found to reduce glial coverage of hippocampal mossy fibre synapses (Ali et al. 2020), and astrocytic coverage of synapses has previously been observed to follow a rhythmic pattern, and increases in response to chronic sleep restriction, particularly of large mature dendritic spines in the mouse cortex (Bellesi et al. 2015). Small changes in the proximity of astrocytic processes to synaptic spines strongly affect extracellular glutamate concentration, thereby regulating the strength and timing of neuronal synaptic activity. Accordingly, the stabilisation of synapses is impaired when astrocyte motility is reduced (Nishida and Okabe 2007). The importance of regulating extracellular glutamate levels is also exemplified by astrocyte-mediated switching of cortical neurons to a slow-oscillation state via glutamatergic signalling, which may be important in sleep modulation (Poskanzer and Yuste 2016). The benefit of daily remodelling of astrocytes is unclear, however it has been hypothesised that reducing synaptic efficacy in the dark phase may lower energy consumption over extended time periods. Alternatively, the change may allow the local network to switch between modes that facilitate improved processing of different types of information at different times (McCauley et al. 2020). Interestingly, computational modelling showed that lower NMDA receptor expression and longer recovery from AMPA receptor activation makes hippocampal activity in the high gamma frequency range more susceptible to circadian modulation. Indeed, behaviours such as learning and novel environment exploration, which rely on activity in the high gamma range, show a higher circadian variation than lower-frequency behaviours, such as memory retrieval and navigation in familiar environments (Colgin et al. 2009; Zheng et al. 2016; McCauley et al. 2020).

In addition to daily structural remodelling, astrocytic BMAL1 has also been found to be important for correct functioning of astrocytes at synapses. Upon finding short- and long-term memory impairments, a global reduction of BMAL1 levels in the cortex, as well as dampening in the rhythmic expression of BMAL1 and PER2 in mice with astrocyte-specific conditional deletion (Bmal1cKO), Barca-Mayo et al. (2017) investigated the cell-autonomous rhythmic entrainment of cortical neurons by astrocytes in vitro. Synchronous astrocytes in a physically separated co-culture with asynchronous primary cortical neurons induced rhythmic clock gene expression in neurons. siRNA-mediated Bmal1 knockdown in astrocytes abolished their ability to synchronise rhythmic transcript expression in neurons, suggesting that an intact astrocytic clock is required to entrain rhythmicity in cortical neurons via exchange of extracellular signals. GABA was sufficient to entrain primary cortical neurons, while inhibition of the GABAA receptor prevented entrainment by astrocytes. In line with this finding, GABA transporters GAT1 and GAT3 were significantly reduced in the cortex of Bmal1cKO, and GABA uptake was severely impaired in arrhythmic cortical astrocytes. Accordingly, GABAA receptor antagonist treatment restored the memory impairment in Bmal1cKO mice. This suggests that astrocytic BMAL1 is required to maintain extracellular GABA levels. Astrocytic GABAergic signalling in the cerebellum has previously been shown to be involved in tonic inhibition of neurons (Lee et al. 2010), which has been implicated in LTP and hippocampus-dependent memory (Martin et al. 2010). Thus, loss of astrocytic BMAL1 and the resulting reduction in GABA reuptake may cause over-inhibition of cortical and hippocampal circuits involved in learning and memory. It is not clear however whether the role of astrocytes in maintenance of extracellular GABA level depends on static or oscillating BMAL1 function. Interestingly, deleting *Bmal1* in Nestin-expressing pericytes (which line the vascular system) causes not only astroglial activation in regions as far-reaching as the cortex and hippocampus, but also novelty-induced hyperactivity (Nakazato et al. 2017), which could be a consequence of astrocyte activation in brain regions required for cognitive function.

6.4.2 Astrocyte Timekeeping and Sleep

The connection between SCN timekeeping, local clock gene expression, and their role in regulating sleep is complex (Franken 2013; Deboer 2020) and is discussed elsewhere in this book (Chap. 5). Here we will review evidence specifically involving astrocytes in sleep, with a particular focus on discerning available information regarding their role in homeostatic and/or circadian regulation of sleep. Astrocytes of the prefrontal cortex show dynamic changes in Ca^{2+} across the sleep/wake cycle (Ingiosi et al., 2020). Astrocytic Ca^{2+} was also found to change in accordance with NREM short wave activity, which decreases throughout the rest period. More in-depth analysis showed a highly dynamic spatiotemporal pattern of Ca^{2+} fluctuations across vigilance states, with a bias towards higher Ca^{2+} levels in astrocyte processes during wake and REM sleep. In response to sleep deprivation, astrocytic Ca^{2+} signalling increased and decreased during recovery sleep, suggesting a role in sleep homeostasis. In addition, the synchrony of Ca^{2+} signals across neuronal networks, which is particularly high during NREM sleep, increased further after sleep deprivation, while Ca^{2+} rhythm synchrony across the astrocyte network was not only lower than neuronal synchrony during NREM sleep, but was further reduced in response to sleep deprivation. This indicates that astrocytic Ca^{2+} rhythms behave independently from neuronal Ca^{2+} rhythms and are not just responding to neuronal signals. To determine whether astrocytic Ca^{2+} contributes to sleep homeostasis (Ingiosi et al. 2020), they used a mouse model with inducible impairment of astrocytic store-operated Ca^{2+} entry (SOCE), which relies on the action of stromal interaction molecule 1 (STIM1) and is important for increasing intracellular calcium levels. Conditional KO of STIM1 in astrocytes showed normal daily patterns of sleep/wake behaviour and body temperature, as well as normal measures of sleep. However, NREM short wave activity and sleep time was altered following sleep deprivation, suggesting that astrocytic Ca^{2+} contributes to sleep homeostasis. In line with these findings, Bojarskaite et al. (2020) found that astrocytic Ca^{2+} increases precede transitions from slow wave sleep to wakefulness. They employed genetic ablation of Itpr2, which mediates inositol triphosphate (IP_3)-dependent astrocytic Ca^{2+} signalling, to show that impaired astrocytic Ca^{2+} signalling results in more fragmented sleep, abnormal brain rhythms, increased NREM sleep transitions, and an increased frequency of sleep spindles. They also observed astrocytic silencing prior to the transition from NREM to REM sleep, suggesting that astrocytic silencing is required for this transition, or that astrocytic Ca^{2+} may be important to maintain slow wave sleep. Adenosine could be mediating the observed effects of altered astrocytic activation on sleep homeostasis (Lazarus et al. 2019). Despite its central role in sleep homeostasis, the exact source of adenosine that regulates sleep is unclear. Neurons can induce adenosine production via activity-dependent synaptic release of ATP, which is hydrolysed to adenosine extracellularly (Wall and Dale 2013). In contrast, astrocyte-mediated adenosine release generally occurs over more prolonged timescales, making them ideal mediators for the slow changes in adenosine levels that modulate sleep (Haydon 2017). Indeed, mice with impaired astrocytic exocytosis exhibit deficits in sleep homeostasis, slow wave activity, and compensatory sleep changes in response to sleep deprivation (Halassa et al. 2009), consistent with effects detected in mice with impaired IP_3-dependent astrocytic Ca^{2+} signalling (Bojarskaite et al. 2020). The same phenotype could be recapitulated in wild-type mice by intracerebroventricular infusion of A1 adenosine receptor antagonists (which did not alter sleep in mice with impaired astrocytic exocytosis), and astrocytic adenosine has previously been found to induce tonic presynaptic inhibition via A1 receptor binding (Pascual et al. 2005). Further studies using a variety of molecular genetic approaches have provided more evidence for astrocytic adenosine in the regulation of sleep. For instance, optogenetic stimulation of astrocytes in the posterior hypothalamus increases time spent asleep (Pelluru et al. 2016), consistent with previous studies showing that microdialysis perfusion of an A1 receptor agonist into the rat posterior-lateral hypothalamus promoted sleep (Alam et al. 2009). Interestingly, Jagannath et al. (2021) have recently identified signalling pathways downstream of adenosine receptors A1 and A2A that regulate circadian entrainment by modulating the phase-shifting effect of light on the clock. The proposed mechanism thus enables sleep/wake history and light to interact in entrainment and suggests an additional pathway through which the astrocyte-derived adenosine may affect circadian function.

Probably the most popular model implicating astrocytes in the regulation of sleep/wake cycles is of them playing a key role in the regulation of the brain interstitial space and the clearance of brain waste, via the glymphatic system, via regulation of the astrocytic water channel aquaporin 4 (AQP4) (Xie et al. 2013). Waste removal by the glymphatic system is more effective during sleep when interstitial space volume increases, as observed through cranial windows in living animals (Xie et al. 2013; DiNuzzo and Nedergaard 2017; Lundgaard et al. 2017). Regulation of brain clearance is, however, part of a far more complex regulation, and to be understood in the wider context of the daily regulation of fluid homeostasis in the brain and body, a process that is deeply affected by circadian timekeeping as also suggested by evidence that brain clearance follows endogenous daily rhythms (Hablitz et al. 2020b) that persist in constant light conditions. Circadian aspects of glymphatic clearance also emerge when Nestin-expressing cells lack *Bmal1*—cortical AQP4 levels then decrease, while mouse brain water content and vascular permeability rise (Nakazato et al. 2017). AQP4 localisation at astrocytic endfeet changes in a circadian fashion and peaks in the perivascular areas during the rest phase in mice (Hablitz et al. 2020b), a phenomenon accompanied by increased clearance during daytime. Drainage of cerebrospinal fluid (CSF) to the lymph nodes also follows a circadian variation. Daily variations of both glymphatic clearance and lymphatic drainage are reduced in *Aqp4* KO mice (constitutive KO). While these findings are of certain interest, whether and how timekeeping processes are directly involved in daily clearance processes remains to be elucidated. CSF production exhibits circadian variations (Nilsson et al. 1992), and the SCN is important to preserve brain-wide daily fluctuations of interstitial space: *Bmal1* deletions sparing the SCN do not affect them, whereas if *Bmal1* is also deleted in the SCN, they are impaired (Kress et al. 2018), thus suggesting that central timekeeping may play a key part in maintaining daily variations of solute levels in the brain interstitial space. Moreover, CSF is refreshed daily by the choroid plexus (CP), a structure consisting of glial cells and responsible for 80% of interstitial fluid production. The CP shows synchronous circadian rhythms of clock gene expression, which are weakened by gap junction inhibitors (Myung et al. 2018). The CP can provide input to the SCN in vitro and in vivo (Myung et al. 2018): in tissue culture, the *Per2* expression period of the SCN accelerates in response to a co-cultured CP, while CP *Per2* expression remains unchanged. Further, *Bmal1* deletion in the CP lengthens the locomotor period in mice, suggesting that the CP can alter the SCN rhythmic patterns to some degree (Myung et al. 2018). Nevertheless, the role of circadian expression of aquaporins in the SCN and CP is poorly understood, as is the interplay among SCN, CP, and local astrocytic timekeepers in brain clearance of solutes.

6.5 Disruption of Astrocyte Circadian Timekeeping in Neurodegeneration

6.5.1 Circadian Disruption in Alzheimer's Disease

The circadian system changes with age, leading to altered rhythms of behaviour and physiology, such as advances in circadian phase and changes in rhythms of body temperature (Duffy et al.

2015). Circadian dysfunction has also been associated with numerous age-related neurodegenerative diseases, including Alzheimer's disease (AD). AD is characterised by significant and progressive cognitive decline and is primarily diagnosed based on the presence of extracellular amyloid β plaques and intracellular accumulation of hyperphosphorylated tau protein in the form of neurofibrillary tangles (Jagust 2018; Long and Holtzman 2019). However, AD has a highly heterogeneous pathophysiology and an extensive pre-symptomatic stage involving aberrations in various molecular processes. People with AD show impaired daily rhythms of rest-activity and temperature, as well as increased evening/nighttime agitation (sundowning) (Volicer et al. 2001; Harper et al. 2005). Consistent with clinical evidence, hypothalamic and brain stem areas mediating circadian and sleep behaviour are heavily affected in AD. Amongst these, the SCN experiences inflammation, reactive gliosis, amyloid and tau pathology, and neuronal loss (Swaab et al. 1985; Stopa et al. 1999; Harper et al. 2008; Cermakian et al. 2011; Roy et al. 2019). Recent large longitudinal studies show that pre-clinical AD is associated with fragmented rest-activity cycles and that circadian disruption is present up to 15 years before disease onset, confirming previous findings in elderly women (Tranah et al. 2011; Musiek et al. 2018; Li et al. 2020). Chronic circadian misalignment due to shift work has a dramatic effect on cognitive performance (Chellappa et al. 2019) and is associated with heightened risk of developing AD and associated increase in mortality rates (Jørgensen et al. 2017; Bokenberger et al. 2018).

Astrocytes themselves have been implicated in neurodegeneration, as they are key regulators of numerous processes involved in pathology, including synaptic homeostasis and neuroinflammation. GFAP CSF and serum levels are increased in patients with AD and other neurodegenerative diseases, including Parkinson's disease and Lewy body dementia, and correlate with cognitive impairment (Oeckl et al. 2019). Plasma level of GFAP is also significantly elevated in cognitively normal elderly adults with a brain amyloid β load that indicates risk of AD (Chatterjee et al. 2021). This suggests that reactive changes in astrocyte activities are present in the pre-symptomatic stage of the disease and are associated with amyloid β burden, strongly implicating them in the pathogenesis of AD. The astrocytic clock regulates key functions of astrocytes, which provides several potential mechanisms through which its disruption may contribute to neurodegeneration. Recent studies have directly linked the astrocytic clock to key pathological hallmarks of AD, such as amyloid β accumulation (Lananna et al. 2020). In the next section, we will review available evidence implicating astrocyte timekeeping mechanisms in neurodegeneration. While circadian involvement is present in several forms of dementia (Leng et al. 2019), specific evidence for altered timekeeping in astrocytes is stronger for AD. We will first focus on cellular mechanisms directly impaired by altered intracellular circadian timekeeping (neuroinflammation and oxidative stress) and then focus on indirect mechanisms by which altered astrocyte timekeeping may affect circuits underpinning cognitive and sleep function, indirectly increasing brain vulnerability in AD.

A schematic overview of the processes linking disruption of astrocyte timekeepers to neurodegeneration in AD is provided in Fig. 6.2.

6.5.2 Neuroinflammation and Astrocyte Timekeeping in Alzheimer's Disease

Astrocytes play a key role in neuroinflammation and their inflammation-induced activation triggers strong structural and function changes (Sun and Jakobs 2012; Linnerbauer et al. 2020). Astrocyte-mediated inflammation may be protective in neurodegenerative diseases or promote neurodegeneration depending on context (Linnerbauer et al. 2020). Importantly, circadian clocks are in a complex bi-directional relationship with inflammatory responses. On the one hand, daily expression of pro-inflammatory cytokines is time-of-day dependent (Fonken et al. 2015); on the other hand, inflammation directly inhibits clock function in the SCN by chronically

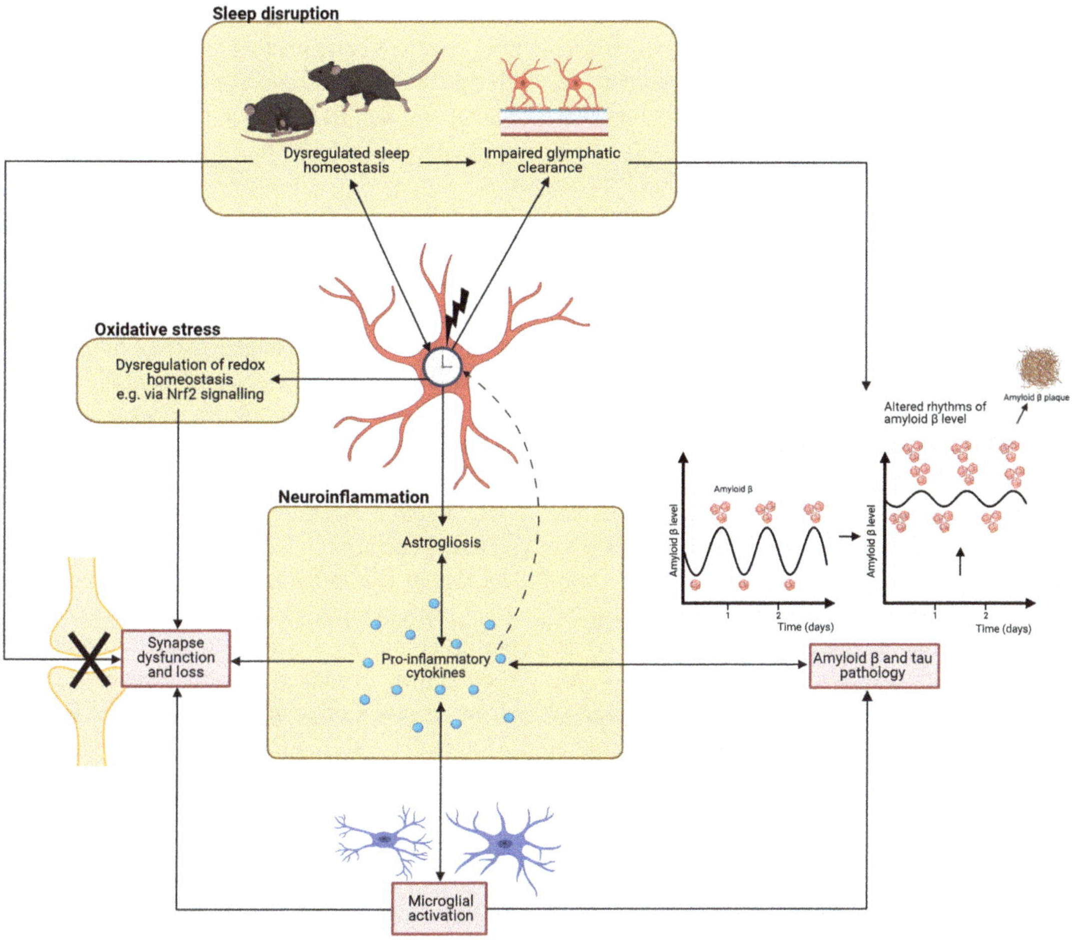

Fig. 6.2 Schematic overview of pathways through which disrupted astrocytic timekeeping contributes to neurodegeneration in AD. The astrocytic clock has been shown to regulate redox homeostasis, neuroinflammation, and numerous processes associated with sleep. Astrocytic clock disruption may impair normal regulation of redox homeostasis, via altered Nrf2 signalling (Baeza-Raja et al. 2013; Ishii et al. 2019), leading to an increase in oxidative stress inducing altered synapse function and loss (González-Reyes et al. 2017; Tönnies and Trushina 2017). Alteration of daily balance of pro-inflammatory pathways by disrupted astrocyte timekeepers may also contribute to synaptic dysfunction and microglial activation, and lead to increased amyloid β and tau loads (Musiek et al. 2013; Rajendran and Paolicelli 2018; Lananna et al. 2018). Other avenues through which an impaired astrocytic clock may contribute to neurodegeneration include reduced clearance of toxic brain waste and subsequent accumulation of amyloid β (Iranzo 2016; Holth et al. 2017; Rasmussen et al. 2018). Illustration created with BioRender.com

suppressing central timekeeping (Cermakian et al. 2014). Therefore, circadian clocks in astrocytes may be key to preserving physiological daily variations of pro-inflammatory/repair brain states and their alterations following toxic protein aggregation may result in further circadian impairment and reduction in clearance.

Brain-wide *Bmal1* or dual *Clock* and *Npas2* deletions age-dependently induce widespread astrogliosis, chronic inflammation, and synapse degeneration (Musiek et al. 2013). In this context, the astrocytic release of pro-inflammatory cytokines that exacerbate neurodegeneration, such as cyclooxygenase-2 (Teismann et al.

2003), is increased. Interestingly, astrocyte-specific *Bmal1* KO is sufficient to induce gliosis, suggesting this inflammatory phenotype is under cell-autonomous circadian control (Lananna et al. 2018). NF-κB-mediated inflammation in particular may be regulated by the astrocytic clock, as RNAi-mediated knockdown of *Per1* was found to induce NF-κB activation and CCL2 and interleukin 6 production in spinal astrocytes in vitro (Sugimoto et al. 2014).

A recent study has revealed a direct link between amyloid β phagocytosis and amyloid plaque deposition and an astrocytic clock-controlled transcript in mouse models of AD. YKL-40, a secreted glycoprotein encoded by the *Chi3l1* gene, is a human CSF biomarker of neuroinflammation that is elevated in neurodegenerative disease, including AD, multiple sclerosis, amyotrophic lateral sclerosis, and frontotemporal degeneration. Lananna et al. (2020) found that crossing an APP/PS1 AD mouse model with *Chi3l1* KO mice reduced amyloid plaque burden and led to increased microglial CD68 expression in vivo. Moreover, siRNA-mediated knockdown of *Chi3l1* in primary cell cultures enhanced the phagocytosis of amyloid β by astrocytes and microglia in vitro. *Chi3l1* mRNA is downregulated in *Bmal1*-null and *Clock/Npas2*-null mice, suggesting that BMAL1 directly regulates its expression. Despite clock regulation, however, *Chi3l1* mRNA does not show circadian oscillations, likely due to its long half-life. Interestingly, LPS- and amyloid β-mediated induction of *Chi3l1* in primary astrocytes was found to be highly phase-dependent and closely mirrored the expression pattern of other genes dependent on *Bmal1* transcriptional activity, suggesting that *Bmal1* activity gates the induction of *Chi3l1* in astrocytes. While this suggests a role of the astrocytic clock in the regulation of neuroinflammation in the context of neurodegeneration, cell-autonomous effects in microglia could not be discounted because of the absence of cell-type specific *Chi3l1* manipulations in in vivo experiments. Interestingly, *Chi3l1* KO induced inflammatory signalling in response to LPS but not amyloid plaques. In fact, some previous studies suggest that *Chi3l1* may have anti-inflammatory roles that are protective in traumatic brain injury and experimental autoimmune encephalomyelitis (Bonneh-Barkay et al. 2012; Wiley et al. 2015). In an induced amyloidosis rat model, inhibition of *Chi3l1* was protective by reducing amyloidogenesis (Choi et al. 2018). As such, the effect of *Chi3l1* in pathology may be context-dependent, in line with glial activation having been found to be neuroprotective by promotion of amyloid β and tau phagocytosis, while in other contexts it is pathology-enhancing through excessive inflammation leading to synapse loss and increased plaque accumulation in AD.

In addition to clock effects on the inflammatory function of astrocytes in the context of neurodegeneration, there is also evidence suggesting that inflammation induces changes in the astrocytic clock. While numerous pro-inflammatory cytokines have previously been shown to induce circadian phase shifts in clock gene expression and behaviour in vivo (Motzkus et al. 2002; Cavadini et al. 2007; Kwak et al. 2008), more recent evidence suggests that SCN astrocytes modulate their clock gene expression in response to cytokine TNF-α in vitro (Duhart et al. 2013). In addition, some inflammatory signals that induce molecular phase shifts in the SCN in vivo primarily act on astrocytes, such as CCL2 (Duhart et al. 2016), suggesting that the astrocytic clock may contribute to mediating their circadian effects. As such, inflammation may also modulate other astrocytic circadian rhythms, which could have widespread effects on circadian regulation at the circuit level. Interestingly, TNF-α is rhythmically expressed in the hypothalamus in vivo (Floyd and Krueger 1997) and in primary microglia in vitro, and its expression in response to immune challenge in primary microglia is attenuated after treatment with a Rev-Erbα agonist (Wolff et al. 2020), suggesting that it may mediate coordination between microglial and astrocyte timekeepers. A recent study in human iPSC-derived neurons from patients with familial AD showed that TNF-α application accelerates the production of protein aggregates, including amyloid β (Whiten et al. 2020), supporting a role for this cytokine in

AD pathogenesis and emphasising the importance of understanding the interactions between the circadian clock and cytokines such as TNF-α, and their effect on glial function.

6.5.3 Oxidative Stress and Astrocyte Timekeeping

Increased production of reactive oxygen species and reduced antioxidant defence occur during aging and have been implicated in numerous age-related neurodegenerative diseases, including AD (Butterfield and Halliwell 2019). Oxidative stress has far-reaching consequences on numerous processes, including intracellular Ca^{2+} levels, mitochondrial function, neuroinflammatory pathways, glutamate uptake, and the production and accumulation of amyloid β, and may ultimately induce cognitive dysfunction as a result of altered synaptic function and neurotransmission (González-Reyes et al. 2017; Tönnies and Trushina 2017). Astrocytes mitigate oxidative stress in neurons via the expression of redox-protective enzymes and their regulation of neuronal glutathione levels (Baxter and Hardingham 2016). This process is under circadian control, as astrocyte-specific *Bmal1* KO reduces the expression of glutathione-S-transferase enzymes in astrocytes and increases oxidative damage in the brain (Lananna et al. 2018). Circadian oscillations in the astrocytic expression of p75 neurotrophin receptor (p75NTR), which is under direct control of the CLOCK-BMAL1 transcription factor complex (Baeza-Raja et al. 2013), have been hypothesised to mediate clock-control of antioxidant function via regulating the nuclear factor erythroid 2–related factor 2 (Nrf2) signalling pathway and subsequently the expression of antioxidant enzymes (Ishii et al. 2019). Interestingly, circadian control over redox states modulates K^{+} channel conductances in SCN neurons, thereby affecting neuronal excitability (Wang et al. 2012). Musiek et al. (2013) have previously shown that circadian expression of *Bmal1* plays a critical role in neuronal redox homeostasis, as diminished *Bmal1* expression exacerbates amyloidosis-related neurodegeneration in vivo. They also performed in vitro *Bmal1* knockdown experiments in neuron-enriched cultures and in primary astrocyte cultures, in which they showed that only neuronal *Bmal1* reduction induced oxidative damage and inflammation. They concluded that *Bmal1* expression in neurons, not astrocytes, is the primary driver of oxidative injury that would eventually lead to the diminished brain-wide connectivity observed in vivo in the *Bmal1* KO model. However, as astrocytes are involved in the formation and elimination of synapses (Eroglu and Barres 2010; Risher et al. 2014), they could potentially contribute to this phenotype through impaired circadian regulation of synaptic remodelling.

Moreover, disruption of clock function in neurons may trigger Nrf2 activation in astrocytes, as previously mentioned. Nrf2 signalling is key for the production of antioxidant enzymes and protection of neurons against ferroptosis-like cell death (Ishii et al. 2019), as astrocytes can store high levels of iron, a process which is Nrf2-dependent (Kerins and Ooi 2017). As such, it is possible that the astrocytic clock contributes substantially to redox homeostasis and that a dysregulation of this function may contribute to neurodegeneration primarily through its negative effects on the mitigation of oxidative stress in neurons, hence explaining the lack of harmful effects on astrocytes observed by Musiek et al. (2013). Further studies of neuronal and astrocyte-specific alterations of clock function will aid disentangling the complex neuron-astrocyte interplay, whose disruption may underpin establishment of a sustained oxidative state and eventually lead to altered synaptic connectivity and neurodegeneration.

6.5.4 Sleep Disruption and Altered Daily Clearance of Toxic Brain Waste by Astrocytes

Neurodegeneration is often associated with astrocyte pathology, and a growing body of evidence shows how circadian dysfunction, interstitial space regulation, and amyloid β clearance represent key contributors to neurodegeneration. Astrocytes regulate cerebral blood flow, inflammatory and oxidative stress responses—all in a circadian manner, as outlined above. Specifically, the daily regulation of glymphatic clearance by astrocytes links these processes by orchestrating bulk fluid flow and solute exchange between the CSF and intracellular regions, where impaired clearance can lead to a build-up of neurotoxic molecules like excess amyloid β and tau protein (Iliff et al. 2014; Da Mesquita et al. 2018; Mestre et al. 2018), followed by synapse degeneration and eventually, cognitive decline (Mawuenyega et al. 2010; Wildsmith et al. 2013).

Astrocyte-mediated clearance processes are subject to daily rhythms. Interestingly, amyloid β levels rise in the interstitial space of sleep-deprived mice and in the CSF of sleep-deprived humans, respectively (Kang et al. 2009; Lucey et al. 2018). Similarly, sleep loss in mice can raise levels of tau protein and pathology (Holth et al. 2019). This illustrates how lifestyle (abnormal sleep patterns) alters astrocyte-mediated clearance outcomes with direct implications for neurodegenerative pathology.

In health, robust influx and efflux of glymphatic flow requires AQP4 expression by astrocytes (Mestre et al. 2018), but AD patients show reduced perivascular AQP4 localisation at astrocytic endfeet (despite overall raised AQP4 levels), which correlates with AD severity (Zeppenfeld et al. 2017). In line with this, mice with AQP4-deficient astrocytes show reduced CSF influx through the glymphatic system, and a drastically reduced clearance of interstitial solutes, including amyloid β and tau proteins (Iliff et al. 2012, 2014). In humans, different *Aqp4* variants represent genetic factors correlating with distinct sleep quality, latency, and duration, which in turn influence amyloid β clearance (Rainey-Smith et al. 2018). Further, several groups have reported altered learning and memory performance in AQP4-depleted mice, where astrocytic glutamate transporters may be key regulators of neuronal function by clearing excess extracellular glutamate (Lan et al. 2016). Moreover, astrocytes can also clear amyloid β via phagocytosis and interact with microglia to mediate extracellular amyloid β clearance (Lian et al. 2016; Liddelow et al. 2017), where internal clock cues and gene expression in both glia likely influence their phagocytic activity.

Sleep disturbances have been extensively implicated in the pathogenesis of neurodegenerative diseases, including AD, Parkinson's disease, and Huntington's disease (Iranzo 2016; Holth et al. 2017; Owen and Veasey 2020). Further evidence comes from human studies, in which CSF biomarkers of neuronal damage and neuroinflammation are increased in sleep-deprived individuals (Frey et al. 2007; Benedict et al. 2014). Astrocytic clock disturbances may thus contribute to neurodegeneration through their role in sleep homeostasis. Indeed, amyloid β induces calcium elevations in astrocytes via increasing intracellular Ca^{2+} release and store-operated Ca^{2+} entry (Grolla et al. 2013; Ronco et al. 2014). As there is substantial evidence implicating astrocytic Ca^{2+} in the regulation of sleep, impaired astrocytic Ca^{2+} homeostasis may negatively affect sleep, thereby contributing to neurodegeneration. Furthermore, astrocytic phagocytosis of synapses in the mouse frontal cortex has been found to increase in response to acute and chronic sleep deprivation, which may contribute to synaptic loss as a result of sleep and/or circadian disturbances (Bellesi et al. 2017; Owen and Veasey 2020). Finally, another potential link between the astrocytic clock and neurodegeneration may stem from circadian regulation of is the astrocyte fatty acid binding protein 7 (*Fabp7*), which is rhythmically expressed (Gerstner et al. 2008) and directly regulated by Rev-Erbα (Schnell et al. 2014; Gerstner and Paschos 2020) in astrocytes. Mutations in *Fabp7* cause fragmented sleep in humans and animal models (Gerstner et al. 2017), a phenotype associated with increased risk of developing AD (Lim et al. 2013).

6.6 Conclusions

Our grasping of how astrocytes affect neuronal circuits and behaviour is still rudimentary. This is partly due to a poor understanding of the organising principles by which glia encode information, as opposed to the well-established, universal principles of action potential propagation, synapse connectivity, and circuit integration underpinning neuronal function. The discovery of universal circadian molecular clockwork affecting behaviour is, in fact, one of the most elegant demonstrations of how information flows from genes to rhythmic neuronal activities to patterns of organised 24-h behaviour. In contrast, while there has never before been a greater array of tools for the investigation of astrocyte function, some scepticism remains regarding whether the experimental observations that follow astrocyte manipulations reflect a true physiological role for astrocytes in the encoding of information in the brain, or just a generic interference with their metabolic activities. Evidence showing that selective de novo expression of clock genes in astrocytes can 'kick-start' and support daily patterns of circadian behaviour with a periodicity that is distinguishable from that imposed by neurons demonstrates a cell-autonomous role for astrocytes in encoding (circadian) information in the brain (Brancaccio et al. 2019). However, these effects are not observed in a 'neuronal vacuum'; neurons are still present and needed for the specification of circadian time: their ablation, as well astrocyte ablation, dampens the molecular 'ticking' of the circadian clockwork, as shown by cell-type specific caspase expression in SCN tissue (Brancaccio et al. 2017). This suggests that neurons and astrocytes are perhaps best understood as a temporal unit, and their reciprocal activities should therefore not be discounted *a priori* in experiments of causal neuroscience. Undoubtedly, understanding how information flows between astrocytes and neurons acting over different temporal domains remains a fundamental challenge, however, the presence of a common 24 h temporal domain provides a testbed in which the principles of astrocyte-neuronal interplay can be more easily investigated and understood. This may also have great relevance in neurodegeneration, as the transition from homeostatic to reactive astrocyte behaviour may not only deprive neurons of metabolic support, but may also deteriorate temporal information within neuronal circuits, well ahead of any neuronal loss, and have wide-ranging consequences for disease progression. Future experiments should thus ideally focus on co-detecting neuronal and astrocyte reciprocal activities, and their selective manipulation, in freely behaving animals. The events accompanying the astrocyte transition from elements of temporal resilience to mediators of early brain vulnerability in neurodegeneration will need to be mapped out to understand whether and how information is degraded and with what consequences for early pathology.

Acknowledgements Figures created with BioRender.com. This work is supported by the UK Dementia Research Institute which receives its funding from UK DRI Ltd, funded by the UK Medical Research Council, Alzheimer's Society and Alzheimer's Research UK. NN was also supported by an Imperial College President's PhD Scholarship.

References

Acuna-Goycolea C, Obrietan K, van den Pol AN (2010) Cannabinoids excite circadian clock neurons. J Neurosci 30:10061–10066. https://doi.org/10.1523/JNEUROSCI.5838-09.2010

Alam MN, Kumar S, Rai S et al (2009) Role of adenosine A1 receptor in the perifornical–lateral hypothalamic area in sleep–wake regulation in rats. Brain Res 1304: 96–104. https://doi.org/10.1016/j.brainres.2009.09.066

Ali AAH, Stahr A, Ingenwerth M et al (2019) Connexin30 and Connexin43 show a time-of-day dependent expression in the mouse suprachiasmatic nucleus and modulate rhythmic locomotor activity in the context of chronodisruption. Cell Commun Signal 17:61. https://doi.org/10.1186/s12964-019-0370-2

Ali AAH, Schwarz-Herzke B, Rollenhagen A et al (2020) Bmal1-deficiency affects glial synaptic coverage of the hippocampal mossy fiber synapse and the actin cytoskeleton in astrocytes. Glia 68:947–962. https://doi.org/10.1002/glia.23754

Amir S, Lamont EW, Robinson B, Stewart J (2004) A circadian rhythm in the expression of PERIOD2 protein reveals a novel SCN-controlled oscillator in the oval nucleus of the bed nucleus of the stria terminalis. J Neurosci 24:781–790. https://doi.org/10.1523/JNEUROSCI.4488-03.2004

Ariznavarreta C, Cardinali DP, Villanúa MA et al (2002) Circadian rhythms in airline pilots submitted to long-haul transmeridian flights. Aviat Space Environ Med 73:445–455

Artiushin G, Sehgal A (2020) The glial perspective on sleep and circadian rhythms. Annu Rev Neurosci 43: 119–140. https://doi.org/10.1146/annurev-neuro-091819-094557

Baeza-Raja B, Eckel-Mahan K, Zhang L et al (2013) p75 neurotrophin receptor is a clock gene that regulates oscillatory components of circadian and metabolic networks. J Neurosci 33:10221–10234. https://doi.org/10.1523/JNEUROSCI.2757-12.2013

Barca-Mayo O, Pons-Espinal M, Follert P et al (2017) Astrocyte deletion of Bmal1 alters daily locomotor activity and cognitive functions via GABA signalling. Nat Commun 8:14336. https://doi.org/10.1038/ncomms14336

Barnes CA, McNaughton BL, Goddard GV et al (1977) Circadian rhythm of synaptic excitability in rat and monkey central nervous system. Science 197:91–92. https://doi.org/10.1126/science.194313

Basu R, Sarma JD (2018) Connexin 43/47 channels are important for astrocyte/ oligodendrocyte cross-talk in myelination and demyelination. J Biosci 43:1055–1068

Baxter PS, Hardingham GE (2016) Adaptive regulation of the brain's antioxidant defences by neurons and astrocytes. Free Radic Biol Med 100:147–152. https://doi.org/10.1016/j.freeradbiomed.2016.06.027

Beaulé C, Swanstrom A, Leone MJ, Herzog ED (2009) Circadian modulation of gene expression, but not glutamate uptake, in mouse and rat cortical astrocytes. PLoS One 4:e7476. https://doi.org/10.1371/journal.pone.0007476

Becquet D, Girardet C, Guillaumond F et al (2008) Ultrastructural plasticity in the rat suprachiasmatic nucleus. Possible involvement in clock entrainment. Glia 56: 294–305. https://doi.org/10.1002/glia.20613

Bellesi M, de Vivo L, Tononi G, Cirelli C (2015) Effects of sleep and wake on astrocytes: clues from molecular and ultrastructural studies. BMC Biol 13:66. https://doi.org/10.1186/s12915-015-0176-7

Bellesi M, de Vivo L, Chini M et al (2017) Sleep loss promotes astrocytic phagocytosis and microglial activation in mouse cerebral cortex. J Neurosci 37:5263–5273. https://doi.org/10.1523/JNEUROSCI.3981-16.2017

Bell-Pedersen D, Cassone VM, Earnest DJ et al (2005) Circadian rhythms from multiple oscillators: lessons from diverse organisms. Nat Rev Genet 6:544–556. https://doi.org/10.1038/nrg1633

Benedict C, Cedernaes J, Giedraitis V et al (2014) Acute sleep deprivation increases serum levels of neuron-specific enolase (NSE) and S100 calcium binding protein B (S-100B) in healthy young men. Sleep 37:195–198. https://doi.org/10.5665/sleep.3336

Bhattacharya A, Vavra V, Svobodova I et al (2013) Potentiation of inhibitory synaptic transmission by extracellular ATP in rat suprachiasmatic nuclei. J Neurosci 33: 8035–8044. https://doi.org/10.1523/JNEUROSCI.4682-12.2013

Bojarskaite L, Bjørnstad DM, Pettersen KH et al (2020) Astrocytic Ca 2+ signaling is reduced during sleep and is involved in the regulation of slow wave sleep. Nat Commun 11:3240. https://doi.org/10.1038/s41467-020-17062-2

Bokenberger K, Sjölander A, Dahl Aslan AK et al (2018) Shift work and risk of incident dementia: a study of two population-based cohorts. Eur J Epidemiol 33: 977–987. https://doi.org/10.1007/s10654-018-0430-8

Bonneh-Barkay D, Wang G, LaFramboise WA et al (2012) Exacerbation of experimental autoimmune encephalomyelitis in the absence of breast regression protein 39/chitinase 3-like 1. J Neuropathol Exp Neurol 71:948–958. https://doi.org/10.1097/NEN.0b013e31826eaee7

Brancaccio M, Maywood ES, Chesham JE et al (2013) A Gq-Ca2+ axis controls circuit-level encoding of circadian time in the suprachiasmatic nucleus. Neuron 78: 714–728. https://doi.org/10.1016/j.neuron.2013.03.011

Brancaccio M, Patton AP, Chesham JE et al (2017) Astrocytes control circadian timekeeping in the suprachiasmatic nucleus via glutamatergic signaling. Neuron 93:1420–1435.e5. https://doi.org/10.1016/j.neuron.2017.02.030

Brancaccio M, Edwards MD, Patton AP et al (2019) Cell-autonomous clock of astrocytes drives circadian behavior in mammals. Science 363:187–192. https://doi.org/10.1126/science.aat4104

Broadhead MJ, Miles GB (2021) A common role for astrocytes in rhythmic behaviours? Prog Neurobiol. https://doi.org/10.1016/j.pneurobio.2021.102052

Burkeen JF, Womac AD, Earnest DJ, Zoran MJ (2011) Mitochondrial calcium signaling mediates rhythmic extracellular ATP accumulation in suprachiasmatic nucleus astrocytes. J Neurosci 31:8432–8440. https://doi.org/10.1523/JNEUROSCI.6576-10.2011

Butterfield DA, Halliwell B (2019) Oxidative stress, dysfunctional glucose metabolism and Alzheimer disease. Nat Rev Neurosci 20:148–160. https://doi.org/10.1038/s41583-019-0132-6

Cagampang FR, Rattray M, Powell JF et al (1996) Circadian variation of EAAC1 glutamate transporter messenger RNA in the rat suprachiasmatic nuclei. Brain Res Mol Brain Res 35:190–196. https://doi.org/10.1016/0169-328x(95)00203-5

Carney RSE (2020) Astrocytes function as an intermediate for retrograde endocannabinoid signaling in the suprachiasmatic nucleus to influence circadian clock timing. eNeuro 7. https://doi.org/10.1523/ENEURO.0323-20.2020

Cavadini G, Petrzilka S, Kohler P et al (2007) TNF-alpha suppresses the expression of clock genes by interfering with E-box-mediated transcription. Proc Natl Acad Sci U S A 104:12843–12848. https://doi.org/10.1073/pnas.0701466104

Cermakian N, Waddington Lamont E, Boudreau P, Boivin DB (2011) Circadian clock gene expression in brain regions of Alzheimer's disease patients and control subjects. J Biol Rhythms 26:160–170. https://doi.org/10.1177/0748730410395732

Cermakian N, Westfall S, Kiessling S (2014) Circadian clocks and inflammation: reciprocal regulation and shared mediators. Arch Immunol Ther Exp 62:303–318. https://doi.org/10.1007/s00005-014-0286-x

Chatterjee P, Pedrini S, Stoops E et al (2021) Plasma glial fibrillary acidic protein is elevated in cognitively normal older adults at risk of Alzheimer's disease. Transl Psychiatry 11:1–10. https://doi.org/10.1038/s41398-020-01137-1

Chaudhury D, Colwell CS (2002) Circadian modulation of learning and memory in fear-conditioned mice. Behav Brain Res 133:95–108. https://doi.org/10.1016/s0166-4328(01)00471-5

Chellappa SL, Morris CJ, Scheer FAJL (2019) Effects of circadian misalignment on cognition in chronic shift workers. Sci Rep 9:699. https://doi.org/10.1038/s41598-018-36762-w

Cho K (2001) Chronic "jet lag" produces temporal lobe atrophy and spatial cognitive deficits. Nat Neurosci 4: 567–568. https://doi.org/10.1038/88384

Choi JY, Yeo IJ, Kim KC et al (2018) K284-6111 prevents the amyloid beta-induced neuroinflammation and impairment of recognition memory through inhibition of NF-κB-mediated CHI3L1 expression. J Neuroinflamm 15:224. https://doi.org/10.1186/s12974-018-1269-3

Colgin LL, Denninger T, Fyhn M et al (2009) Frequency of gamma oscillations routes flow of information in the hippocampus. Nature 462:353–357. https://doi.org/10.1038/nature08573

Colwell CS, Ghiani CA (2020) Potential circadian rhythms in oligodendrocytes? working together through time. Neurochem Res 45:591–605. https://doi.org/10.1007/s11064-019-02778-5

Condorelli DF, Belluardo N, Trovato-Salinaro A, Mudò G (2000) Expression of Cx36 in mammalian neurons. Brain Res Brain Res Rev 32:72–85. https://doi.org/10.1016/s0165-0173(99)00068-5

Cotrina ML, Gao Q, Lin JH, Nedergaard M (2001) Expression and function of astrocytic gap junctions in aging. Brain Res 901:55–61. https://doi.org/10.1016/s0006-8993(01)02258-2

Da Mesquita S, Louveau A, Vaccari A et al (2018) Functional aspects of meningeal lymphatics in ageing and Alzheimer's disease. Nature 560:185–191. https://doi.org/10.1038/s41586-018-0368-8

Deboer T (2020) Circadian regulation of sleep in mammals. Curr Opin Physiol 15:89–95. https://doi.org/10.1016/j.cophys.2019.12.015

Debski KJ, Ceglia N, Ghestem A et al (2020) The circadian dynamics of the hippocampal transcriptome and proteome is altered in experimental temporal lobe epilepsy. Sci Adv 6. https://doi.org/10.1126/sciadv.aat5979

DiNuzzo M, Nedergaard M (2017) Brain energetics during the sleep-wake cycle. Curr Opin Neurobiol 47:65–72. https://doi.org/10.1016/j.conb.2017.09.010

Duffy JF, Zitting K-M, Chinoy ED (2015) Aging and circadian rhythms. Sleep Med Clin 10:423–434. https://doi.org/10.1016/j.jsmc.2015.08.002

Duhart JM, Leone MJ, Paladino N et al (2013) Suprachiasmatic astrocytes modulate the circadian clock in response to TNF-α. J Immunol 191:4656–4664. https://doi.org/10.4049/jimmunol.1300450

Duhart JM, Brocardo L, Mul Fedele ML et al (2016) CCL2 mediates the circadian response to low dose endotoxin. Neuropharmacology 108:373–381. https://doi.org/10.1016/j.neuropharm.2016.05.005

Eroglu C, Barres BA (2010) Regulation of synaptic connectivity by glia. Nature 468:223–231. https://doi.org/10.1038/nature09612

Fasciani I, Pluta P, González-Nieto D et al (2018) Directional coupling of oligodendrocyte connexin-47 and astrocyte connexin-43 gap junctions. Glia 66:2340–2352. https://doi.org/10.1002/glia.23471

Floyd RA, Krueger JM (1997) Diurnal variation of TNFα in the rat brain. NeuroReport 8:915–918

Fonken LK, Frank MG, Kitt MM et al (2015) Microglia inflammatory responses are controlled by an intrinsic circadian clock. Brain Behav Immun 45:171–179. https://doi.org/10.1016/j.bbi.2014.11.009

Franken P (2013) A role for clock genes in sleep homeostasis. Curr Opin Neurobiol 23:864–872. https://doi.org/10.1016/j.conb.2013.05.002

Frey DJ, Fleshner M, Wright KP (2007) The effects of 40 hours of total sleep deprivation on inflammatory markers in healthy young adults. Brain Behav Immun 21:1050–1057. https://doi.org/10.1016/j.bbi.2007.04.003

Gerstner JR, Paschos GK (2020) Circadian expression of Fabp7 mRNA is disrupted in Bmal1 KO mice. Mol Brain 13:26. https://doi.org/10.1186/s13041-020-00568-7

Gerstner JR, Bremer QZ, Heyden WMV et al (2008) Brain fatty acid binding protein (Fabp7) is diurnally regulated in astrocytes and hippocampal granule cell precursors in adult rodent brain. PLoS One 3:e1631. https://doi.org/10.1371/journal.pone.0001631

Gerstner JR, Perron IJ, Riedy SM et al (2017) Normal sleep requires the astrocyte brain-type fatty acid binding protein FABP7. Sci Adv 3:e1602663. https://doi.org/10.1126/sciadv.1602663

Giaume C, Koulakoff A, Roux L et al (2010) Astroglial networks: a step further in neuroglial and gliovascular interactions. Nat Rev Neurosci 11:87–99. https://doi.org/10.1038/nrn2757

Giaume C, Naus CC, Sáez JC, Leybaert L (2020) Glial connexins and pannexins in the healthy and diseased brain. Physiol Rev 101:93–145. https://doi.org/10.1152/physrev.00043.2018

González-Reyes RE, Nava-Mesa MO, Vargas-Sánchez K et al (2017) Involvement of astrocytes in Alzheimer's disease from a neuroinflammatory and oxidative stress perspective. Front Mol Neurosci 10. https://doi.org/10.3389/fnmol.2017.00427

Griffin P, Sheehan PW, Dimitry JM et al (2020) REV-ERBα mediates complement expression and diurnal regulation of microglial synaptic phagocytosis. eLife 9:e58765. https://doi.org/10.7554/eLife.58765

Grolla AA, Fakhfouri G, Balzaretti G et al (2013) Aβ leads to Ca^{2+} signaling alterations and transcriptional changes in glial cells. Neurobiol Aging 34:511–522. https://doi.org/10.1016/j.neurobiolaging.2012.05.005

Hablitz LM, Gunesch AN, Cravetchi O et al (2020a) Cannabinoid signaling recruits astrocytes to modulate presynaptic function in the suprachiasmatic nucleus. eNeuro 7. https://doi.org/10.1523/ENEURO.0081-19.2020

Hablitz LM, Plá V, Giannetto M et al (2020b) Circadian control of brain glymphatic and lymphatic fluid flow. Nat Commun 11:4411. https://doi.org/10.1038/s41467-020-18115-2

Halassa MM, Florian C, Fellin T et al (2009) Astrocytic modulation of sleep homeostasis and cognitive consequences of sleep loss. Neuron 61:213–219. https://doi.org/10.1016/j.neuron.2008.11.024

Harper DG, Volicer L, Stopa EG et al (2005) Disturbance of endogenous circadian rhythm in aging and Alzheimer disease. Am J Geriatr Psychiatry 13:359–368. https://doi.org/10.1176/appi.ajgp.13.5.359

Harper DG, Stopa EG, Kuo-Leblanc V et al (2008) Dorsomedial SCN neuronal subpopulations subserve different functions in human dementia. Brain 131:1609–1617. https://doi.org/10.1093/brain/awn049

Harris AL (2007) Connexin channel permeability to cytoplasmic molecules. Prog Biophys Mol Biol 94:120–143. https://doi.org/10.1016/j.pbiomolbio.2007.03.011

Hastings MH, Maywood ES, Brancaccio M (2018) Generation of circadian rhythms in the suprachiasmatic nucleus. Nat Rev Neurosci 19:453. https://doi.org/10.1038/s41583-018-0026-z

Hastings MH, Maywood ES, Brancaccio M (2019) The mammalian circadian timing system and the suprachiasmatic nucleus as its pacemaker. Biology 8:13. https://doi.org/10.3390/biology8010013

Haydon PG (2017) Astrocytes and the modulation of sleep. Curr Opin Neurobiol 44:28–33. https://doi.org/10.1016/j.conb.2017.02.008

Hegyi Z, Oláh T, Kőszeghy Á et al (2018) CB 1 receptor activation induces intracellular Ca 2+ mobilization and 2-arachidonoylglycerol release in rodent spinal cord astrocytes. Sci Rep 8:10562. https://doi.org/10.1038/s41598-018-28763-6

Holth J, Patel T, Holtzman DM (2017) Sleep in Alzheimer's disease—beyond amyloid. Neurobiol Sleep Circadian Rhythms 2:4–14. https://doi.org/10.1016/j.nbscr.2016.08.002

Holth JK, Fritschi SK, Wang C et al (2019) The sleep-wake cycle regulates brain interstitial fluid tau in mice and CSF tau in humans. Science 363:880–884. https://doi.org/10.1126/science.aav2546

Houades V, Koulakoff A, Ezan P et al (2008) Gap junction-mediated astrocytic networks in the mouse barrel cortex. J Neurosci 28:5207–5217. https://doi.org/10.1523/JNEUROSCI.5100-07.2008

Iliff JJ, Wang M, Liao Y et al (2012) A paravascular pathway facilitates CSF flow through the brain parenchyma and the clearance of interstitial solutes, including amyloid β. Sci Transl Med 4:147ra111. https://doi.org/10.1126/scitranslmed.3003748

Iliff JJ, Chen MJ, Plog BA et al (2014) Impairment of glymphatic pathway function promotes tau pathology after traumatic brain injury. J Neurosci 34:16180–16193. https://doi.org/10.1523/JNEUROSCI.3020-14.2014

Ingiosi AM, Hayworth CR, Harvey DO et al (2020) A role for astroglial calcium in mammalian sleep and sleep regulation. Curr Biol 30:4373–4383.e7. https://doi.org/10.1016/j.cub.2020.08.052

Iranzo A (2016) Sleep in neurodegenerative diseases. Sleep Med Clin 11:1–18. https://doi.org/10.1016/j.jsmc.2015.10.011

Ishii T, Warabi E, Mann GE (2019) Circadian control of BDNF-mediated Nrf2 activation in astrocytes protects dopaminergic neurons from ferroptosis. Free Radic Biol Med 133:169–178. https://doi.org/10.1016/j.freeradbiomed.2018.09.002

Itri J, Michel S, Waschek JA, Colwell CS (2004) Circadian rhythm in inhibitory synaptic transmission in the mouse suprachiasmatic nucleus. J Neurophysiol 92:311–319. https://doi.org/10.1152/jn.01078.2003

Jagannath A, Varga N, Dallmann R et al (2021) Adenosine integrates light and sleep signalling for the regulation of circadian timing in mice. Nat Commun 12:2113. https://doi.org/10.1038/s41467-021-22179-z

Jagust W (2018) Imaging the evolution and pathophysiology of Alzheimer disease. Nat Rev Neurosci 19:687–700. https://doi.org/10.1038/s41583-018-0067-3

Jørgensen JT, Karlsen S, Stayner L et al (2017) Shift work and overall and cause-specific mortality in the Danish nurse cohort. Scand J Work Environ Health 43:117–126. https://doi.org/10.5271/sjweh.3612

Kamasawa N, Sik A, Morita M et al (2005) Connexin-47 and connexin-32 in gap junctions of oligodendrocyte somata, myelin sheaths, paranodal loops and Schmidt-Lanterman incisures: implications for ionic homeostasis and potassium siphoning. Neuroscience 136:65–86. https://doi.org/10.1016/j.neuroscience.2005.08.027

Kang J-E, Lim MM, Bateman RJ et al (2009) Amyloid-beta dynamics are regulated by orexin and the sleep-wake cycle. Science 326:1005–1007. https://doi.org/10.1126/science.1180962

Karatsoreos IN, Bhagat S, Bloss EB et al (2011) Disruption of circadian clocks has ramifications for metabolism, brain, and behavior. PNAS 108:1657–1662. https://doi.org/10.1073/pnas.1018375108

Kay CWP, Ursu D, Sher E, King AE (2016) The role of Cx36 and Cx43 in 4-aminopyridine-induced rhythmic activity in the spinal nociceptive dorsal horn: an electrophysiological study in vitro. Physiol Rep 4. https://doi.org/10.14814/phy2.12852

Kerins MJ, Ooi A (2017) The roles of NRF2 in modulating cellular iron homeostasis. Antioxid Redox Signal 29: 1756–1773. https://doi.org/10.1089/ars.2017.7176

Kress GJ, Liao F, Dimitry J et al (2018) Regulation of amyloid-β dynamics and pathology by the circadian clock. J Exp Med 215:1059–1068. https://doi.org/10.1084/jem.20172347

Kwak Y, Lundkvist GB, Brask J et al (2008) Interferon-γ alters electrical activity and clock gene expression in suprachiasmatic nucleus neurons. J Biol Rhythms 23: 150–159. https://doi.org/10.1177/0748730407313355

Lan Y-L, Zhao J, Ma T, Li S (2016) The potential roles of aquaporin 4 in Alzheimer's disease. Mol Neurobiol 53: 5300–5309. https://doi.org/10.1007/s12035-015-9446-1

Lananna BV, Nadarajah CJ, Izumo M et al (2018) Cell-autonomous regulation of astrocyte activation by the circadian clock protein BMAL1. Cell Rep 25:1–9.e5. https://doi.org/10.1016/j.celrep.2018.09.015

Lananna BV, McKee CA, King MW et al (2020) Chi3l1/YKL-40 is controlled by the astrocyte circadian clock and regulates neuroinflammation and Alzheimer's disease pathogenesis. Sci Transl Med 12. https://doi.org/10.1126/scitranslmed.aax3519

Lavialle M, Servière J (1993) Circadian fluctuations in GFAP distribution in the Syrian hamster suprachiasmatic nucleus. Neuroreport 4:1243–1246. https://doi.org/10.1097/00001756-199309000-00008

Lazarus M, Chen J-F, Huang Z-L et al (2019) Adenosine and sleep. In: Landolt H-P, Dijk D-J (eds) Sleep-wake neurobiology and pharmacology. Springer, Cham, pp 359–381

Lee S, Yoon B-E, Berglund K et al (2010) Channel-mediated tonic GABA release from glia. Science 330: 790–796. https://doi.org/10.1126/science.1184334

Leng Y, Musiek ES, Hu K et al (2019) Association between circadian rhythms and neurodegenerative diseases. Lancet Neurol 18:307–318. https://doi.org/10.1016/S1474-4422(18)30461-7

Li P, Gao L, Gaba A et al (2020) Circadian disturbances in Alzheimer's disease progression: a prospective observational cohort study of community-based older adults. Lancet Healthy Longev 1:e96–e105. https://doi.org/10.1016/S2666-7568(20)30015-5

Lian H, Litvinchuk A, Chiang AC-A et al (2016) Astrocyte-microglia cross talk through complement activation modulates amyloid pathology in mouse models of Alzheimer's disease. J Neurosci 36:577–589. https://doi.org/10.1523/JNEUROSCI.2117-15.2016

Liddelow SA, Guttenplan KA, Clarke LE et al (2017) Neurotoxic reactive astrocytes are induced by activated microglia. Nature 541:481–487. https://doi.org/10.1038/nature21029

Liedhegner ES, Sasman A, Hillard CJ (2014) Brain region-specific changes in N-acylethanolamine contents with time of day. J Neurochem 128:491–506. https://doi.org/10.1111/jnc.12495

Lim ASP, Kowgier M, Yu L et al (2013) Sleep fragmentation and the risk of incident Alzheimer's disease and cognitive decline in older persons. Sleep 36:1027–1032. https://doi.org/10.5665/sleep.2802

Linnerbauer M, Wheeler MA, Quintana FJ (2020) Astrocyte crosstalk in CNS inflammation. Neuron 108:608–622. https://doi.org/10.1016/j.neuron.2020.08.012

Long JM, Holtzman DM (2019) Alzheimer disease: an update on pathobiology and treatment strategies. Cell 179:312–339. https://doi.org/10.1016/j.cell.2019.09.001

Lucey BP, Hicks TJ, McLeland JS et al (2018) Effect of sleep on overnight cerebrospinal fluid amyloid β kinetics. Ann Neurol 83:197–204. https://doi.org/10.1002/ana.25117

Lundgaard I, Lu ML, Yang E et al (2017) Glymphatic clearance controls state-dependent changes in brain lactate concentration. J Cereb Blood Flow Metab 37: 2112–2124. https://doi.org/10.1177/0271678X16661202

Marpegan L, Krall TJ, Herzog ED (2009) Vasoactive intestinal polypeptide entrains circadian rhythms in astrocytes. J Biol Rhythms 24:135–143. https://doi.org/10.1177/0748730409332042

Marpegan L, Swanstrom AE, Chung K et al (2011) Circadian regulation of ATP release in astrocytes. J Neurosci 31:8342–8350. https://doi.org/10.1523/JNEUROSCI.6537-10.2011

Marsicano G, Goodenough S, Monory K et al (2003) CB1 cannabinoid receptors and on-demand defense against excitotoxicity. Science 302:84–88. https://doi.org/10.1126/science.1088208

Martin LJ, Zurek AA, MacDonald JF et al (2010) α5GABAA receptor activity sets the threshold for long-term potentiation and constrains hippocampus-dependent memory. J Neurosci 30:5269–5282. https://doi.org/10.1523/JNEUROSCI.4209-09.2010

Marvin JS, Borghuis BG, Tian L et al (2013) An optimized fluorescent probe for visualizing glutamate neurotransmission. Nat Methods 10:162–170. https://doi.org/10.1038/nmeth.2333

Mawuenyega KG, Sigurdson W, Ovod V et al (2010) Decreased clearance of CNS beta-amyloid in Alzheimer's disease. Science 330:1774. https://doi.org/10.1126/science.1197623

McCauley JP, Petroccione MA, D'Brant LY et al (2020) Circadian modulation of neurons and astrocytes controls synaptic plasticity in hippocampal area CA1. Cell Rep 33:108255. https://doi.org/10.1016/j.celrep.2020.108255

Meng Q-J, Logunova L, Maywood ES et al (2008) Setting clock speed in mammals: the CK1 epsilon tau mutation in mice accelerates circadian pacemakers by selectively destabilizing PERIOD proteins. Neuron 58:78–88. https://doi.org/10.1016/j.neuron.2008.01.019

Menichella DM, Majdan M, Awatramani R et al (2006) Genetic and physiological evidence that oligodendrocyte gap junctions contribute to spatial buffering of potassium released during neuronal activity. J Neurosci

26:10984–10991. https://doi.org/10.1523/JNEUROSCI.0304-06.2006

Mestre H, Hablitz LM, Xavier AL et al (2018) Aquaporin-4-dependent glymphatic solute transport in the rodent brain. eLife 7:e40070. https://doi.org/10.7554/eLife.40070

Motzkus D, Albrecht U, Maronde E (2002) The human PER1 gene is inducible by interleukin-6. J Mol Neurosci 18:105–109. https://doi.org/10.1385/JMN:18:1-2:105

Murillo-Rodriguez E, Désarnaud F, Prospéro-García O (2006) Diurnal variation of arachidonoylethanolamine, palmitoylethanolamide and oleoylethanolamide in the brain of the rat. Life Sci 79:30–37. https://doi.org/10.1016/j.lfs.2005.12.028

Musiek ES, Lim MM, Yang G et al (2013) Circadian clock proteins regulate neuronal redox homeostasis and neurodegeneration. J Clin Invest 123:5389–5400. https://doi.org/10.1172/JCI70317

Musiek ES, Bhimasani M, Zangrilli MA et al (2018) Circadian rest-activity pattern changes in aging and preclinical Alzheimer disease. JAMA Neurol 75:582–590. https://doi.org/10.1001/jamaneurol.2017.4719

Myung J, Schmal C, Hong S et al (2018) The choroid plexus is an important circadian clock component. Nat Commun 9:1062. https://doi.org/10.1038/s41467-018-03507-2

Nagai J, Yu X, Papouin T et al (2021) Behaviorally consequential astrocytic regulation of neural circuits. Neuron 109:576–596. https://doi.org/10.1016/j.neuron.2020.12.008

Nagy JI, Patel D, Ochalski PA, Stelmack GL (1999) Connexin30 in rodent, cat and human brain: selective expression in gray matter astrocytes, co-localization with connexin43 at gap junctions and late developmental appearance. Neuroscience 88:447–468. https://doi.org/10.1016/s0306-4522(98)00191-2

Nagy JI, Ionescu A-V, Lynn BD, Rash JE (2003) Coupling of astrocyte connexins Cx26, Cx30, Cx43 to oligodendrocyte Cx29, Cx32, Cx47: implications from normal and connexin32 knockout mice. Glia 44:205–218. https://doi.org/10.1002/glia.10278

Nagy JI, Dudek FE, Rash JE (2004) Update on connexins and gap junctions in neurons and glia in the mammalian nervous system. Brain Res Rev 47:191–215. https://doi.org/10.1016/j.brainresrev.2004.05.005

Nakanishi H, Ni J, Nonaka S, Hayashi Y (2021) Microglial circadian clock regulation of microglial structural complexity, dendritic spine density and inflammatory response. Neurochem Int 142:104905. https://doi.org/10.1016/j.neuint.2020.104905

Nakazato R, Kawabe K, Yamada D et al (2017) Disruption of Bmal1 impairs blood-brain barrier integrity via pericyte dysfunction. J Neurosci 37:10052–10062. https://doi.org/10.1523/JNEUROSCI.3639-16.2017

Navarrete M, Araque A (2008) Endocannabinoids mediate neuron-astrocyte communication. Neuron 57:883–893. https://doi.org/10.1016/j.neuron.2008.01.029

Nedergaard M, Goldman SA (2020) Glymphatic failure as a final common pathway to dementia. Science 370:50–56. https://doi.org/10.1126/science.abb8739

Nilsson C, Ståhlberg F, Thomsen C et al (1992) Circadian variation in human cerebrospinal fluid production measured by magnetic resonance imaging. Am J Physiol 262:R20–R24. https://doi.org/10.1152/ajpregu.1992.262.1.R20

Nishida H, Okabe S (2007) Direct astrocytic contacts regulate local maturation of dendritic spines. J Neurosci 27:331–340. https://doi.org/10.1523/JNEUROSCI.4466-06.2007

Oeckl P, Halbgebauer S, Anderl-Straub S et al (2019) Glial fibrillary acidic protein in serum is increased in Alzheimer's disease and correlates with cognitive impairment. J Alzheimers Dis 67:481–488. https://doi.org/10.3233/JAD-180325

Owen JE, Veasey SC (2020) Impact of sleep disturbances on neurodegeneration: insight from studies in animal models. Neurobiol Dis 139:104820. https://doi.org/10.1016/j.nbd.2020.104820

Pannasch U, Vargová L, Reingruber J et al (2011) Astroglial networks scale synaptic activity and plasticity. Proc Natl Acad Sci U S A 108:8467–8472. https://doi.org/10.1073/pnas.1016650108

Papaneophytou C, Georgiou E, Kleopa KA (2019) The role of oligodendrocyte gap junctions in neuroinflammation. Channels (Austin) 13:247–263. https://doi.org/10.1080/19336950.2019.1631107

Parys B, Côté A, Gallo V et al (2010) Intercellular calcium signaling between astrocytes and oligodendrocytes via gap junctions in culture. Neuroscience 167:1032–1043. https://doi.org/10.1016/j.neuroscience.2010.03.004

Pascual O, Casper KB, Kubera C et al (2005) Astrocytic purinergic signaling coordinates synaptic networks. Science 310:113–116. https://doi.org/10.1126/science.1116916

Pelluru D, Konadhode RR, Bhat NR, Shiromani PJ (2016) Optogenetic stimulation of astrocytes in the posterior hypothalamus increases sleep at night in C57BL/6J mice. Eur J Neurosci 43:1298–1306. https://doi.org/10.1111/ejn.13074

Poskanzer KE, Yuste R (2016) Astrocytes regulate cortical state switching in vivo. Proc Natl Acad Sci U S A 113:E2675–E2684. https://doi.org/10.1073/pnas.1520759113

Prolo LM, Takahashi JS, Herzog ED (2005) Circadian rhythm generation and entrainment in astrocytes. J Neurosci 25:404–408. https://doi.org/10.1523/JNEUROSCI.4133-04.2005

Raghavan AV, Horowitz JM, Fuller CA (1999) Diurnal modulation of long-term potentiation in the hamster hippocampal slice. Brain Research 833:311–314. https://doi.org/10.1016/S0006-8993(99)01523-1

Rainey-Smith SR, Mazzucchelli GN, Villemagne VL et al (2018) Genetic variation in Aquaporin-4 moderates the relationship between sleep and brain Aβ-amyloid burden. Transl Psychiatry 8:47. https://doi.org/10.1038/s41398-018-0094-x

Rajendran L, Paolicelli RC (2018) Microglia-mediated synapse loss in Alzheimer's disease. J Neurosci 38: 2911–2919. https://doi.org/10.1523/JNEUROSCI.1136-17.2017

Rasmussen MK, Mestre H, Nedergaard M (2018) The glymphatic pathway in neurological disorders. Lancet Neurol 17:1016–1024. https://doi.org/10.1016/S1474-4422(18)30318-1

Rinholm JE, Hamilton NB, Kessaris N et al (2011) Regulation of oligodendrocyte development and myelination by glucose and lactate. J Neurosci 31: 538–548. https://doi.org/10.1523/JNEUROSCI.3516-10.2011

Rinholm JE, Vervaeke K, Tadross MR et al (2016) Movement and structure of mitochondria in oligodendrocytes and their myelin sheaths. Glia 64:810–825. https://doi.org/10.1002/glia.22965

Risher WC, Patel S, Kim IH et al (2014) Astrocytes refine cortical connectivity at dendritic spines. eLife 3: e04047. https://doi.org/10.7554/eLife.04047

Ronco V, Grolla AA, Glasnov TN et al (2014) Differential deregulation of astrocytic calcium signalling by amyloid-β, TNFα, IL-1β and LPS. Cell Calcium 55:219–229. https://doi.org/10.1016/j.ceca.2014.02.016

Rouach N, Koulakoff A, Abudara V et al (2008) Astroglial metabolic networks sustain hippocampal synaptic transmission. Science 322:1551–1555. https://doi.org/10.1126/science.1164022

Roy U, Heredia-Muñoz MT, Stute L et al (2019) Degeneration of the suprachiasmatic nucleus in an Alzheimer's disease mouse model monitored by in vivo magnetic resonance relaxation measurements and immunohistochemistry. J Alzheimers Dis 69:363–375. https://doi.org/10.3233/JAD-190037

Schnell A, Chappuis S, Schmutz I et al (2014) The nuclear receptor REV-ERBα regulates Fabp7 and modulates adult hippocampal neurogenesis. PLoS One 9:e99883. https://doi.org/10.1371/journal.pone.0099883

Shinozaki Y, Nomura M, Iwatsuki K et al (2014) Microglia trigger astrocyte-mediated neuroprotection via purinergic gliotransmission. Sci Rep 4:4329. https://doi.org/10.1038/srep04329

Spanagel R, Pendyala G, Abarca C et al (2005) The clock gene Per2 influences the glutamatergic system and modulates alcohol consumption. Nat Med 11:35–42. https://doi.org/10.1038/nm1163

Stopa EG, Volicer L, Kuo-Leblanc V et al (1999) Pathologic evaluation of the human suprachiasmatic nucleus in severe dementia. J Neuropathol Exp Neurol 58:29–39

Sugimoto T, Morioka N, Zhang FF et al (2014) Clock gene Per1 regulates the production of CCL2 and interleukin-6 through p38, JNK1 and NF-κB activation in spinal astrocytes. Mol Cell Neurosci 59:37–46. https://doi.org/10.1016/j.mcn.2014.01.003

Sun D, Jakobs TC (2012) Structural remodeling of astrocytes in the injured CNS. Neuroscientist 18:567–588. https://doi.org/10.1177/1073858411423441

Svobodova I, Bhattaracharya A, Ivetic M et al (2018) Circadian ATP release in organotypic cultures of the rat suprachiasmatic nucleus is dependent on P2X7 and P2Y receptors. Front Pharmacol 9. https://doi.org/10.3389/fphar.2018.00192

Swaab DF, Fliers E, Partiman TS (1985) The suprachiasmatic nucleus of the human brain in relation to sex, age and senile dementia. Brain Res 342:37–44. https://doi.org/10.1016/0006-8993(85)91350-2

Takahashi JS (2017) Transcriptional architecture of the mammalian circadian clock. Nat Rev Genet 18:164–179. https://doi.org/10.1038/nrg.2016.150

Teismann P, Tieu K, Choi D-K et al (2003) Cyclooxygenase-2 is instrumental in Parkinson's disease neurodegeneration. PNAS 100:5473–5478. https://doi.org/10.1073/pnas.0837397100

Tönnies E, Trushina E (2017) Oxidative stress, synaptic dysfunction, and Alzheimer's disease. J Alzheimers Dis 57:1105–1121. https://doi.org/10.3233/JAD-161088

Tranah GJ, Blackwell T, Stone KL et al (2011) Circadian activity rhythms and risk of incident dementia and mild cognitive impairment in older women. Ann Neurol 70: 722–732. https://doi.org/10.1002/ana.22468

Tso CF, Simon T, Greenlaw AC et al (2017) Astrocytes regulate daily rhythms in the suprachiasmatic nucleus and behavior. Curr Biol 27:1055–1061. https://doi.org/10.1016/j.cub.2017.02.037

Vainchtein ID, Molofsky AV (2020) Astrocytes and microglia: in sickness and in health. Trends Neurosci 43:144–154. https://doi.org/10.1016/j.tins.2020.01.003

Vainchtein ID, Chin G, Cho FS et al (2018) Astrocyte-derived interleukin-33 promotes microglial synapse engulfment and neural circuit development. Science 359:1269–1273. https://doi.org/10.1126/science.aal3589

Valenti M, Viganò D, Casico MG et al (2004) Differential diurnal variations of anandamide and 2-arachidonoyl-glycerol levels in rat brain. Cell Mol Life Sci 61:945–950. https://doi.org/10.1007/s00018-003-3453-5

van der Horst GT, Muijtjens M, Kobayashi K et al (1999) Mammalian Cry1 and Cry2 are essential for maintenance of circadian rhythms. Nature 398:627–630. https://doi.org/10.1038/19323

Volicer L, Harper DG, Manning BC et al (2001) Sundowning and circadian rhythms in Alzheimer's disease. AJP 158:704–711. https://doi.org/10.1176/appi.ajp.158.5.704

Wagner S, Castel M, Gainer H, Yarom Y (1997) GABA in the mammalian suprachiasmatic nucleus and its role in diurnal rhythmicity. Nature 387:598–603. https://doi.org/10.1038/42468

Wall MJ, Dale N (2013) Neuronal transporter and astrocytic ATP exocytosis underlie activity-dependent adenosine release in the hippocampus. J Physiol 591:

3853–3871. https://doi.org/10.1113/jphysiol.2013.253450
Walter L, Stella N (2003) Endothelin-1 increases 2-arachidonoyl glycerol (2-AG) production in astrocytes. Glia 44:85–90. https://doi.org/10.1002/glia.10270
Walter L, Dinh T, Stella N (2004) ATP induces a rapid and pronounced increase in 2-arachidonoylglycerol production by astrocytes, a response limited by monoacylglycerol lipase. J Neurosci 24:8068–8074. https://doi.org/10.1523/JNEUROSCI.2419-04.2004
Wang TA, Yu YV, Govindaiah G et al (2012) Circadian rhythm of redox state regulates excitability in suprachiasmatic nucleus neurons. Science 337:839–842. https://doi.org/10.1126/science.1222826
Wang N, De Bock M, Decrock E et al (2013) Paracrine signaling through plasma membrane hemichannels. Biochimica et Biophysica Acta (BBA) Biomembr 1828:35–50. https://doi.org/10.1016/j.bbamem.2012.07.002
Wardlaw SM, Phan TX, Saraf A et al (2014) Genetic disruption of the core circadian clock impairs hippocampus-dependent memory. Learn Mem 21:417–423. https://doi.org/10.1101/lm.035451.114
Whiten DR, Brownjohn PW, Moore S et al (2020) Tumour necrosis factor induces increased production of extracellular amyloid-β- and α-synuclein-containing aggregates by human Alzheimer's disease neurons. Brain Commun 2. https://doi.org/10.1093/braincomms/fcaa146
Wildsmith KR, Holley M, Savage JC et al (2013) Evidence for impaired amyloid β clearance in Alzheimer's disease. Alzheimers Res Ther 5:33. https://doi.org/10.1186/alzrt187
Wiley CA, Bonneh-Barkay D, Dixon CE et al (2015) Role for mammalian chitinase 3-like protein 1 in traumatic brain injury. Neuropathology 35:95–106. https://doi.org/10.1111/neup.12158
Woelfle MA, Ouyang Y, Phanvijhitsiri K, Johnson CH (2004) The adaptive value of circadian clocks: an experimental assessment in cyanobacteria. Curr Biol 14:1481–1486. https://doi.org/10.1016/j.cub.2004.08.023
Wolff SEC, Wang X-L, Jiao H et al (2020) The effect of Rev-erbα agonist SR9011 on the immune response and cell metabolism of microglia. Front Immunol 11. https://doi.org/10.3389/fimmu.2020.550145
Womac AD, Burkeen JF, Neuendorff N et al (2009) Circadian rhythms of extracellular ATP accumulation in suprachiasmatic nucleus cells and cultured astrocytes. Eur J Neurosci 30:869–876. https://doi.org/10.1111/j.1460-9568.2009.06874.x
Xie L, Kang H, Xu Q et al (2013) Sleep drives metabolite clearance from the adult brain. Science 342:373–377. https://doi.org/10.1126/science.1241224
Yamazaki S, Ishida Y, Inouye S (1994) Circadian rhythms of adenosine triphosphate contents in the suprachiasmatic nucleus, anterior hypothalamic area and caudate putamen of the rat—negative correlation with electrical activity. Brain Res 664:237–240. https://doi.org/10.1016/0006-8993(94)91978-x
Ye Z-C, Wyeth MS, Baltan-Tekkok S, Ransom BR (2003) Functional hemichannels in astrocytes: a novel mechanism of glutamate release. J Neurosci 23:3588–3596. https://doi.org/10.1523/JNEUROSCI.23-09-03588.2003
Yum SW, Zhang J, Valiunas V et al (2007) Human connexin26 and connexin30 form functional heteromeric and heterotypic channels. Am J Physiol Cell Physiol 293:C1032–C1048. https://doi.org/10.1152/ajpcell.00011.2007
Zeppenfeld DM, Simon M, Haswell JD et al (2017) Association of perivascular localization of aquaporin-4 with cognition and Alzheimer disease in aging brains. JAMA Neurol 74:91–99. https://doi.org/10.1001/jamaneurol.2016.4370
Zheng C, Bieri KW, Hwaun E, Colgin LL (2016) Fast gamma rhythms in the hippocampus promote encoding of novel object–place pairings. eNeuro 3. https://doi.org/10.1523/ENEURO.0001-16.2016

Part III

Clock and Mental Illness

The Role of the Circadian System in Attention Deficit Hyperactivity Disorder

7

Upasana Bondopadhyay, Unai Diaz-Orueta, and Andrew N. Coogan

Abstract

Attention deficit hyperactivity disorder (ADHD) is a common neurodevelopmental condition characterised by the core symptoms of inattention, impulsivity and hyperactivity. Similar to many other neuropsychiatric conditions, ADHD is associated with very high levels of sleep disturbance. However, it is not clear whether such sleep disturbances are precursors to, or symptoms of, ADHD. Neither is it clear through which mechanisms sleep and ADHD are linked. One possible link is via modulation of circadian rhythms. In this chapter we overview the evidence that ADHD is associated with alterations in circadian processes, manifesting as later chronotype and delayed sleep phase in ADHD, and examine some mechanisms that may lead to such changes. We also interrogate how the circadian clock may be a substrate for therapeutic intervention in ADHD (chronotherapy) and highlight important new questions to be addressed to move the field forward.

Keywords

ADHD · Circadian · Sleep · Attention · Impulsivity

7.1 Clinical Vignette

Frank is a cheerful 9-year-old boy, currently enrolled in primary school, who seems to be imaginative and humorous. In a consultation with a psychologist, his parents reported being concerned about their son's academic performance and certain aspects of his behaviour. As conveyed by his school teachers, staff and by his mother, Frank exhibits certain disruptive behaviours within the classroom/the school premises and sometimes in social gatherings. These behaviours are not resultant from any apparent emotional concerns such as anger, but are for the most times playfully or accidentally occurring due to lack of control in Frank over his own reactions. He also has difficulty sitting still in the class or paying attention during lessons and has a tendency to move around and experiences some difficulties with interpersonal relationships among his peers. Further, Frank has been facing some academic concerns in certain subjects, especially English and Maths. Following careful psychological assessment, it is apparent that Frank possesses an average intellectual ability with a below average academic functioning

Disclosures: The authors have no potential conflicts of interest to disclose.

U. Bondopadhyay · U. Diaz-Orueta · A. N. Coogan (✉)
Department of Psychology, Maynooth University, National University of Ireland, Maynooth, County Kildare, Ireland
e-mail: andrew.coogan@mu.ie

O. Engmann, M. Brancaccio (eds.), *Circadian Clock in Brain Health and Disease*, Advances in Experimental Medicine and Biology 1344, https://doi.org/10.1007/978-3-030-81147-1_7

level in the domain of Mathematics, coupled with clinically significant difficulty in paying selective attention and lack of age appropriate behavioural inhibition, that are leading to substantial levels of concerns in his school performance and functioning in social environments. Frank fulfils the criteria to warrant a diagnosis of attention deficit hyperactivity disorder (ADHD).

Although Frank's primary difficulties were directly related to inattention and difficulty controlling his impulses, upon careful evaluation of his functioning a majority of his disturbances could be interlinked to his day to day functioning and especially to his sleep. Detailed clinical interview revealed that at home Frank would generally start doing his homework late after dinner and would consistently refuse to go to bed at the parentally-designated time; this refusal had led his parents to design most of the work at home towards the later part of the night. Frank generally goes to bed around midnight on schooldays and is woken (with difficulty) at 7 am in the morning to leave for school. Most days, in the morning Frank misses his breakfast and forgets to bring his schoolbooks. At school he generally gives the reason that he was asking a classmate for a text/note book during a lesson, when he is found chatting with others and moving around in the class. By break time, Frank partially finishes his lunch to join his friends for free-play. By the later part of the afternoon, teachers report Frank seems irritable and sleepy. On a near daily basis Frank ends up in a physical fight with the student standing in front or behind him in the school yard line. This picture demonstrates how clinically significant inattention and hyperactivity impact on Frank's day-to-day functioning and associate with particular temporal patterns in his psychological and physiological functioning. Frank's presentation is typical for many children with a diagnosis of ADHD and illustrates how the circadian system has a crucial involvement in forming daily patterns of behaviour that manifest as the clinical presentation of ADHD symptoms. This chapter attempts to elucidate the various alterations in the physiological, endocrinological, behavioural and cognitive functioning observed in ADHD that represents disruption in the individual's circadian systems.

7.2 Neuropsychological Presentation of Attention Deficit Hyperactivity Disorder

Attention deficit hyperactivity disorder (ADHD) is a neurodevelopmental disorder which usually emerges in childhood and continues through adulthood in 30–50% of cases (Noggle et al. 2012). ADHD is characterised by the core psychopathological features of inattention, impulsivity and hyperactivity and may be associated with detrimental consequences for the individual as well as for the community as a whole (McGough 2014). ADHD is associated with impairment across various spheres of functioning, including academic achievement, age appropriate interpersonal/social skills, personal health/safety, parenting and occupational outcome (Noggle et al. 2012). According to the DSM-V (APA 2013), ADHD is characterised by a persistent pattern of inattention and/or hyperactivity-impulsivity that interferes with functioning or development, for at least 6 months to a degree that is inconsistent with the developmental level and that negatively impacts directly on social and academic/occupational activities (5th ed.; DSM–5; American Psychiatric Association 2013) (Table 7.1).

In addition, the diagnostic criteria state that the presence of the above symptoms (as per criteria) should be demonstrated prior to age 12 years and that the symptoms should occur in two or more settings (at home, school or work; with friends or relatives or during other activities). Lastly, symptoms should not occur exclusively during the course of a psychotic disorder and cannot be better explained by another mental disorder (APA 2013).

7.3 ADHD Aetiology

Attention deficit hyperactivity disorder (ADHD) has been a subject of investigation since the

Table 7.1 In the clinical presentation of ADHD in children and adolescents, the presence of either/both the following symptoms (6 or more in each category) represents the diagnostic Criteria (APA 2013)

Symptoms	Inattention	Hyperactivity
A.	Often fails to give close attention to details or makes careless mistakes in schoolwork, at work, or during other activities (e.g., overlooks or misses details, work is inaccurate)	Often fidgets with or taps hands or feet or squirms in seat
B.	Often has difficulty sustaining attention in tasks or play activities (e.g., has difficulty remaining focused during lectures, conversations, or lengthy reading)	Often leaves seat in situations when remaining seated is expected (e.g., leaves his or her place in the classroom, in the office or other workplace, or in other situations that require remaining in place)
C.	Often does not seem to listen when spoken to directly (e.g., mind seems elsewhere, even in the absence of any obvious distraction)	Often runs about or climbs in situations where it is inappropriate. (Note: In adolescents or adults, may be limited to feeling restless.)
D.	Often does not follow through on instructions and fails to finish schoolwork, chores, or duties in the workplace (e.g., starts tasks but quickly loses focus and is easily side-tracked)	Often unable to play or engage in leisure activities quietly
E.	Often has difficulty organising tasks and activities (e.g., difficulty managing sequential tasks; difficulty keeping materials and belongings in order; messy, disorganised work; has poor time management; fails to meet deadlines)	Is often 'on the go', acting as if 'driven by a motor' (e.g., is unable to be or uncomfortable being still for extended time, as in restaurants, meetings; may be experienced by others as being restless or difficult to keep up with)
F.	Often avoids, dislikes, or is reluctant to engage in tasks that require sustained mental effort (e.g., schoolwork or homework; for older adolescents and adults, preparing reports, completing forms, reviewing lengthy papers)	Often talks excessively
G.	Often loses things necessary for tasks or activities (e.g., school materials, pencils, books, tools, wallets, keys, paperwork, eyeglasses, and mobile telephones)	Often blurts out an answer before a question has been completed (e.g., completes people's sentences; cannot wait for turn in conversation)
H.	Is often easily distracted by extraneous stimuli (for older adolescents and adults, may include unrelated thoughts)	Often has difficulty waiting his or her turn (e.g., while waiting in line)
I.	Is often forgetful in daily activities (e.g., doing chores, running errands; for older adolescents and adults, returning calls, paying bills, keeping appointments)	Often interrupts or intrudes on others (e.g., butts into conversations, games, or activities; may start using other people's things without asking or receiving permission; for adolescents and adults, may intrude into or take over what others are doing)

beginning of the twentieth century, when Still (1902) described an 11-year-old hyperactive boy who displayed 'loss of moral control' and 'psychical disturbance'. A detailed description of symptoms including hyperkinetic behaviour and loss of motor inhibition was included in his lecture. The core symptoms characterising ADHD (inattention, hyperactivity and impulsivity) have been presented in research and clinical practice guidelines (such as ICD-10 and DSM-V) with relative consistency, with only minor alterations in recent editions of contemporary guidelines. However, the search of insight into the aetiology of ADHD has seen plenty of turns and novel outlooks over the past number of decades. Mattes in 1980s argued that the dysfunctions in the frontal lobe regions are related to ADHD, and Niedermeyer and Naidu (1997) further validated that disinhibition of motor activity and features of inattention observed in ADHD resulted from a 'lazy' rather than a damaged frontal lobe. ADHD symptomatology has strongly pointed towards a lack of executive control. Executive dysfunction in the ADHD clinical picture is most clearly manifested among other deficits, in the form of a lack of inhibitory control. Inhibitory control can be described as processes that affect information selection at the time of attentional processing and choosing between conflicting situations (or actions) that require the individual to suppress less dominant incongruent information (Bob and Konicarova 2018). For example, in a neuropsychological test (NEPSY-II) of inhibitory control the child must suppress the urge to consider saying the actual colour of the target shape, rather than the test trial rule, where they must say the opposite colour (black for white or white for black). Further studies have shown that ADHD features are mainly concerned with two types of deficits in executive control, the cognitive deficits, related to the attentional concerns and the affective deficits, related to altered emotional reactions. Here the role of an excitable limbic system has been postulated (Antonini et al. 2015; Castellanos et al. 2006; Martinez et al. 2016; Toplak et al. 2005). Therefore, ADHD symptoms may be a result of not only the executive function concerns of the frontal cortex, but also disrupted functioning and structural features of the subcortical brain regions such as those of the limbic system.

Another line of investigations has focussed on the role of dysfunctional catecholamine secretion in ADHD. Animal studies have demonstrated abnormality of the dopaminergic (DA) and norepinephrinergic (NE) systems in ADHD. Rodent studies involving the spontaneous hypertensive rat (SHR) showed reduced level of dopamine in the prenatal SHR mid brain regions and heightened dopamine transporter activity in the adult SHR (Leo et al. 2003; Russell 2007; Watanabe et al. 1997). These studies also emphasise on whether ADHD pathophysiology is a result of a hyperdopaminergic or a hypodopaminergic tendency. In 2003, Viggiano and his colleagues demonstrated low motor activity in gene-knockout model mice that lacked the DA transporter, tyrosine hydroxylase. These mice were observed to be more hyperactive in novel situations and appeared less active on administration of stimulants. This study inclines towards the hyperdopaminergic model of ADHD. The same authors also studying the Naples high excitability (NHE) rat (showing signs of hyperactivity and poor working memory) found increased staining of the DA transporter in the prefrontal cortex. On the other hand, the SHR model displayed low levels of DA in the ventral tegmental, substantia nigra and the caudate nucleus areas and more release of DA in the prefrontal cortex and nucleus accumbens (Adriani et al. 2003). In addition to an inclination towards hypo DA hypothesis, this study suggested the possibility of an impaired DA storage function implicated in ADHD.

Such hypotheses spurred the investigation of neuronal and brain substrates of ADHD through neuroimaging and cognitive neuroscience paradigms. Recent meta-analysis of nearly two decades of work in the cognitive neuroscience of ADHD has revealed that ADHD is associated with executive function domain-specific alterations in a number of brain networks, including impaired deactivation of the default mode network (Rubia 2018). Another important area of investigation has been in the genetics of ADHD; similar to other neurodevelopmental disorders, ADHD is highly heritable (~75%; Faraone and Larsson 2018). A recent genome-wide meta-analysis of ADHD revealed for the first time 12 loci associated with ADHD that are statistically significant at the genome-wide level (Demontis et al. 2019). Although the level of variance explained by these individual genetic associations remains very modest, these recent findings do support the hypothesis that ADHD symptoms represent extremes of heritable traits expressed in the general population. Unfortunately, both cognitive neuroscience and genetic studies of ADHD have yet to be translated into widespread clinical practice, although there are some promising areas with an accumulating evidence base (e.g., neurofeedback; Enriquez-Geppert et al. 2019).

7.4 Sleep in ADHD

Children diagnosed with ADHD often have disturbed lifestyles, with a host of disorganised behaviours constantly tinting their daily lives.

Although these behaviours may be goal directed, they frequently end up being incomplete or filled with errors, resulting in detrimental implications for their mental health and occupational endeavours. Within the facets of their disturbed lifestyle, individuals with ADHD often report experiencing sleep related concerns and disturbances in considerable excess compared to matched control groups (Owens 2009). Co-occurrence of sleep concerns and ADHD reflect their prevalence in a pronounced way, with 70% of individuals with ADHD having sleep related concerns (Yoon et al. 2012) as opposed to 20–30% in the general population (Quach et al. 2012). Such sleep related concerns reported include sleep onset problems manifesting in long sleep latency, sleep phase delay syndrome, increased periodic limb movements during sleep, daytime sleepiness and altered total sleep (Corkum et al. 1998; Cortese et al. 2006; Konofal et al. 2001, 2010; Mayes et al. 2009). A number of the above concerns are expressed as difficulty in initiating and maintaining sleep (Ball et al. 1997) and the severity of sleep problems may be associated with the severity of ADHD symptoms and may be useful therapeutic targets for ADHD symptom management (Sciberras et al. 2019). Further, in the general population sleep disturbances may be associated with inattention and hyperactivity; for example, Sung et al. (2008) demonstrated that children with behavioural sleep problems also experienced more ADHD-like symptoms, lower quality of life and daily functioning. As such, dysfunctional behavioural traits associated with ADHD might be related to impaired sleep functioning, which in turn acts as a maintaining factor for the condition, fuelling the core characteristic features of the disorder (Coogan et al. 2016a). Indeed, some authors have recently questioned whether ADHD should be considered primarily as a sleep disorder (Bijlenga et al. 2019). However, there are a number of central questions that remain unanswered in relation to the sleep-ADHD associations, including questions of causality, directionality and mechanism (Raman and Coogan 2019).

7.5 The Circadian System and ADHD

Both objective and subjective assessment tools examining the rest and activity cycles among individuals with ADHD (children, adolescents and adults) have shown significant variations in daily rhythms in a number of behavioural, cognitive, endocrine, physiological and molecular parameters when compared to typically functioning individuals (Korman et al. 2019). An important behavioural and psychological manifestation of variations in circadian functioning is chronotype/diurnal preference (Adan et al. 2012). Chronotype/diurnal preference can be broadly conceptualised as inter-individual differences in actual or preferred timing of sleep/wake behaviours, and psychometric and other instruments such as Morningness/Eveningness Questionnaire (Horne and Ostberg 1976), Composite Scale of Morningness (CSM, Smith et al. 1989) and Munich ChronoType Questionnaire (MCTQ, Roenneberg et al. 2003) can be used for its assessment. Both children and adults with ADHD have been found to have a significant preference for eveningness/later chronotype, characterised as later bedtime and wake time or an increased psychological preference for such later timings of sleep behaviours (Baird et al. 2012; Rybak et al. 2007; Voinescu et al. 2012). In a systematic review (Coogan and McGowan 2017), nine target studies were found to demonstrate diurnal preferences among ADHD samples. In a 2007 study, Rybak and his colleagues showed an association between eveningness and ADHD symptoms such as poorer sustained attention among 29 diagnosed adults. More recently, Durmus et al. (2017) found significantly greater eveningness preference among 7–12-year-old children mostly with ADHD-Combined type, when compared to the control group. Further in this study a positive correlation was found between the eveningness scores and total score on resistance to sleep time. Gamble et al. (2013) reported a positive correlation between delay in rest/activity cycle and the severity of ADHD symptoms among adults. Rybak et al. (2007)

also documented a correlation between greater eveningness with increased inattention and impulsivity. These findings point towards a direct relationship between chronotype preferences and major ADHD characteristics.

Examining rhythms of sleep/wake behaviours in non-clinical samples aids in delineating which ADHD symptoms may be associated with later chronotype/evening preference. For example, McGowan in 2016 demonstrated among 396 adults that the construct of Social Jetlag (the mismatch of the time of midsleep on work versus 'free' days indicative of tension between circadian and behavioural cycles), rather than chronotype is a predictor of ADHD symptom of impulsivity. Similar association has also been demonstrated through negative correlations between scores on 'morningness' and symptoms of inattention when using subjective questionnaires (Caci et al. 2009). Later chronotype is associated with greater social jetlag (e.g., Roenneberg et al. 2012), and as such it may be that social jetlag, and not chronotype per se, is the factor associated with impulsivity and other ADHD symptoms. Other studies have linked later and more variable sleep timing with trait impulsivity (McGowan and Coogan 2018), and have also linked later chronotype with other behavioural manifestations of impulsivity, including sensation seeking and response inhibition (Kang et al. 2015). Ottoni et al. (2012) report that eveningness is associated with being emotionally volatile and the behavioural traits of ADHD such as apathetic and disinhibited temperamental inclinations. Therefore, there may be widespread influence of the circadian system on the individual's cognitive, affective, behavioural domains that are pertinent for ADHD symptomatology.

As illustrated in the clinical vignette case of Frank, his behavioural symptoms governing lack of inhibitory control revealed through psychological test findings as well as subjective reports are related perhaps to his altered sleep timings, which in turn may be the result of inclination towards later chronotype. Therefore, Frank's temperamental reactions with peers or with others in his social environments and his frequent tendency to get into unpleasant physical interactions with peers may in part reflect chronobiological traits.

7.6 Delayed Sleep Phase and Sleep Onset Insomnia in ADHD

Gruber et al. (2000), using both subjective reports of sleep log and objective findings from actigraphic recordings from 7- to 11-year-old children diagnosed with ADHD, demonstrated that ADHD was associated with more variable and later sleep onset time compared to controls. Subsequently, similar findings were reported both for younger children and older adolescents (Hvolby et al. 2008; Hysing et al. 2016). Van der Heijden et al. (2005) reported that the presence of sleep onset insomnia in children with ADHD is associated with significantly longer sleep latencies and delayed dim-light melatonin onset (DLMO) compared to children with ADHD but no sleep onset insomnia. Bron et al. (2016) found that adults with lifetime depression/anxiety (LDA) in addition to high ADHD symptoms were found to have the least favourable sleep characteristics as compared to the individuals in the depression/anxiety and control group, and that the LDA and ADHD group were found to have the highest results for Delayed Sleep Phase Syndrome, shortest sleep duration and extremely late chronotype. Van Veen et al. (2010) found that, out of 40 ADHD adults, 31 reported the presence of sleep onset insomnia and were also found to have a delayed DLMO, indicated a significantly delayed phase of circadian entrainment. In a large retrospective cohort of 9338 adolescents, a significant association was found between delayed sleep phase disorder (DSPD) and inattentive and hyperactive symptoms (Sivertsen et al. 2015), indicating potential for delayed sleep phase to feed into ADHD symptoms.

The pineal hormone melatonin's synthesis in the pinealocytes from the precursor tryptophan exhibits a clear circadian rhythm, with peak plasma levels usually between 2 and 3 am, and the master pacemaker in the suprachiasmatic nucleus (SCN, located in the anterior ventral hypothalamus) has indirect efferent projections to the pineal gland crucial for synchronising the circadian rhythm of melatonin to the light-dark cycle and maintaining its persistence (Arendt 2005a). As such, delays in DLMO observed in ADHD may be indicative of altered entrainment of the SCN master clock to environmental zeitgebers. However, to date there has been no direct examination of SCN function in ADHD (for example, in post mortem tissue), and other work to date has focussed on the use of peripheral (and accessible) circadian oscillators as proxies for circadian function in ADHD (e.g., Baird et al. 2012; Coogan et al. 2019).

Psychopharmacological factors might also influence the presence of longer sleep latency and other insomnia symptoms sleep in ADHD. In a study by Boonstra et al. (2007), administration of methylphenidate (MPH, the frontline psychostimulant used in the pharmacological management of ADHD) in adults diagnosed with ADHD has been linked to anomalies in sleep parameters, such as concerns with sleep latency and sleep duration. However, the same study also documented evidence of lesser nocturnal awakenings and hence a more consolidated sleep because of MPH (Boonstra et al. 2007). Among the younger ADHD populations, methylphenidate has been linked to a shorter total sleep time and later sleep onset times in several investigations (Lee et al. 2012; Sangal et al. 2006; Snitselaar et al. 2013; Tirosh et al. 1993). Another recent study demonstrated that pharmacotherapy of ADHD was associated with alterations in circadian and sleep function in adults with ADHD when compared to treatment-naïve ADHD patients (Coogan et al. 2019). For example, treatment was associated with longer, but not more frequent, wake bouts during the night in medicated patients. As such, it is important to delineate the associations of ADHD itself with sleep and circadian changes from ADHD-treatment effects.

Returning to the case of Frank, his consistently later bedtimes (and the resultant short sleep duration, as he wakes up by 7 am) could be attributed to the possible presence of a delayed sleep phase (although no sleep measures were used in his assessment particularly). This delayed sleep phase may be directly amplifying Frank's manifestation of inattention or hyperactivity. Alternatively, or additionally, shortened sleep duration may be important as this has been associated with hyperactivity and inattention in a study with a large adolescent cohort (Gau et al. 2007). Therefore, functionally, later bedtimes could be a crucial influencing factor for Frank's ADHD symptoms.

7.7 Genetic and Environmental Factors That May Link ADHD and the Circadian System

As mentioned earlier, ADHD is a highly heritable disorder. Circadian traits are also reported to be heritable; for example, chronotype appears to be strongly heritable (Inderkum and Tarokh 2018) and genome-wide studies have revealed significant associations with polymorphisms in clock genes and other loci (e.g., Jones and Jane 2019). As such, there is the possibility for shared genetic risk between ADHD and circadian traits associated with ADHD. Although at present no association between ADHD and clock genes has been reported at the genome-wide level, there are some interesting reports from candidate gene studies (notwithstanding the recognised weaknesses of such approaches; Duncan et al. 2019). *PER1* has been associated with ADHD in children and adolescents (Lasky-Su et al. 2008) and *PER2* polymorphisms has also been associated with ADHD (Brookes et al. 2006), although these associations did not reach genome-wide statistical significance. A single nucleotide polymorphism in the CLOCK gene has been associated with adult ADHD symptoms in three separate studies (Jeong et al. 2014; Kissling et al. 2008; Xu et al. 2010). In a recent study, Carpena et al. (2019) explored the association between ADHD and *CLOCK*, using haplotype analysis, and demonstrated an association between *CLOCK* haplotype and ADHD status, further implicating *CLOCK* in the aetiology of ADHD. There have also been a number of studies that examined the diurnal rhythms in expression of clock genes in different tissues derived from ADHD patients. Baird et al. (2012) reported that, in adults, ADHD is associated with blunting of rhythms in the expression profiles of the clock genes *BMAL1* and *PER2*. Coogan et al. (2019) also reported ADHD-related changes in clock gene expression profiles in ex-vivo cultures of fibroblasts derived from patients with ADHD (and with or without ADHD medication). These data indicate that there may be alterations in the core molecular circadian cycle associated with ADHD, and warrants further study.

As light is the most important environmental zeitgeber that determines circadian phase, it is possible that geographical variations in the timing and levels of exposure to sunlight might influence ADHD prevalence. Arns et al. (2013) tested this hypothesis when they analysed the record of solar intensity from ten countries, reporting associations between higher solar intensities and lower ADHD prevalence. The proposed mechanism that may underpin this association is that bright morning light would phase advance the circadian clock and decrease the association between delayed phase and ADHD symptoms. In 2015, Huber and his co-workers stated that higher altitude geographical regions have lower prevalence of ADHD, based their rationale on the association of higher altitude with hypobaric hypoxia and hence increased levels of dopamine. Based on the hypodopaminergic model of ADHD, such an increase in the dopamine would mitigate against ADHD. However, Arns et al. (2015) found that the association of altitude with ADHD prevalence is actually related to the solar intensity levels rather than the altitude level, which must have been the confounding factor detected previously. The influence of natural sunlight, mediated through the biological clock, on ADHD symptoms may offer therapeutic opportunities through increasing exposure to daylight.

7.8 Neuroendocrine and Autonomic Features of ADHD: Circadian Aspects

Turner-Cobb (2005) showed that the hypothalamus pituitary adrenal (HPA) axis plays an important role in regulating neurobehavioural domains such as attention, emotion, learning, memory and movement. Under a stressful condition, the neurons of the hypothalamus release the corticotropic hormone which in turn leads to the secretion of the adrenocorticotropic hormone and the subsequent secretion and release of cortisol from the adrenal cortex. In a study by Musser in 2011 it was found that there were no significant sympathetic variations found in ADHD children when they were shown emotional stimuli as opposed to typically functioning individuals, supporting the hypothesis of autonomic dysregulation in individuals with ADHD. This under-reactivity of the HPA was also found to correlate with the neurocognitive performance of the ADHD individual and was further demonstrated empirically by measuring low levels of cortisol for ADHD patients as compared to the control group in a study by Ma and colleagues in 2011. Abnormal diurnal cortisol rhythms have been linked with hyperactivity manifested in childhood features of ADHD (Blomqvist et al. 2007; Kaneko et al. 1993) and this finding is crucial as the HPA axis is under strong circadian control (Nicolaides et al. 2014). Further, the diurnal profile of cortisol expressed relative to habitual wake time appears to be phase-delayed in adults with ADHD compared to controls (Baird et al. 2012), whilst children with ADHD show morning hypo-arousal of the HPA axis as assessed via salivary cortisol levels (Imeraj et al. 2012). Dysfunction in arousal mechanisms can be viewed as the causal factor for ADHD, with motor hyperactivity being considered a reaction to the hypo-arousal condition that is required to counteract somnolence (Lecendreux et al. 2000). Hence, hypo-reactivity of the HPA axis in stressful condition has been linked to the symptoms of hyperactivity and impulsivity in ADHD (Blomqvist et al. 2007; Hong et al. 2003; Moss et al. 1995; Virkkunen

1985). This behavioural hyperactivity and impulsiveness is expressed in the form of a significant lack of behavioural inhibition characterising their psychosocial reactions. Hong et al. (2003) postulated therefore that an abnormal HPA-axis response to stress should be considered as an attributional factor for the dysfunctional behavioural inhibition observed in ADHD. A corollary of this hypothesis is that given the HPA operates under strong circadian influence, that dysfunction of HPA function in ADHD may be associated with altered rhythms in HPA processes.

The possible dysfunction of the HPA axis has been linked to the major defining characteristics found in ADHD which are cognitive, affective and behavioural in nature. Lack of behavioural inhibition has been linked to the problematic behaviours observed in ADHD (Lackschewitz et al. 2008), and such impairing behaviours should be considered in line with the abnormal HPA-axis response to stress (Hong et al. 2003). Low levels of cortisol has been associated with lack of age appropriate cognitive performance (Hong et al. 2003), maladaptive behaviours as well as variations in levels of anxiety among ADHD children (Hastings et al. 2009); these findings have also been replicated in adults with ADHD (Lackschewitz et al. 2008). Certain studies have shed light upon specific relationship between the ADHD subtype and under-reactivity of HPA axis in response to stress. Both hyperactivity/impulsive subtype and the inattentive subtype have been linked to dysfunctional HPA activation (Hong et al. 2003; Moss et al. 1995; Randazzo et al. 2008; Virkkunen 1985). However these findings are not completely consistent; for example, Van West et al. (2009) did not find a relation between low cortisol responsivity and psychosocial stress. These discrepancies however could be attributed to the study sample, design, comorbidity or treatment effects.

7.9 Chronotherapy for ADHD

If circadian rhythms are indeed altered in ADHD, and these alterations are linked to specific features of the condition, then interventions to re-synchronise the circadian cycles might then be particularly effective in alleviating these symptoms. A number of studies have examined the effect of melatonin treatment on ADHD symptoms and sleep related outcomes. The nature of the current evidence base for the utility of melatonin in ADHD management include randomised, placebo controlled and double-blind trials to longitudinal investigations on samples of ADHD or typically developing individuals (Coogan and McGowan 2017). Systematic administration of melatonin has been associated with significant decrease in sleep onset latency and increase in sleep duration among the participants. For example, in a recent study by Masi et al. (2019), ADHD children and adolescents treated with methylphenidate were administered melatonin for a period of 1–12 months, and both younger children and adolescents reported improvements in sleep concerns. Other studies over the past decade have been more specific with regard to sleep outcomes as a result of melatonin use. Tjon Pian Gi et al. (2003) reported a rapid decrease in sleep onset latency upon prior to bedtime administration of melatonin among children with ADHD and insomnia. Another earlier study even reported a decrease of sleep onset latency from 60 to 30 min as a result of melatonin treatment for 4 weeks (Smits et al. 2001). Different studies have also argued regarding the best possible design and co-treatment that might lead to best sleep related outcome among the patients. Weiss et al. (2006) demonstrated that maximum reduction in insomnia symptoms took place when melatonin administration was coupled with sleep hygiene training. Similarly a decrease in sleep latency resulted from the co-administration of melatonin with methylphenidate, accompanied by an increase in the height and weight of the participating children (Mostafavi et al. 2012). Further, another study following a similar design of co-administering melatonin and methylphenidate showed decrease in sleep onset latency, but did not result in improvements of core ADHD symptoms (Mohammadi et al. 2012). Therefore, the above studies point towards the positive effects of melatonin in improving sleep onset among ADHD; however, the treatment has not shown direct effect on the symptoms of inattention or hyperactivity/impulsivity.

With regard to delayed sleep phase, melatonin administration has shown some positive effects. A meta-analysis (Van Geijlswijk et al. 2010) reported that exogenous intake of melatonin leads to the advancement of the endogenous melatonin onset in both children and adolescents. In 2006, Szeinburg showed that a 6 months intake of melatonin resulted in shorter sleep latency and longer sleep duration among a group of children and adolescents diagnosed with delayed sleep phase syndrome. Melatonin treatment has shown advancement of DLMO, indicating melatonin's utility in correcting delayed circadian phase (Van der Heijden et al. 2007). One question of importance regarding melatonin in these studies is whether melatonin is deployed primarily as a somnolent or as a chronobiotic; the doses and timing of optimal treatments will differ accordingly, and as such this clear distinction should be made at the conception of studies (Arendt 2005b).

Behavioural chronotherapeutic interventions for treatment of circadian and/or sleep related concerns in ADHD populations over the last decade have shown some promise. Mullane and Corkum (2006) investigated the effect of behavioural intervention for sleep in three unmedicated children over a 5-week treatment period and found that the children's sleep improved and were maintained over a 3-month follow-up period. Corkum et al. (2009) validated the above findings on a larger randomised control trial (RCT), where the ADHD children were found to have significantly improved sleep as compared to typically functioning control group. The intervention included facets such as psychoeducation about basic sleep physiology and the different types of sleep problems/disorders, sleep hygiene and bedtime routines, implementing a faded bedtime strategy and reward program (Mullane and Corkum 2006). Over the following period other well-designed RCTs also supported the effectiveness of behavioural interventions for the improvement of sleep concerns in ADHD sample of participants (Hiskock et al. 2015; Keshavarzi et al. 2014). Hiskock et al. (2015) found significantly improved ADHD symptoms (through parent and teacher ADHD rating scale), sleep problems (parent reported severity through children's sleep habits questionnaire and actigraphy), behaviour and daily functioning (measured through strengths and difficulties questionnaire) and working memory (working memory test battery for children) because of the behavioural sleep intervention-controlled trial. Similar results were also highlighted by Peppers et al. (2016) where sleep hygiene program led to better sleep quality and improved ADHD symptoms.

The use of morning bright light therapy among ADHD children has been shown to result in correction of delayed sleep phase and also improvements in ADHD ratings (Gruber et al. 2007). Similar findings (alleviation of core ADHD symptoms and improvements in affective symptoms) have also been demonstrated in case of adults that demonstrated that shifts towards earlier circadian preference were associated with improvement in the overall subjective and objective ADHD symptom ratings (Rybak et al. 2006). Other preliminary pilot findings further support the principle that light therapy as a chronotherapeutic may be useful in the management of ADHD symptoms: Fargason et al. (2017) report that morning bright light, coupled with the minimisation of evening bright light, advanced DLMO and was associated with decreased ADHD scores. Given the promise of light therapy in other areas of psychiatry and psychology (Cunningham et al. 2019), further study of this strategy in ADHD is warranted.

Agomelatine is a licenced antidepressant in the European Union and Australia, which is an agonist of MT1 and MT2 melatonin receptor and an antagonist of 5-HT2c and 5-HT2B serotonin receptors that at least partially functions as a chronobiotic (Guardiola-Lemaitre et al. 2014). Preliminary evidence has suggested that agomelatine treatments lead to decreased ADHD-related symptoms (Niederhofer 2012). Salardini et al. (2016) reported in a study with a small group of ADHD children and adolescents that outcomes related to ADHD symptoms were not significantly better for the agomelatine-treated group than those treated with methylphenidate; however, the former reported less concerns associated with insomnia as a treatment outcome. Although these findings reveal some promise, concerns around hepatotoxicity with agomelatine and the lack of licensing in major jurisdictions are likely to curtail interest in further exploring its use in ADHD, and future studies may rather focus on other melatoninergic agonists (Comai et al. 2019).

In the clinical vignette case of Frank, his habitually late bedtime could be reinforcing a delayed circadian and sleep phase through increased exposure to evening light, which will delay the circadian clock (Raman and Coogan 2019). As such, there is potential for a 'vicious cycle' to develop, though which late bedtime reinforces late bedtime. Equally however, there may be an opportunity here to break this cycle through increasing morning light exposure (coupled with decreased evening light exposure), phase advancing the clock and helping establish more socially-congruent sleep/wake behavioural cycles. The utility of such approaches may vary strongly between individuals, as recent evidence indicates very high levels of inter-individual variability in responses to light (Phillips et al. 2019).

7.10 Future Directions

The current literature leaves a number of important questions still unanswered in relation to ADHD and altered circadian rhythms. Chief amongst these is whether circadian changes are symptoms of ADHD, or precursors to ADHD, or risk factors for ADHD, or a mix of these? There are some suggestions from recent longitudinal studies that suggest sleep disturbances preface and predict ADHD diagnosis in young children (Soehner et al. 2019; Tso et al. 2019). Indeed, there have been recent suggestions that ADHD might be reasonably reconceptualised as a sleep disorder, with ADHD features emerging (in part) as a neurocognitive result of chronic sleep disturbances and circadian changes (Bijlenga et al. 2019).

Certainly, it would appear that cognitive and behavioural strategies to improve sleep and/or circadian function in ADHD might be expected to yield benefit. Inspiration for those of us working in the field of ADHD comes from findings in major depression that indicated that Cognitive Behavioural Therapy for insomnia (CBTi) results in significant and durable antidepressant effects (e.g., Kalmbach et al. 2019). Indeed, large trials of internet-delivered interventions indicate that CBTi results in improvements in psychopathology and quality of life (Espie et al. 2019). Thus, trials of CBTi in ADHD appear to be strongly warranted. Chronotherapy incorporating bright light therapy may also be a useful treatment modality and has shown much promise in other diagnoses (Cunningham et al. 2019). Further, circadian principles may be deployed in the pharmacotherapy of ADHD: either through the use of melatonin or melatoninergic agonists as chronobiotics, or through the use of circadian medicine to identify time-of-day for optimal effectiveness of psychostimulants used routinely in the treatment of ADHD (Ruben et al. 2019).

Incorporating perspectives from chronobiology into ADHD research and practice may lead to advances in our understanding of ADHD, and improvements in the lives of patients like Frank, as well as their families and carers.

References

Adan A, Archer SN, Hidalgo MP et al (2012) Circadian typology: a comprehensive review. Chronobiol Int 29 (9):1153–1175

Adriani W, Caprioli A, Granstrem O et al (2003) The spontaneously hypertensive-rat as an animal model of ADHD: evidence for impulsive and non-impulsive subpopulations. Neurosci Biobehav Rev 27:639–651

American Psychiatric Association (2013) Neurodevelopmental disorders. In Diagnostic and statistical manual of mental disorders (5th ed.). https://doi.org/10.1176/appi.books.9780890425596.dsm01

Antonini TN, Becker SP, Tamm L, Epstein JN (2015) Hot and cool executive functions in children with attention-deficit/hyperactivity disorder and comorbid

oppositional defiant disorder. J Int Neuropsychol Soc 21(8):584–595
Arendt J (2005a) Melatonin in humans: it's about time. J Neuroendocrinol 17:537–538
Arendt J (2005b) Melatonin: characteristics, concerns, and prospects. J Biol Rhythm 20:291–303
Arns M, van der Heijden KB, Arnold LE, Kenemans JL (2013) Geographic variation in the prevalence of attention-deficit/hyperactivity disorder: the sunny perspective. Biol Psychiatry 74:585–590
Arns M, Swanson JM, Arnold LE (2015) ADHD prevalence: altitude or sunlight? Better understanding the interrelations of dopamine and the circadian system. J Atten Disord 22(2):163–166
Baird AL, Coogan AN, Siddiqui A, Donev RM, Thome J (2012) Adult attention-deficit hyperactivity disorder is associated with alterations in circadian rhythms at the behavioural endocrine and molecular levels. Mol Psychiatry 17(10):988–995
Ball J, Tiernan M, Janusz J, Furr A (1997) Sleep patterns among children with attention-deficit hyperactivity disorder: a reexamination of parent perceptions. J Pediatr Psychol 22:389–398
Bijlenga D, Vollebregt MA, Kooij JJS, Arns M (2019) The role of the circadian system in the etiology and pathophysiology of ADHD: time to redefine ADHD? Atten Defic Hyperact Disord 11(1):5–19
Blomqvist M, Holmberg K, Lindblad F, Fernell E, Ek U, Dahllöf G (2007) Salivary cortisol levels and dental anxiety in children with attention deficit hyperactivity disorder. Eur J Oral Sci 115:1–6
Bob P, Konicarova J (2018) ADHD, stress, and development. Springer, Cham
Boonstra AM, Kooij JJ, Oosterlaan J, Sergeant JA, Buitelaar JK, Van Someren EJ (2007) Hyperactive night and day? Actigraphy studies in adult ADHD: a baseline comparison and the effect of methylphenidate. Sleep 30:433–442
Bron T, Bijlenga D, Kooij JJ et al (2016) Attention-deficit hyperactivity disorder symptoms add risk to circadian rhythm sleep problems in depression and anxiety. J Affect Disord 200:74–81
Brookes K, Chen XXW, Zhou K et al (2006) The analysis of 51 genes in DSM-IV combined type attention deficit hyperactivity disorder: association signals in DRD4, DAT1 and 16 other genes. Mol Psychiatry 11:53–934
Caci H, Bouchez J, Bayle FJ (2009) Inattentive symptoms of ADHD are related to evening orientation. J Atten Disord 13:36–41
Carpena MX, Hutz MH, Salatino-Oliveira A et al (2019) CLOCK polymorphisms in attention-deficit/hyperactivity disorder (ADHD): further evidence linking sleep and circadian disturbances and ADHD. Genes (Basel) 10(2):E88
Castellanos FX, Sonuga-Barke EJ, Milham MP, Tannock R (2006) Characterizing cognition in ADHD: beyond executive dysfunction. Trends Cogn Sci 10:117–123
Comai S, Bertazzo A, Vachon J et al (2019) Trace elements among a sample of prisoners with mental and personality disorders and aggression: correlation with impulsivity and ADHD indices. J Trace Elem Med Biol 51:123–129
Coogan AN, McGowan NM (2017) A systematic review of circadian function, chronotype and chronotherapy in attention deficit hyperactivity disorder. Atten Defic Hyperact Disord 9:129–147
Coogan AN, Baird AL, Popa-Wagner A, Thome J (2016a) Circadian rhythms and attention deficit hyperactivity disorder: the what, the when and the why. Prog Neuropsychopharmacol Biol Psychiatry 67:74–81
Coogan AN, Schenk M, Palm D et al (2019) Impact of adult attention deficit hyperactivity disorder and medication status on sleep/wake behavior and molecular circadian rhythms. Neuropsychopharmacol 44:1198–1206
Corkum P, Tannock R, Moldofsky H (1998) Sleep disturbance in children with attention-deficit/hyperactivity disorder. J Am Acad Child Adolesc Psychiatry 37:637–646
Corkum P, Davidson F, Lingley-Pottie P et al (2009) Behavioral treatment for diagnosed pediatric insomnia in school-aged children. In: Fifth annual conference on pediatric sleep medicine (symposium), Denver, CO
Cortese S, Konofal E, Yareman N, Mouren MC, Lecendreux M (2006) Sleep and alertness in children with attention deficit/hyperactivity disorder: a systematic review of the literature. Sleep 29:504–511
Cunningham JEA, Stamp JA, Shapiro CM (2019) Sleep and major depressive disorder: a review of non-pharmacological chronotherapeutic treatments for unipolar depression. Sleep Med. https://doi.org/10.1016/j.sleep.2019.04.012
Demontis D, Walters RK, Martin J et al (2019) Discovery of the first genome-wide significant risk loci for attention deficit/hyperactivity disorder. Nat Genet 51 (1):63–75
Duncan LE, Ostacher M, Ballon J (2019) How genome-wide association studies (GWAS) made traditional candidate gene studies obsolete. Neuropsychopharmacology 44(9):1518–1523
Durmus FB, Armaan AR, Ayaaz AB (2017) Chronotype and its relationship with sleep disorders in children with attention deficit hyperactivity disorder. Chronobiol Int 34(7):886–894
Enriquez-Geppert S, Smit D, Pimenta MG et al (2019) Neurofeedback as a treatment intervention in ADHD: current evidence and practice. Curr Psychiatry Rep 21:46
Espie CA, Emsley R, Kyle SD et al (2019) Effect of digital cognitive behavioral therapy for insomnia on health, psychological well-being, and sleep-related quality of life: a randomized clinical trial. JAMA Psychiatry 76 (1):21–30
Faraone SF, Larsson H (2018) Genetics of attention deficit hyperactivity disorder. Mol Psychiatry 24:562–575
Fargason RE, Fobian AD, Hablitz LM et al (2017) Correcting delayed circadian phase with bright light therapy predicts improvement in ADHD symptoms: a pilot study. J Psychiatr Res 91:105–110

Gamble KL, May RS, Besing RC, Tankersly AP, Fargason RE (2013) Delayed sleep timing and symptoms in adults with attention-deficit/hyperactivity disorder: a controlled actigraphy study. Chronobiol Int 30 (4):598–606
Gau SS, Kessler RC, Tseng WL, Wu YY, Chiu YN, Yeh CB et al (2007) Association between sleep problems and symptoms of attention-deficit/hyperactivity disorder in young adults. Sleep 30:195–201
Gruber R, Sadeh A, Raviv A (2000) Instability of sleep patterns in children with attention-deficit/hyperactivity disorder. J Am Acad Child Adolesc Psychiatry 39 (4):495–501
Gruber R, Grizenko N, Joober R (2007) Delayed sleep phase syndrome, ADHD, and bright light therapy. J Clin Psychiatry 68(2):337–338
Guardiola-Lemaitre B et al (2014) Agomelatine: mechanism of action and pharmacological profile in relation to antidepressant properties. Br J Pharmacol 171 (15):3604–3619
Hastings PD, Fortier I, Utendale WT, Simard LR, Robaey P (2009) Adrenocortical functioning in boys with attention-deficit/hyperactivity disorder: examining subtypes of ADHD and associated comorbid conditions. J Abnorm Child Psychol 37(4):565–578
Hiskock H, Sciberras E, Mensah F et al (2015) Impact of a behavioural sleep intervention on symptoms and sleep in children with attention deficit hyperactivity disorder, and parental mental health: randomised controlled trial. BMJ 350(h68):1–14
Hong HJ, Shin DW, Lee EH, Oh YH, Noh KS (2003) Hypothalamic–pituitary–adrenal reactivity in boys with attention deficit hyperactivity disorder. Yonsei Med J 44:608–614
Horne JA, Ostberg O (1976) A self-assessment questionnaire to determine morningness eveningness in human circadian rhythms. Int J Chronobiol 4(2):97–110
Huber RS, Kim TS, Kim N, Kuykendall MD, Sherwood SN, Renshaw PF, Kondo DG (2015) Association between altitude and regional variation of ADHD in youth. J Atten Disord 22(14):1299–1306
Hvolby A, Jørgensen J, Bilenberg N (2008) Actigraphic and parental reports of sleep difficulties in children with attention-deficit/hyperactivity disorder. Arch Pediatr Adolesc Med 162(4):323–329
Hysing M, Lundervold AJ, Posserud MB, Sivertsen B (2016) Association between sleep problems and symptoms of attention deficit hyperactivity disorder in adolescence: results from a large population-based study. Behav Sleep Med 14(5):550–564
Imeraj L, Antrop I, Roeyers H, Deschepper E, Bal S, Deboutte D (2012) Time-of-day effects in arousal: disrupted diurnal cortisol profiles in children with ADHD. J Child Psychol Psychiatry 53(7):782–789
Inderkum AP, Tarokh L (2018) High heritability of adolescent sleep–wake behavior on free, but not school days: a long-term twin study. Sleep 41(3):1–9
Jeong SH, Yu JC, Lee C, Choi KS, Choi JE, Kim SJ, Joo EJ (2014) Human CLOCK gene-associated attention deficit hyperactivity disorder-related features in healthy adults: quantitative association study using Wender Utah Rating Scale. Eur Arch Psychiatry Clin Neurosci 264:71–81
Jones SE, Jane JN (2019) Genome-wide association analyses of chronotype in 697,828 individuals provides insights into circadian rhythms. Nat Commun 10:343
Kalmbach DA, Cheng P, Arnedt T et al (2019) Improving daytime functioning, work performance, and quality of life in postmenopausal women with insomnia: comparing cognitive behavioral therapy for insomnia, sleep restriction therapy, and sleep hygiene education. J Clin Sleep Med 15(7):999–1010
Kaneko M, Hoshino Y, Hashimoto S, Okano T, Kumashiro H (1993) Hypothalamic–pituitary–adrenal axis function in children with attention-deficit hyperactivity disorder. J Autism Dev Disord 23:59–65
Kang JI, Park CI, Sohn SY, Kim HW, Namkoong K, Kim SJ (2015) Circadian preference and trait impulsivity, sensation-seeking and response inhibition in healthy young adults. Chronobiol Int 32:235–241
Keshavarzi Z, Bajoghli H, Mohamadi MR et al (2014) In a randomized case-control trial with 10-years olds suffering from attention deficit/hyperactivity disorder (ADHD) sleep and psychological functioning improved during a 12-week sleep-training program. World J Biol Psychiatry 15(8):609–619
Kissling C, Retz W, Wiemann S, Coogan AN et al (2008) A polymorphism at the 3′-untranslated region of the CLOCK gene is associated with adult attention-deficit hyperactivity disorder. Am J Med Genet B Neuropsychiatr Genet 147:333–338
Konofal E, Lecendreux M, Bouvard MP, Mouren-Simeoni M (2001) High levels of nocturnal activity in children with attention-deficit hyperactivity disorder: a video analysis. Psychiatry Clin Neurosci 55:97–103
Konofal E, Lecendreux M, Cortese S (2010) Sleep and ADHD. Sleep med 11(7):652–658
Korman M, Palm D, Uzoni A, Faltraco F, Tucha O, Thome J et al (2019) ADHD 24/7: circadian clock genes, chronotherapy and sleep/wake cycle insufficiencies in ADHD. World J Biol Psychiatry 5:1–16
Lackschewitz H, Hüther G, Kröner-Herwig B (2008) Physiological and psychological stress responses in adults with attention-deficit/hyperactivity disorder (ADHD). Psychoneuroendocrinology 33:612–624
Lasky-Su J, Neale BM, Franke B et al (2008) Genome-wide association scan of quantitative traits for attention deficit hyperactivity disorder identifies novel associations and confirms candidate gene associations. Am J Med Genet B Neuropsychiatr Genet 147B:1345–1354
Lecendreux M, Konofal E, Bouvard M, Falissard MC et al (2000) Sleep and alertness in children with ADHD. J Child Psychol Psychiatry 41:803–812
Lee SH, Seo WS, Sung HM et al (2012) Effect of methylphenidate on sleep parameters in children with ADHD. Psychiatry Invest 9(4):384–390
Leo D, Sorrentino E, Volpicelli F et al (2003) Altered midbrain dopaminergic neurotransmission during development in an animal model of ADHD. Neurosci Biobehav Rev 27:661–669

Martinez L, Prada E, Satler C, Tavares MC, Tomaz C (2016) Executive dysfunctions: the role in attention deficit hyperactivity and post-traumatic stress neuropsychiatric disorders. Front Psychol 7:1230
Masi G, Fantozzi P, Villafranca A et al (2019) Effects of melatonin in children with attention-deficit/hyperactivity disorder with sleep disorders after methylphenidate treatment. Neuropsychiatr Dis Treat 15:663–667
Mattes JA (1980) The role of frontal lobe dysfunction in childhood hyperkinesis. Compr Psychiatry 21:358–369
Mayes SD, Calhoun SL, Bixler ED et al (2009) ADHD subtypes and comorbid anxiety, depression, andoppositional-defiant disorder: differences in sleep problems. J Pediatr Psychol 34:328–337
McGough J (2014) ADHD: ADHD. Retrieved from https://ebookcentral.proquest.com
McGowan NM, Coogan AN (2018) Sleep and circadian rhythm function and trait impulsivity: an actigraphy study. Psychiatry Res 268:251–256
McGowan NM, Voinescu BI, Coogan AN (2016) Sleep quality, chronotype and social jetlag differentially associate with symptoms of attention deficit hyperactivity disorder in adults. Chronobiol Int 33 (10):1433–1443
Mohammadi MR, Mostafavi SA, Keshavarz SA et al (2012) Melatonin effects in methylphenidate treated children with attention deficit hyperactivity disorder: a randomized double blind clinical trial. Iran J Psychiatry 7(2):87–92
Moss HB, Vanyukov MM, Martin CS (1995) Salivary cortisol responses and the risk for substance abuse in prepubertal boys. Biol Psychiatry 3:547–555
Mostafavi SA, Mohammadi MR, Hosseinzadeh P et al (2012) Dietary intake, growth and development of children with ADHD in a randomized clinical trial of Ritalin and Melatonin co-administration: through circadian cycle modification or appetite enhancement? Iran J Psychiatry 7(3):114–119
Mullane J, Corkum P (2006) Case series: evaluation of a behavioural sleep intervention for three children with attention-deficit/hyperactivity disorder and dyssomnia. J Atten Disord 10(2):217–227
Musser ED, Backs R, Schmitt CF, Ablow JC et al (2011) Emotion regulation via the autonomic nervous system in children with attention-deficit/hyperactivity disorder (ADHD). J Abnorm Child Psychol 39(6):841–852
Nicolaides NC, Charmandari E, Chrousos GP, Kino T (2014) Circadian endocrine rhythms: the hypothalamic–pituitary–adrenal axis and its actions. Ann N Y Acad Sci 1318:71–80
Niederhofer H (2012) Treating ADHD with agomelatine. J Atten Disord 16(4):346–348
Niedermeyer E, Naidu SB (1997) Attention-deficit hyperactivity disorder (ADHD) and frontal-motor cortex disconnection. Clin Electroencephal 28:130–136
Noggle C, Dean R, Horton A (2012) The encyclopedia of neuropsychological disorders. Springer, New York
Ottoni GL, Antoniolli E, Lara DR (2012) Circadian preference is associated with emotional and affective temperaments. Chronobiol Int 29:786–793
Owens JA (2009) A clinical overview of sleep and attention-deficit/hyperactivity disorder in children and adolescents. J Can Acad Child Adolesc Psychiatry 18 (2):92–102
Peppers KH, Eisbach S, Atkins S et al (2016) An intervention to promote sleep and reduce ADHD symptoms. J Pediatr Health Care 30:e43–e48
Phillips AJK, Vidafar P, Burns AC et al (2019) High sensitivity and interindividual variability in the response of the human circadian system to evening light. Proc Natl Acad Sci U S A 116(24):12019–12024
Quach J, Hiscock H, Wake M (2012) Sleep problems and mental health in primary school new entrants: cross-sectional community-based study. J Paediatr Child Health 48:1076–1081
Raman S, Coogan AN (2019) Closing the loop between circadian rhythms, sleep, and attention deficit hyperactivity disorder. In: Dringenberg HC (ed) Handbook of behavioral neuroscience, vol 30. Elsevier, pp 707–716
Randazzo WT, Dockray S, Susman EJ (2008) The stress response in adolescents with inattentive type ADHD symptoms. Child Psychiatry Hum Dev 39:27–38
Roenneberg T, Allebrandt KV, Merrow M, Vetter C (2012) Social jetlag and obesity. Curr Biol 22 (10):939–943
Roenneberg T, Wirz-Justice A, Merrow M (2003) Life between clocks: daily temporal patterns of human chronotypes. J Biol Rhythms 18(1):80–90
Ruben MD, Hogenesch JB, Smith DF (2019) Sleep and circadian medicine—time of day in the neurologic clinic. Neurol Clin 37(3):615–629
Rubia K (2018) Cognitive neuroscience of attention deficit hyperactivity disorder (ADHD) and its clinical translation. Front Hum Neurosci 12:100
Russell VA (2007) Neurobiology of animal models of attention-deficit hyperactivity disorder. J Neurosci Methods 161:185–198
Rybak YE, McNeely HE, Mackenzie BE, Jain UR, Levitan RD (2006) An open trial of light therapy in adult attention-deficit/hyperactivity disorder. J Clin Psychiatry 67:1527–1535
Rybak YE, McNeely HE, Mackenzie BE, Jain UR, Levitan RD (2007) Seasonality and circadian preference in adult attention-deficit/hyperactivity disorder: clinical and neuropsychological correlates. Compr Psychiatry 48:562–571
Salardini E, Zeinoddini A, Kohi A, Mohammadi MR, Mohammadinejad P, Khiabany M, Shahriari M, Akhondzadeh S (2016) Agomelatine as a treatment for attention-deficit/hyperactivity disorder in children and adolescents: a double-blind, randomized clinical trial. J Child Adolesc Psychopharmacol 26(6):513–519
Sangal RB, Owens J, Allen AJ, Sutton V, Schuh K, Kelsey D (2006) Effects of atomoxetine and methylphenidate on sleep in children with ADHD. Sleep 29 (12):1573–1585

Sciberras E, Mulraney M, Mensah F, Oberklaid F, Efron D, Hiscock H (2019) Sustained impact of a sleep intervention and moderators of treatment outcome for children with ADHD: a randomised controlled trial. Psychol Med 18:1–10

Sivertsen B, Harvey AG, Pallesen S, Hysing M (2015) Mental health problems in adolescents with delayed sleep phase: results from a large population-based study in Norway. J Sleep Res 24(1):11–18

Smith CS, Reilly C, Midkiff K (1989) Evaluation of three circadian rhythm questionnaires with suggestion for an improved measure of morningness. J Appl Psychol 75:728–738

Smits MG, Nagtegaal EE, van der Heijden J, Coenen AM, Kerkof GA (2001) Melatonin for chronic sleep onset insomnia in children: a randomized placebo-controlled trial. J Child Neurol 16(2):86–92

Snitselaar MA, Smits MG, van der Heijden KB et al (2013) Influence of methylphenidate on circadian rhythmicity and sleep in adult attention-deficit/hyperactivity disorder. Sleep Biol Rhythms 11:282–285

Soehner A, Bertocci MA, Levenson JC et al (2019) Longitudinal associations between sleep patterns and psychiatric symptom severity in high-risk and community comparison youth. J Am Acad Child Adolesc Psychiatry 58–6:608–617

Still G (1902) The Goulstonian lectures on some abnormal psychical conditions in children, Lecture 1. Lancet 1:1008–1012

Sung V, Hiscock H, Sciberras E, Efron D (2008) Sleep problems in children with attention-deficit/hyperactivity disorder: prevalence and the effect on the child and family. Arch Pediatr Adolesc Med 162:336–342

Szeinberg A, Borodkin K, Dagan Y (2006) Melatonin treatment in adolescents with delayed sleep phase syndrome. Clin Pediatr (Phila) 45(9):809–818

Tirosh A, Sadeh R, Munvez P et al (1993) Effects of methylphenidate on sleep in children with attention-deficient hyperactivity disorder An activity monitor study. Am J Dis Child 147:1313–1315

Tjon Pian Gi CV, Broeren JP, Starreveld JS, Versteegh FG (2003) Melatonin for treatment of sleeping disorders in children with attention deficit/hyperactivity disorder: a preliminary open label study. Eur J Pediatr 162 (7–8):554–555

Toplak ME, Jain U, Tannock R (2005) Executive and motivational processes in adolescents with attention deficit-hyperactivity disorder (ADHD). Behav Brain Funct 1(1):8

Tso W, Chan M, Ho FK et al (2019) Early sleep deprivation and attention-deficit/hyperactivity disorder. Pediatr Res 85(4):449–455

Turner-Cobb JM (2005) Psychological and stress hormone correlates in early life: a key to HPA-axis dysregulation and normalisation. Stress 8:47–57

Van der Heijden KB, Smits MG, van Someren EJ, Gunning WB (2005) Idiopathic chronic sleep onset insomnia in attention-deficit/hyperactivity disorder: a circadian rhythm sleep disorder. Chronobiol Int. 22 (3):559–570

Van der Heijden KB, Smits MG, Van Someren EJ, Ridderinkhof KR, Gunning WB (2007) Effect of melatonin on sleep, behavior, and cognition in ADHD and chronic sleep-onset insomnia. J Am Acad Child Adolesc Psychiatry 46(2):233–241

Van Geijlswijk IM, Korzilius HP, Smits MG (2010) The use of exogenous melatonin in delayed sleep phase disorder: a meta-analysis. Sleep 33(12):1605–1614

Van Veen MM, Kooij JJ, Boonstra AM, Gordijn MC, Van Someren EJ (2010) Delayed circadian rhythm in adults with attention-deficit/hyperactivity disorder and chronic sleep-onset insomnia. Biol Psychiatry 67 (11):1091–1096

Van West D, Claes S, Deboutte D (2009) Differences in hypothalamic–pituitary–adrenal axis functioning among children with ADHD predominantly inattentive and combined types. Eur Child Adolesc Psychiatry 18:543–553

Viggiano D, Ruocco LA, Sadile AG (2003) Dopamine phenotype and behaviour in animal models: in relation to attention deficit hyperactivity disorder. Neurosci Biobehav Rev 27:623–637

Virkkunen M (1985) Urinary free cortisol secretion in habitually violent offenders. Acta Psychiatr Scand 72:40–44

Voinescu BI, Szentagotai A, David D (2012) Sleep disturbance, circadian preference and symptoms of adult attention deficit hyperactivity disorder (ADHD). J Neural Transm (Vienna) 119(10):1195–1204

Watanabe Y, Fujita M, Ito Y, Okada T, Kusuoka H, Nishimura T (1997) Brain dopamine transporter in spontaneously hypertensive rats. J Nucl Med 38:470–474

Weiss MD, Wasdell MB, Bomben MM, Rea KJ, Freeman RD (2006) Sleep hygiene and melatonin treatment for children and adolescents with ADHD and initial insomnia. J Am Acad Child Adolesc Psychiatry 45(5):512–519

Xu X, Breen G, Chen CK, Huang YS, Wu YY, Asherson P (2010) Association study between a polymorphism at the 3′-untranslated region of CLOCK gene and attention deficit hyperactivity disorder. Behav Brain Funct 6:48

Yoon SYR, Jain U, Shapiro C (2012) Sleep in attention-deficit/hyperactivity disorder in children and adults: past, present, and future. Sleep Med Rev 16:371–388

How Psychoactive Drugs and the Circadian Clock Are Enlightening One Another

8

Olivia Engmann

Abstract

Psychoactive drugs are a popular way to induce pleasant feelings, but also to modify wakefulness and sleep. In turn, insomnia and circadian often impact on drug-taking behavior. This book chapter explores the interplay between drugs and the circadian system. The reader will be introduced to the main classes of psychoactive drugs and the role they play in circadian pathways and behaviors. The importance of circadian interventions on drug-taking and implications for our society are discussed.

Keywords

Psychoactive drugs · Circadian system · Sleep · Psychedelics · Psychostimulants

8.1 Eat–Prey–Love: A Biological Explanation for Drug Use

Why do humans consume psychoactive drugs? The general answer is: drugs stimulate pathways that feel rewarding or in other ways agreeable. The brain reward system, and other neural circuits that mediate pleasant experiences, evolved to positively reinforce behaviors, which contribute to survival. These systems can be hijacked with drugs to elicit similarly positive feelings without the environmental counterpart that would normally trigger underlying pathways. However, during addiction, drug use itself can become reinforced and drug-taking may eventually become independent of the agreeable experiences they at first evoked (Hyman et al. 2006).

A few basic conserved behaviors, that have contributed to survival in the animal kingdom early on, are:

1. Ingestion of nutritious or health-promoting substances (eat)
2. Fighting off or hiding from predators (prey)
3. Reproduction (love)

Importantly, the availability of food sources, enemies, and mates fluctuates around the clock as animals evolved in a 24 h rhythm of light and darkness. Hence, to maximize success, all three behavioral groups are strongly regulated by circadian rhythms. The circadian regulation of reinforced behaviors is both, determined genetically (mutations in major circadian regulators can disrupt these behaviors)—and epigenetically (through the entrainment by Zeitgebers from the environment including light and, as we will see, drugs).

In this book chapter, I will elucidate how the major classes of psychoactive drugs, despite very different modes of action, are tightly linked to the regulation of the circadian system. In this context,

O. Engmann (✉)
Institute of Human Genetics, University Hospital Jena, Friedrich Schiller University, Jena, Germany
e-mail: olivia.engmann@med.uni-jena.de

O. Engmann, M. Brancaccio (eds.), *Circadian Clock in Brain Health and Disease*, Advances in Experimental Medicine and Biology 1344, https://doi.org/10.1007/978-3-030-81147-1_8

I define a psychoactive drug as any non-vital substance that causes mind-altering changes.

One of the many artifacts of modern society is a deviation from our natural circadian rhythms. These include chronic shifts in the circadian period (getting up earlier or later than usual), altered lengths of active and inactive phases (e.g., sleeping less) as well as rapid and transient alterations in circadian time (often leading to jetlag). I will review data on how these changes in circadian rhythms can affect drug-taking behavior and discuss circadian interventions to reduce drug abuse issues. Lastly, I will make a case that research on psychoactive drugs, in particular, those still understudied, may provide highly valuable tools to improve our understanding of how the circadian system is functioning in the brain.

8.2 Psychoactive Drugs Target Circadian Rhythms Via Reinforcing Brain Circuits

Evidence that drugs and circadian rhythms are directly linked can be found nearly every day: Many of us start our mornings with a cup of coffee and end evenings with a glass of wine. Indeed, most drugs are preferentially used at a certain time of day (Raymond et al. 1992; Logan et al. 2014).

To conceive by what mechanisms drugs interact with circadian signaling—how their consumption is influenced by the circadian phase and alters it in return—it is key to understand what brain systems are most affected by psychoactive drugs. In order to synchronize reinforced natural behaviors with diurnal or circadian changes, these circuits are highly linked to the circadian clock.

8.2.1 Brain Reward System

The reward system of the brain consists of small nuclei in the mesolimbic area, namely the ventral tegmental area, the nucleus accumbens, the striatum, and the habenula (Robison and Nestler 2011; Lecca et al. 2014). They are all connected through the neurotransmitter dopamine. Dopamine is a highly conserved monoamine and can be found in virtually all animals, where it regulates locomotor activity, in order to seek pleasurable stimuli such as mates and food (Baird and Gauvin 2000). As these exhibit rhythmic patterns, dopamine levels undergo circadian oscillations as well (Ferris et al. 2014). However, if highly rewarding stimuli are present, the circadian regulation can be overridden to pursue these rewarding subjects.

The brain reward system can be synchronized with the body's circadian rhythms by melatonin (Uz et al. 2005a) and through regulation by the suprachiasmatic nucleus (SCN). At the same time, the brain reward system contains its own circadian clock machinery (see Chap. 4).

Psychostimulant drugs (e.g., cocaine and amphetamine) induce wakefulness and locomotor performance by increasing synaptic dopamine levels. These drugs simulate rewarding stimuli and can therefore alter circadian signaling in the brain reward system. This effect is so pronounced that it can reinstate circadian rhythms in the absence of SCN activity (Tataroğlu et al. 2006). In consequence, the brain reward system can affect circadian rhythms and sleep as a partially independent circadian regulator (Logan et al. 2014; Baño-Otálora and Piggins 2017).

8.2.2 Opioid System

The brain produces endogenous opioids including beta-endorphin to help relieve pain during strenuous activities that are necessary to survival. These include: exercise while foraging for food (eat) or running from predators (prey), sex, orgasm, or giving birth (love). There are several types of opioid receptors that are located all over the brain, including the amygdala, the cortex, the hypothalamus, the locus ceruleus, and the brain reward system. Both, beta-endorphin and the sleep hormone melatonin are produced by the pituitary gland. However, in contrast to melatonin, beta-endorphin suppresses sleep to increase performance during eat–prey–love (King et al.

1981). The endogenous opioid system can be stimulated by the uptake of opioids such as morphine and heroin.

8.2.3 Raphe Nuclei

The Raphe nuclei are small clusters of cells in the brain stem. They are characterized by synthesizing serotonin and project widely throughout the brain, including to the SCN, where they provide the main non-photic inputs (Glass et al. 2003). Serotonin is an essential neurotransmitter and present in all bilateral animals as well as some plants and fungi. Although best known for its positive effects on mood, serotonin has many functions in the brain. It regulates behaviors as diverse as reward, libido, sociability, and sleep (Siegel and Crockett 2013; Snoeren et al. 2014). Interestingly, serotonin receptors are also linked to the effects of psychedelic drugs including lysergic acid diethylamide-25 (LSD), psilocybin, and mescalin. Despite the anatomical connections between raphe nuclei and the SCN, the effect of serotonergic psychoactive drugs on circadian rhythms is not well investigated.

8.2.4 Cholinergic and Endocannabinoid Systems

Signaling through the neurotransmitter acetylcholine modulates the neural excitability of a variety of brain areas. It, therefore, has broad effects on arousal, reward, attention, cognition, alertness, and sleep (Hobson et al. 1993; Picciotto et al. 2012). Acetylcholine is produced by small nuclei in the brain stem and the basal forebrain. Its targets include the nucleus accumbens and the SCN (O'Hara et al. 1998; Picciotto et al. 2012), which are both involved in circadian regulation. The most popular drug to stimulate the cholinergic system is nicotine.

The endocannabinoid system, which can be altered through the use of cannabis, is spread widely across brain areas including the basal ganglia, the cerebellum, the SCN (Wittmann et al. 2007a; Sanford et al. 2008), and the raphe nuclei (Moldrich and Wenger 2000; Häring et al. 2007; Sanford et al. 2008). It regulates various physiological mechanisms including mood, memory, pain sensation, stress, body temperature, reward, and appetite. Both, endocannabinoids and their induced effects, exhibit circadian rhythms (Santucci et al. 1996; Sanford et al. 2008; Vaughn et al. 2010). In return, endocannabinoids can affect circadian rhythms through melatonin synthesis (Koch et al. 2006, 2008).

8.3 Circadian Patterns Affect Drug Use

8.3.1 Circadian Rhythm Changes May Promote Drug Use

As we have seen, the brain systems regulated by psychoactive drugs are strongly intertwined with the circadian system. It, therefore, comes as no surprise that circadian rhythm changes also affect drug use.

For instance, sleep deprivation can increase nicotine use in humans (Hamidovic and de Wit 2009), shift work can increase the use of alcohol and nicotine (Trinkoff and Storr 1998; Bildt and Michélsen 2002), and patients with insomnia tend to generally use more drugs (Breslau et al. 1996). This may be in part because individuals self-medicate with drugs to induce sleep (sedatives), or to promote wakefulness (psychostimulants) (Teplin et al. 2006). Since drug use is regulated by melatonin in rodents (Brick et al. 1984; Akhisaroglu et al. 2004; Garmabi et al. 2016), altered melatonin levels might mediate an increased drug use in populations with altered circadian rhythms. More studies will be necessary to rule out the indirect effects of stress on drug use, which is also caused by circadian shifts.

8.3.2 The Intoxicated Night Owl: Chronotypes and Drug Use

As is the case for most biological processes, human circadian rhythmicity occurs on a spectrum. Early chronotypes, also called "early birds," rise early in the morning and go to bed early as well. In contrast, late chronotypes or "night owls" (Logan et al. 2014) get up late and fall asleep late. Chronotypes can change during development and aging. A child may start with an early chronotype, exhibit a late chronotype peak during adolescence and then stabilize around age 20. Hence, there are individual as well as population differences in chronotypes. Interestingly, studies consistently show that night-owl chronotypes exhibit a higher intake of a variety of drugs including alcohol, caffeine, and nicotine (Adan 1994; Broms et al. 2011). This may be in part because, in our society, work and learning schedules are adapted to early chronotypes. Adhering to an early-bird schedule can induce a sleep debt in night owls ("social jet lag hypothesis") and it may be this sleep deprivation that puts night owls at a higher risk for drug abuse, and certain mental illnesses (Logan et al. 2014). It has been proposed that social jet lag equals "flying overseas and back each week," inducing profound circadian rhythm alterations on the body and potentially causing chronic stress (Logan et al. 2014). However, future studies on adults should rule out artifacts in the correlation between night-owl chronotypes and drug taking due to societal conventions, as alcohol and other psychoactive drugs are often consumed during the after-work hours and this time window is larger for night-owl chronotypes.

On the other hand, the impact of social jetlag on drug-taking has been extensively studied in adolescents as well, which, as a population, have a later chronotype, yet are usually forced to adhere to a similar school schedule as younger children (Colrain and Baker 2011). More than 10% of adolescents have insomnia (Johnson 2006). Additionally, adolescents are still "naïve" to chronic drug use or are at least socially confined in the use of such, and are therefore ideal models to investigate.

These studies suggest that sleep problems precede the onset of alcohol abuse in adolescents (Shibley et al. 2008; Roberts et al. 2009; Hasler and Clark 2013). In fact, sleep disturbances in early childhood may predict an early onset of alcohol and marijuana use in adolescence (Wong et al. 2004), indicating that sleep and/or circadian rhythm changes are facilitators of drug use. Additionally, the use of alcohol may aggravate sleep problems (Johnson and Breslau 2001), potentially leading to a vicious cycle of sleep deprivation and drug use in adolescents.

8.4 Rodent Studies Mechanistically Link Drugs and Circadian Rhythms

In humans, there is a significant relationship between the use of various drugs and altered circadian rhythms (Vescovi et al. 1992; Drummond et al. 1998; Morgan et al. 2006; Webb et al. 2009; Fakier and Wild 2011). However, human studies are often confounded by the fact that common situations such as stress can promote both insomnia and self-medication with psychoactive drugs. Furthermore, character traits such as impulsivity or comorbid illnesses, e.g., mood disorders, may affect both sleep/circadian rhythms and drug use. In contrast, rodent studies are a simple model to measure the effect of acute or chronic drug administration on circadian rhythms: They enable targeted molecular manipulations that causally link genetic predispositions (e.g., mutations in clock genes) with altered drug-taking behavior. Additionally, animal studies ethically allow drug administration in order to study consequences on circadian systems and behaviors. These studies have provided ample evidence that drugs can affect circadian rhythms. Unfortunately, the field has been biased toward a few drug types including cocaine, alpha-methylphenethylamine (amphetamine), and alcohol. However, in this limited research field, abundant information suggests that these drugs can be regarded as Zeitgeber stimuli that entrain the body clock (Shibata et al. 2010).

In the following, we will explore individual classes of psychoactive drugs in greater depth:

8.5 Psychostimulants Are Zeitgebers

Psychostimulant drugs all hijack the dopamine system to promote wakefulness, euphoria, and locomotor activity. The brain reward system is able to quickly adapt to extreme or chronic stimuli and persistent reward memories can be formed from single events. Hence, drugs that regulate the brain reward system can be habit forming and addictive. Addiction refers to "compulsive seeking and taking of drugs despite horrendous consequences or loss of control over drug use" (Nestler 2013) and is characterized by long-lasting, drug-induced changes. Since the propensity for addiction is only approximately 50% heritable (Nestler 2013), environmental factors, such as altered circadian rhythms, can play a major role.

Moreover, a main characteristic of psychostimulants is their wakefulness-promoting properties, which are strongly linked to circadian rhythms. For that reason, the moderately addictive psychostimulant drugs cocaine and amphetamine have been studied in great depth in the context of circadian rhythms. However, we will see that certain psychostimulants, among them 3,4-methylene-dioxy-metamphetamine (MDMA) and caffeine, are still relatively understudied. More research into their effects on circadian rhythms may lead to a deeper understanding of the brain's circadian regulation.

8.5.1 Classical Psychostimulants: Cocaine, Amphetamine, and Metamphetamine

Cocaine and amphetamines are among the most consumed illegal psychostimulants worldwide. Cocaine is the active component in Coca leaves (*Erythroxylum coca* var. *coca*), which have been used for thousands of years by South American indigenous populations for their appetite suppressing, performance-boosting properties. Amphetamine was first synthesized in Germany in the late nineteenth century and initially found its medical use in Benzedrine nose decongestants. Metamphetamine was synthesized independently in Japan shortly afterward. Cocaine and amphetamines increase the levels of dopamine, but also serotonin and norepinephrine, by inhibiting their reuptake from the synapse. They are now heavily controlled substances but remain popular for recreational use.

Behavioral studies on rodents show that cocaine intake strongly changes over the course of the day (Uz et al. 2003), indicating that psychostimulant use is affected by the circadian clock. Melatonin, the circadian "hormone of darkness", accumulates toward the end of the active phase, when it reduces the motivation to self-administer cocaine and prevents relapse in rats (Takahashi et al. 2017). Additionally, shorter photoperiods suppress reinstatement of cocaine-seeking (Sorg et al. 2011). Evidence for a molecular link to the circadian system comes from genetic studies, where the presence of a gene variant of the circadian regulator Per2 correlated with an increased vulnerability to cocaine addiction in humans (Shumay et al. 2012). In another study, the expression of the *Per1*-gene positively correlated with cocaine sensitization (Uz et al. 2003). Altogether, the relevance of Clock genes on cocaine-related behaviors has been characterized in great depth (Tables 8.1 and 8.2).

In turn, cocaine and amphetamine also affect the circadian system. On the anatomical level, the SCN is regulated by dopaminergic innervations arriving from the ventral tegmental area (VTA). Additionally, cocaine decreases the serotonin transport in the SCN (Prosser et al. 2008, 2014; Glass et al. 2012), thereby affecting SCN-oscillations.

The SCN is only one brain region in which cocaine (and amphetamine) alter circadian signaling. Circadian genes are also expressed locally within dopamine-receptor-containing neurons of the brain reward system, which are the main targets of psychostimulants (McClung 2007; Sleipness et al. 2007; Luo and Aston-Jones 2009). Indeed, the brain reward system exhibits circadian patterns of neural firing activity and transcription (McClung 2007). In arhythmic animals, such as animals with SCN lesions,

Table 8.1 Clock genes and their impact on drug use

Gene	Model	Drug	Effect	Reference
Per1	*Per1Brdm1*-mutant mice	Cocaine	No CPP	Abarca et al. (2002)
Per1	*Per1Brdm1*-mutant mice	Alcohol	No effect	Zghoul et al. (2007)
Per1	Masspec (Human)	Cannabis	SNP is associated with consumption	Saffroy et al. (2019)
Per2	*Per2Brdm1*-mutant mice	Cocaine (acute)	Increases non-photic phase-advance response at midday	Brager et al. (2013)
Per2	*Per2Brdm1*-mutant mice	Alcohol (acute, chronic)	Deletion abolishes alcohol-induced endorphin release	Agapito et al. (2010)
Per2	*Per2Brdm1*-mutant mice	Alcohol (acute)	Increases alcohol seeking and consumption	Spanagel et al. (2005b); Brager et al. (2011)
Per2	*Per2Brdm1*-mutant mice	Morphine	Reduces tolerance to withdrawal	Perreau-Lenz et al. (2010)
Per2	PCR (human)	Alcohol	Alcohol use in adolescent boys	Comasco et al. (2010)
Per2	PCR (human)	Alcohol	High levels of alcohol drinking in alcoholics	Spanagel et al. (2005b)
Per2	PCR (human)	Alcohol	Alcoholism risk	Kovanen et al. (2010)
Per3	PCR (human)	Heroin	Increases addiction risk	Zou et al. (2008)
Per3	QTL mapping (human)	Alcohol	Per3 variant is associated with increased alcohol intake	Wang et al. (2012)
Clock	*ClockΔ19*-mutant mice	Cocaine	Increases CPP and SA	McClung et al. (2005); Ozburn et al. (2012)
Clock	*ClockΔ19*-mutant mice	Alcohol	Increases alcohol sensitivity and consumption	Ozburn et al. (2013)
Clock	*ClockΔ19*-mutant mice	Nicotine	No effect	Bernardi and Spanagel (2013)
Bmal1	PCR (human)	Alcohol	Alcoholism risk	Kovanen et al. (2010)

CPP, conditioned place preference; SNP, single nucleotide polymorphism; SA, self-administration

metamphetamine can restore circadian locomotor rhythmicity (Tataroğlu et al. 2006). Metamphetamine also enhances the amplitude of locomotor rhythms in a normal L:D setting (Honma et al. 1986), supporting the hypothesis that the dopaminergic system may have some circadian regulatory function beyond the SCN. Chronic administration of psychostimulants is also “speeding up” the entrainment from constant darkness to a daily pulse of 1 min dim light (Stowie et al. 2015) and lengthens the free-running period under constant darkness conditions (Stowie et al. 2015). Furthermore, psychostimulants disrupt sleep (Morgan et al. 2006; Angarita et al. 2016). Taken together, dopaminergic psychostimulants clearly induce effects on circadian rhythms that are, at least in part, independent of the SCN.

Interestingly, the effect of cocaine on the circadian system may persist across generations. In rats, chronic paternal cocaine administration leads to differences in light-induced phase shifts in the offspring in a sex-dependent manner (Yaw et al. 2018).

8.5.2 Caffeine: A Little Investigated Influencer of Mood and the Clock

Caffeine, the stimulant component in coffee beans, is the world’s most consumed psychoactive drug. Approximately 80% of the population habitually uses caffeine, contained in coffee or energy drinks. It is ingested to induce wakefulness and may give a slight feeling of euphoria—hence coffee has been termed “concentrated sunshine” by the discoverer and scientist Alexander von Humboldt (Letter to HC Schumacher 1836). In healthy populations, side effects of caffeine are

Table 8.2 Impact of psychoactive drugs on Clock genes

Drug	Model	Gene	Brain region	Reference
Cocaine (acute)	Mouse	Up: PER1	Striatum	Falcon et al. (2013)
Cocaine (chronic)	Mouse	Up: NPAS2, PER1-3	Striatum, NAc	Falcon et al. (2013)
Cocaine (chronic)	Mouse	Up: Per1, Per2 Down: Bmal1, Cry1	Hippocampus	Uz et al. (2005b)
Cocaine (chronic)	Mouse	Up: Per1, Clock	Striatum	Uz et al. (2005b)
Cocaine (chronic)	Rat	Up: Per2, Cry1, Bmal1, Clock	Dorsal striatum	Lynch et al. (2008)
Amphetamine (chronic)	Rat	Shifted Per1, Per2 phase	Striatum	Wongchitrat et al. (2013)
Metamphetamine (Chronic/acute)	Mouse, Rat	Up: Per1	SCN	Iijima et al. (2002); Sanchis-Segura et al. (2009); Webb et al. 2009; Nikaido et al. (2018)
Metamphetamine (Chronic/acute)	Mouse, rat	Per1, Per2, Bmal1, Reverbα	Striatum	Iijima et al. (2002); Sanchis-Segura et al. (2009); Webb et al. (2009); Nikaido et al. (2018)
Caffeine (acute)	Mouse	Altered CLOCK: BMAL1 binding to chromatin	Striatum	Trautmann et al. 2020
MDMA (acute)	Rat	Up: Per1, Per2	SCN	Ogeil et al. (2012b)
CBD (acute)	Mouse	Up: Per1, Bmal1, Cry2 Down: Clock	BV-2 cell line	Lafaye et al. (2018)
Morphine (acute)	Rat	Up: Per1	SCN (dark phase), blood	Pačesová et al. (2015)
Morphine (chronic)	Rat	Up: Per1, Per2	Frontal cortex, striatum	Ammon-Treiber and Höllt (2005)
Morphine (chronic)	Rat	Down: Clock	SCN	Li et al. (2009)
Morphine (withdrawal)	Rat	Up: Per2 mRNA and protein	Forebrain, frontal cortex	Ammon et al. (2003); Ammon-Treiber and Höllt (2005); Hood et al. (2011)
Morphine (withdrawal)	Rat	Down: Per1	Amygdala, NAc core	Li et al. (2009)
		Down: Per2	Amygdala, NAc core, blood, VTA	
		Down: Clock	NAc core, amygdala, blood	
		Up: Per1	Hippocampus	
		Up: Clock	Hippocampus	
Alcohol (chronic)	Rat	Flattened rhythms Per1-3	Arcuate nucleus	Chen et al. (2004)
Alcohol (chronic)	Rat	Reduced Per1-3	SCN	Chen et al. (2004)
Alcohol (chronic)	Mouse	Up: Bmal1, Per2	NAc	Melendez et al. (2012)
Alcohol (chronic)	Mouse	Down: Clock	NAc, VTA	Ozburn et al. (2013)
Alcohol (chronic)	Human	Down: Clock	Blood mononuclear cells	Huang et al. (2010)
LSD	Rat	Up: Per1, Per2	Medial prefrontal cortex	Martin et al. (2014)

mild and may actually be beneficial, since caffeine increases attention, cognitive function (Lara 2010), motivation and self-confidence (Fredholm et al. 1999), might reduce the risk for depression (Smith 2009) and suicide (Klatsky et al. 1993; Kawachi et al. 1996). Interestingly, both early and late chronotypes consume caffeine (Adan 1994; Suh et al. 2017; Treur et al. 2018).

Caffeine acts by antagonizing the receptors for adenosine, especially the adenosine 1- and adenosine 2A-receptors (A1A-R/A2A-R). A1A-R can dimerize with the dopamine-1 receptor and A2A-R with the dopamine-2 receptor. Hence, caffeine effects may be indirectly mediated by dopaminergic signaling as well. Moreover, in the brain reward system, A2A-R can heterodimerize with mGluR5 receptors (Ferre et al. 1991; Lindberg et al. 2018), which are known to affect circadian rhythms (Ahnaou et al. 2015). To a lesser extent, caffeine also has serotonergic and cholinergic effects (Fredholm et al. 1999) and can inhibit phosphodiesterase enzymes (Ribeiro and Sebastio 2010).

Extracellular adenosine accumulates toward the end of the active phase and during sleep deprivation, when it induces drowsiness. Adenosine levels diminish again during sleep. Circadian changes also affect levels of adenosine receptors themselves (Virus et al. 1984).

It is well known that caffeine increases daytime wakefulness and improves work performance during night-shift work or sleep deprivation (Penetar et al. 1993). Caffeine also consistently delays the onset and duration of sleep (Landolt et al. 1995; Clark and Landolt 2017), especially when consumed shortly before going to bed. There is an increased frequency of awakenings and stage-1 sleep earlier in the night as well as reduced stage-3 and 4 sleep. However, data conflicts in whether caffeine consumption increases sleepiness the next day, perhaps because, under normal circumstances (Clark and Landolt 2017), most people avoid consuming caffeine directly before bedtime.

There is good evidence that caffeine affects circadian rhythms: Caffeine can decrease melatonin levels or delay melatonin rhythms in humans when administered in the evening (Shilo et al. 2002; Burke et al. 2015), suggesting that caffeine may affect the circadian phase on a molecular level. In mice, both acute and chronic oral caffeine consumption potentiated photic phase-delays (Ruby et al. 2018). This effect was mimicked by an A1-R but not an A2A-R antagonist (Ruby et al. 2018). Caffeine modulates shifts in sleep cycle (Lindberg et al. 2018), potentiates light-induced phase delays and phase advances (Jha et al. 2017). This may be mediated, at least in part, by activation of A1-R and subsequent inhibition of glutamate into the SCN, which attenuates photic phase-resetting (Watanabe et al. 1996; Elliott et al. 2001; Hallworth et al. 2002; Sigworth and Rea 2003). Caffeine also lengthens the endogenous circadian period in rodents (Oike et al. 2011; van Diepen et al. 2014) and increases light-responsiveness in the SCN (van Diepen et al. 2014). In blind individuals, caffeine did not correct circadian disorders, although it improved daytime alertness (St. Hilaire and Lockley 2015).

Although these data suggest that caffeine affects circadian rhythms, surprisingly little is known about the relationship between caffeine and the circadian clock system on a molecular level. In particular, a clear link between the A1-R, A2A-R, and clock genes is largely missing in vivo, although data from cell culture systems show, that caffeine is capable of increasing the circadian period of clock genes (Burke et al. 2015).

However, recent evidence shows that caffeine may affect circadian rhythms in the striatum via a key molecule of the brain reward system, Dopamine and cAMP-regulated phosphoprotein 32kD (DARPP-32) (Trautmann et al. 2020): CLOCK and DARPP-32 directly bind to one another, and this interaction disrupts CLOCK:BMAL1 binding to chromatin. Specifically, caffeine induced the phosphorylation of DARPP-32 on threonine 75 (Lindskog et al. 2002), which reduced CLOCK:BMAL1 binding. Furthermore, we observed that this effect was blocked in a diurnal manner and in mice carrying the T75A-variant of DARPP-32.

Furthermore, caffeine's rapid mood-elevating effects and gene expression changes in the striatum were similarly affected by the light-phase and T75A-DARPP-32 mutation. This suggests a diurnal regulation of the CLOCK: BMAL1 complex by caffeine via DARPP-32. However, more research is necessary to untangle acute, psychostimulant effects and chronic actions of caffeine on the circadian system.

8.5.3 MDMA May Disrupt Circadian Rhythms Via Serotonin

MDMA, or ecstasy, was first synthesized in Germany in the early twentieth century in search of a blood coagulant drug. As in the case of amphetamine and LSD, its psychoactive properties were later discovered by serendipity. MDMA increases the levels of dopamine, norepinephrine, and serotonin. The serotonergic effects of MDMA are stronger than for the other psychostimulant drugs. Besides its psychostimulant properties, MDMA also induces feelings of sociability, disinhibition, and relaxation.

Regular ecstasy users tend to show impaired sleep patterns (Allen et al. 1993; McCann et al. 2000; Parks and Kennedy 2004; Montgomery et al. 2007; Carhart-Harris et al. 2009; Fisk and Montgomery 2009; Ogeil et al. 2012a). For instance, MDMA users report reductions in total sleep and non-REM sleep (Allen et al. 1993), less stage-2 sleep and more stage-1 sleep than controls (McCann et al. 2007). This may be in part due to dopaminergic effects, and partially because MDMA is a neurotoxin for serotonergic axons projecting from the raphe nuclei to the SCN (Mamounas et al. 1991; McCann and Ricaurte 2007), so that SCN signaling, and in consequence, sleep regulation may be impaired (Prosser et al. 1993).

The effects of MDMA on the circadian system are only beginning to be understood. In rodents, an acute dose of MDMA increased motor activity, wakefulness, and sleep latency for at least 6 hours. Some changes in locomotor activity persisted for up to several days. Increased activity was measured during the resting phase and decreased during the active phase, suggesting that the amplitude of circadian locomotor activity rhythms may have been reduced (Balogh et al. 2004). Certain sleep alterations persisted for up to a month after MDMA treatment. In constant darkness, acute MDMA administration increased the duration of the active phase for at least 1 week (Ogeil et al. 2010). Furthermore, chronic MDMA intake may reduce the ability of serotonin to phase-shift the circadian clock (Biello and Dafters 2001; Colbron et al. 2002; Dafters and Biello 2003; Gardani et al. 2005), perhaps through damaging serotonergic projections. While these data delineate that MDMA does affect circadian rhythms, more studies will be necessary to establish precisely, how this affects the health of MDMA users.

8.6 Opioids Dampen Circadian Rhythms

Opioids were originally extracted from the opium poppy *Papaver somniferum* and may have been used for their anesthetic and sedative properties for more than 7000 years (Kislev et al. 2004). They have also been abused for at least several hundred years since the East India Company promoted opium use in Asia. The most consumed opioids today are morphine and heroin, which not only provide relief from pain and induce euphoria but are also highly addictive and can be lethal when overdosed.

Opioids tap into the endogenous opioid system, which regulates pain, nociception, and stress responses. They indirectly regulate GABAergic, glutamatergic as well as dopaminergic signaling (Mazei-Robison and Nestler 2012). The release of endogenous opioids is regulated by melatonin (Miguel Asai et al. 2007) and follows circadian patterns (Naber et al. 1981; Przewłocki et al. 1983; Asai et al. 1988, 1998).

Opioid use is associated with multiple sleep-wake disturbances (Oyefeso et al. 1997). Besides daytime fatigue (Cao and Javaheri 2018), they also negatively affect sleep architecture, which may be in part due to side effects on the respiratory system (Cheatle and Webster 2015). For instance, opioids can reduce deep sleep and increase stage-2 sleep (Dimsdale et al. 2007).

Opioids also decrease the amplitude of circadian cortisol release in humans (Facchinetti et al. 1985) and disrupt circadian rhythms in rodents. Heroin decreases the amplitude of circadian changes in body temperature (Thronhill et al. 1976). Circadian behaviors are altered by morphine (Stinus et al. 1998; Caillé et al. 2002) and stress-induced effects on sleep and circadian rhythms can be reduced through the administration of a kappa-opioid receptor antagonist (Wells et al. 2017). In turn, mutations in Clock genes appear to be related to opioid use in humans and animal models, suggesting that an altered circadian system may in part underlie the susceptibility for opioid use (Tables 8.1 and 8.2). Taken together, there is some evidence for a relationship between circadian rhythms and opioids, although more in-depth studies are necessary.

8.7 Nicotine Use Can Be Relaxing but Disturbs the Circadian Cycle

Nicotine is the primary psychoactive component in tobacco leaves. It is moderately addictive and can induce relaxation, suppress appetite, improve attention and memory. The main pharmacological target of nicotine is the nicotinic acetylcholine receptor, which is widely distributed in the brain. Nicotine's rewarding properties are mediated by activating VTA to NAc signaling (Pich et al. 1997), while its' cognitive effects are largely mediated by the cerebral cortex (O'Hara et al. 1998).

Cholinergic influences on the circadian system are well established: Acetylcholine levels show strong circadian oscillations, caused largely through changes in the synthesizing enzyme choline acetyltransferase as well as degradation by acetylcholinesterase (Hut and Van der Zee 2011). The SCN is innervated by cholinergic neurons and in some mammals even contains some itself (Hut and Van der Zee 2011). In consequence, nicotine can alter the neural activity in the SCN (Miller et al. 1987). As a result, nicotine may phase-advance SCN firing during the resting phase in mouse brain slices (Trachsel et al. 1995; O'Hara et al. 1998), although another study debates this (Liu and Gillette 2018). Furthermore, nicotine alters the circadian expression of dopamine and serotonin (Pietilä et al. 1995) and may affect melatonin excretion (Ferguson et al. 1999). In rats, chronic nicotine increases the locomotor activity in the light (resting) phase but not in the dark phase (Morley and Garner 1990).

In humans, acute nicotine decreases the total sleep time, sleep efficiency, and amount of REM sleep (Davila et al. 1994; Gillin et al. 1994). Chronic nicotine may delay falling asleep, affect staying asleep, and leads to greater daytime sleepiness (Phillips and Danner 1995). There is a link between the Night owl-chronotype and smoking (Wittmann et al. 2006) and smoking occurs in a diurnal pattern (Jarvik et al. 1993). Interestingly, while no effect of nicotine on clock genes has been demonstrated yet, the expression of *Per2* and *Nr1d1* in the brain is increased through vapors of propylene glycol and glycerol, which occur in e-cigarettes (Lechasseur et al. 2017). These factors need to be taken into consideration, when designing future experiments. Notably, second-hand smoke is associated with insufficient sleep in non-smokers as well, suggesting that the effect is due to tobacco smoke itself and not to comorbidities between sleep and smoking-related behaviors (Sabanayagam and Shankar 2011).

8.8 Cannabinoids Have Mixed Effects on Circadian Rhythms

Cannabis, made from the flowers of the Asian hemp plant *Cannabis sativa*, is the second most inhaled psychoactive drug after nicotine. Cannabis has been harvested throughout recorded history (Abel 1943). Two main psychoactive compounds have been extracted from cannabis, tetrahydrocannabinol (THC) and cannabidiol (CBD). Both components mimic the endocannabinoid system. There are two endocannabinoid receptors: The CB1 receptor is mostly located in the brain, while the CB2 receptor can be found peripherally in the body. THC is a partial CB1/2-receptor agonist, whereas cannabidiol is a mild CB1/2 receptor antagonist. THC and CBD occur in cannabis in varying amounts.

Very few studies have investigated the effect of CBD and THC on circadian rhythms, although it has been shown that chronic THC affects circadian rhythms of the core brain temperature (but not the core body temperature) in rats (Perron et al. 2001). In vitro data suggest a link between circadian genes and acute CBD administration (Lafaye et al. 2018). More studies have assessed the relationship between cannabinoids and sleep:

The effect of cannabis on sleep is mixed, which may in part be due to varying doses and cannabis composition (Babson et al. 2017). However, sleep disturbances are the most consistent symptom of cannabis withdrawal and poor sleep can predict relapse (Babson et al. 2017). Cannabis use is increased in individuals with erratic sleep or a Night owl-chronotype (Kervran et al. 2015; Nguyen-Louie et al. 2018).

Different cannabinoids may have opposing effects on sleep, which may lead to conflicting data on cannabis in human studies: Acute THC decreases REM sleep and increases slow-wave sleep (Vaughn et al. 2010), decreases stage-3 sleep, and may increase sleepiness the next day (Nicholson et al. 2004). However, in patients with posttraumatic stress disorder, chronic THC led to a reduction in the frequency of nightmares (Roitman et al. 2014). There is preliminary evidence that THC improves sleep apnea (Babson et al. 2017). Additionally, dronabinol, the medical term for THC, appears to improve circadian symptoms in dementia patients (Walther et al. 2006).

CBD on the other hand elicits differential effects on sleep, based on its dose, in both, humans and rodents: Low-dose CBD has a stimulating effect, while high-dose CBD has a sedating effect (Nicholson et al. 2004; Zuardi 2008; Carlini and Cunha 2013; Chagas et al. 2013). Similar to THC, CBD can improve sleep disturbances related to posttraumatic stress disorder (Shannon and Opila-Lehman 2015).

In summary, cannabinoids do have an effect on sleep, but this may be, at least to some degree, due to their anxiolytic effects, e.g., reduced nightmares, as well as the improvement of physiological factors such as sleep apnea. Effects of cannabis use and cannabinoids on circadian rhythms and the circadian CLOCK system should be investigated in more depth in the future.

8.9 Alcohol Use and Circadian Rhythm Changes Are Mutually Disruptive

Alcohol is the colloquial term for ethanol, which is typically contained in dilutions between 5 and 50% in alcoholic beverages. Multiple neurotransmitter systems are regulated by alcohol, including glutamate, GABA, dopamine, ATP, beta-endorphin, and adenosine (Lindberg et al. 2018). Alcohol is a sedative, but depending on the concentration and time course it can induce other effects, such as disinhibition, mild mania, aggression, and sentimentality (Colrain et al. 2014). Some effects may be due to the metabolites of ethanol, e.g., aldehydes.

Many studies have explored a connection between alcohol and sleep, as there is strong comorbidity between sleep disturbances and alcoholism (Chakravorty et al. 2016). The vast majority of alcoholics report insomnia (Arnedt et al. 2007), display decreased short wave sleep (Colrain et al. 2014) and other changes in sleep architecture (Landolt and Gillin 2001). An acute dose of alcohol prior to bedtime decreases sleep onset but reduces REM sleep, and increases wakefulness and stage-1 sleep later in the night (Colrain et al. 2014). Hence, a subjective reduction in sleep latency through alcohol may actually go hand in hand with impaired sleep quality. Data suggest that alcohol is sometimes used to self-medicate sleep problems (Arnedt et al. 2007), but these problems may indeed be exacerbated by alcohol use, potentially leading to a vicious cycle.

Besides sleep, chronic alcohol consumption also affects circadian rhythms in humans. Body temperature rhythms are altered by alcohol, both through central and peripheral effects

(Wasielewski and Holloway 2001), and alcohol shifts rhythms in cholesterol, glucose, potassium, and lactic acid (Rajakrishnan et al. 1999). Alcohol reduces the number of neurons expressing arginine vasopressin, vasoactive-intestinal polypeptide, and somatostatin in the SCN (Madeira et al. 1997; Madeira and Paula-Barbosa 1999). Melatonin secretion is inhibited by alcohol in humans (Rupp et al. 2007), while melatonin, in turn, reduces the relapse to alcohol (Vengeliene et al. 2015).

Business travelers and shift workers, who suffer from frequent circadian alterations, are at a higher risk for alcohol abuse (Trinkoff and Storr 1998; Rogers and Reilly 2002). Adults with a Night owl-chronotype have a higher alcohol intake (Wittmann et al. 2006). Insomnia and other sleep problems also interfere with the recovery from alcohol dependence (Arnedt et al. 2007; Brower et al. 2011; Brower 2015). A possible genetic link between altered Clock-signaling and alcohol abuse may partially explain why sleep, circadian rhythm changes, and alcohol use are connected (Table 8.1). Animal models with disturbed circadian cycles increasingly consume ethanol, providing a causal and directional link between both mechanisms (Rosenwasser 2015).

Animal models have provided a strong directional impact of alcohol intake on circadian rhythms as well (Spanagel et al. 2005a): Chronic alcohol alters the circadian phase adaptation to light changes ("photic phase resetting") (Baird et al. 1998; Rajakrishnan et al. 1999; Rosenwasser et al. 2005; Glass et al. 2009). Alcohol intake also shortens the free-running period of mice (Seggio et al. 2009). Additionally, alcohol shifts and shortens circadian rhythms of the core body temperature in rodents (Baird et al. 1998) and disrupts circadian fluctuations in melatonin levels (Kühlwein et al. 2003; Conroy et al. 2012). Furthermore, chronic alcohol dampens circadian changes in plasma corticosterone levels in mice (Kakihana and Moore 1976; Tabakoff et al. 1978). Accordingly, alcohol induces multiple alterations in sleep and circadian rhythms. These effects may in part be mediated by alcohol's effect on astrocyte-mediated glutamatergic signaling (Rossetti and Carboni 1995; Dahchour and De Witte 1999; Nam et al. 2011; Wu et al. 2011), which regulate circadian rhythms (Brancaccio et al. 2017). Additionally, alcohol treatment may indirectly affect phase delays of SCN firing via GABAergic (McElroy et al. 2009; Prosser and Glass 2009) or serotonergic signaling (Prosser et al. 2008).

Taken together, there is a reciprocal connection between circadian rhythms and alcohol. Sleep problems or altered circadian cycles increase the risk for alcoholism, in part because of attempted self-medication. In turn, alcohol consumption impairs circadian/sleep quality further, which can provoke a vicious cycle and, in the worst case, addiction (Rosenwasser 2015).

8.10 Understudied Distorters of Time: Psychedelics

Psychedelic drugs are grouped together because of their hallucinogenic effects. The main psychedelic drugs in use in the Western world are LSD (first synthesized in Switzerland in the mid-twentieth century), psilocybin (the main active compound in magic mushrooms, e.g., *Psilocybe spec.*), and mescaline (the main active ingredient in peyote cactus, *Lophophora williamsii*). Their main target is the serotonin 2A-receptor (5-HT_{2A}-R) (Kovacic and Somanathan 2009). Furthermore, they can have effects on dopaminergic signaling (Ferri et al. 1977; De Gregorio et al. 2016); however, they are deemed non-addictive. While psychedelics may cause adverse reactions such as anxiety, paranoia, and delusions (Passie et al. 2008), they are ranked as one-tenth as harmful as alcohol (Nutt et al. 2010). Approximately 17% of US-Americans of age 26 and older have experimented with hallucinogenic drugs—as many as have tried cocaine (NIDA 2019).

Since both, serotonergic and dopaminergic signaling are involved in circadian rhythms, and this class of drugs is relatively safe, it is surprising how few studies have explored the effect of psychedelic drugs on circadian rhythms. An early study performed on animals suggests that an acute administration of mescaline and LSD may abolish circadian measures in the Y-maze behavioral paradigm in rats (Davies and Redfern 1973). Moreover, LSD may increase wakefulness and decrease slow-wave sleep, both in rats and cats (Depoortere and Loew 1971; Stern et al. 1972). More recently, RNA-sequencing on rat medial prefrontal cortex has demonstrated an upregulation of Per1 and Per2 after a chronic administration of high doses of LSD (Martin et al. 2014).

Psychedelic drugs are also of special interest because they alter subjective measures of time perception, or "interval timing" (Boardman et al. 1957; Aronson et al. 1959). This effect occurs already at low doses of LSD, in which hallucinogenic properties do not occur, so that these effects may be differentiated (Coull et al. 2011; Yanakieva et al. 2018). Both, LSD and psilocybin lead to impaired estimations of long suprasecond intervals (Aronson et al. 1959; Wittmann et al. 2007b).

Interval timing and circadian rhythms have partially overlapping neural circuits (Petersen and Mistlberger 2017): Interval timing is regulated by the prefrontal cortex and basal ganglia, relies on the dopaminergic activity of the corticostriatal circuits and interactions with serotonergic as well as glutamatergic signaling (Cheng et al. 2007a; Coull et al. 2011). Mechanisms of interval timing (short-duration estimation) and circadian free-running behaviors (long-duration estimation) are partially intertwined and this connection is not sufficiently understood. For instance, CLOCK protein is not necessary for interval timing (Cordes and Gallistel 2008) but interval timing is regulated by circadian rhythms, as interval timing differs along the circadian cycle and is completely lost in conditions of constant light as well as in circadian de-synchronizations (Agostino et al. 2011). This could potentially be explained by a 5-HT2A-mediated inhibition of dopamine (Rammsayer 2008; Wackermann et al. 2008; De Gregorio et al. 2016). Moreover, both interval timing and circadian rhythms are affected by cocaine, metamphetamine (Cheng et al. 2006, 2007b; Matell et al. 2006; Williamson et al. 2010; Heilbronner and Meck 2014) and to a small extent by nicotine (Daniels et al. 2015).

It is intriguing that there is a link between the short-term and longer-term timing circuits, similar to the short and long hands of a clock. Psychedelic drugs may be one tool to better understand this connection. Additionally, future studies should investigate, whether psychedelic drugs affect photic phase-shifts, free-running periods, and molecular circadian rhythm markers such as melatonin, corticosterone, and clock genes.

There is an increasing trend to chronically use psychedelic drugs in small, hardly noticeable non-hallucinogenic doses, referred to as microdoses, to self-medicate mood disorders (Rucker et al. 2016) and to increase creativity (Prochazkova et al. 2018; Anderson et al. 2019). In consequence, it will be duly important to understand potential side effects, such as altered sleep and circadian rhythms, which may arise for users.

Furthermore, several studies have suggested that psychedelics may reduce relapse to various drugs of abuse (Nichols et al. 2017). However, more research is necessary to understand the underlying molecular mechanisms, their link to the circadian system, and exact impacts on health.

8.11 Circadian Approaches to Ameliorate Drug Abuse Issues

The strong mutual connection between psychoactive drugs and the circadian system suggests that drug abuse issues may be ameliorated by circadian interventions. Since sleep disturbances are a common symptom during drug withdrawal (McGregor et al. 2005; Syed and Keating 2013; Brower 2015), improved sleep may for instance be beneficial to prevent the relapse to cannabis (Haney 2009). However, to date, few studies specifically treat insomnia and investigate the outcome on cannabis relapse. Furthermore, it has been shown that, while poor sleep in abstinent alcoholics is predictive of relapse (Brower 2015; Dolsen and Harvey 2017), treating insomnia in these patients does not reduce the risk for relapse—it was suggested that insomnia and alcoholism should be regarded as two separate but highly comorbid disorders (Brower 2015). In some cases, pharmacological compounds that improve relapse (for instance, naltrexone for opioid addiction), may even cause insomnia as a side effect (Syed and Keating 2013). Hence, more research needs to be done to understand how sleep/circadian rhythm disturbances may contribute to the risk for relapse (Table 8.3).

Nevertheless, treating insomnia in patients that are recovering from drug dependency is generally recommended (Jones et al. 2003). Importantly, adjusting circadian rhythms to an individual optimum may help prevent drug abuse in the first place, given that sleep problems are predictive of later drug use (Breslau et al. 1996; Wong et al. 2004; Hasler et al. 2016; Miller et al. 2017). Several options are available to improve circadian rhythms and sleep:

Chronotherapy is currently being used in the field of psychiatry to improve symptoms of seasonal affective disorder and depression (Khalifeh 2017). Chronotherapy is a safe manipulation of the circadian system through light exposure, sleep deprivation, or phase shifts. Some advances have been made in patients with circadian sleep- and mood- disorders. However, while the use of chronotherapy in the context of drug addiction has repeatedly been suggested (Arnedt et al. 2007; Brower et al. 2011), to our knowledge, no progress has been made in this field. Given the diversity of drug effects on circadian rhythms and vice versa, chronotherapy would have to be tailored to the type of drug addiction.

Sleep and circadian rhythmicity may be ameliorated by *changes in daily schedules.* Since our schedules are only partially determined by ourselves, this issue needs to be addressed by policy makers as well.

Table 8.3 Summary of interactions between psychoactive drugs and the circadian system

Drug	Sleep/ wakefulness	Chronotype	Known effect on circadian markers	Link with CLOCK genes?
Cocaine/ amphetamine	Wakefulness: up Sleep: down	Unknown	– Meth. restores rhythms in absence of SCN – Coc. facilitates phase entrainment – Coc. lengthens free running period	Yes
Caffeine	Wakefulness: up Sleep: down	All	– Reduces melatonin – Lengthens free running period – Potentiates photic phase shifts	Yes
MDMA	Wakefulness: up Sleep: down	Unknown	– Lengthens active phase – Reduces 5-HT induced phase shifts	Yes
Opioids	Wakefulness: down Sleep: down	Unknown	– Reduces circadian amplitude of cortisol, body temperature, behaviors	Yes
Nicotine	Wakefulness: unclear Sleep: down	Late	– Phase-advances SCN firing – Alters circadian expression of 5-HT, melatonin & dopamine – Increases locomotor activity in resting phase	Unknown
Cannabis	Wakefulness: down Sleep: unclear	Late	– THC alters core brain temperature rhythms	Yes
Alcohol	Wakefulness: down Sleep: down	Late	– Attenuates phase advances – Alters rhythms in body temperature, corticosterone, cholesterol, glucose, potassium & lactic acid	Yes
Psychedelics	Wakefulness: up Sleep: down	Unknown	– Reduces circadian behaviors in Y-maze	Yes

Meth, methamphetamine; coc, cocaine; "Sleep" can refer to sleep quality and/or duration

Adolescents are particularly vulnerable to drug use and show a high level of sleep disturbances. High school start times should be adjusted to accommodate the evening chronotypes of adolescents. Currently, less than 10% of teenagers get the recommended amount of sleep per night (Wheaton et al. 2016). As on average, teenagers tend toward a night-owl chronotype, delaying high school start times is consistently associated with increased sleep time and improved mental health, better performance in school, and fewer risk-taking behaviors such as drug abuse (Wheaton et al. 2016; Wahlstrom and Owens 2017).

Similarly, work schedules for adults should be made more flexible. Social jetlag in night owl-chronotypes may be reduced, which is likely contributing to their increased risk for drug use. Providing periods of stable circadian rhythms for shift workers or frequent long-distance travellers (e.g., switching between ground- and airplane duty for airline personnel) might ameliorate some of the associated mental health risks.

Lifestyle interventions such as Apps for filtering blue light from technical devices in a circadian manner should be installed and light-therapy lamps may be provided in workplaces with a lot of time spent indoors.

Treatment of comorbid illnesses that may contribute to insomnia, including depression and posttraumatic stress disorder, or physical issues such as sleep apnea, may ameliorate sleep symptoms as well as relapse in patients with drug abuse issues (Brower 2015).

Cognitive behavioral therapy may be helpful in treating insomnia in abstinent patients and is preferred to medical approaches, because the results are more long lasting and do not support the habit of modifying mental problems with pharmacological interventions (Brower 2015).

Pharmacological treatment of insomnia should remain a last resort (Jones et al. 2003). Melatonin agonists, sedative antidepressants, sedating antipsychotics, and anticonvulsants have all been tested. In the majority of studies, these drugs have improved insomnia but were not successful in preventing relapse (Brower 2015).

8.12 A Case for the "Psychoactive Toolbox" in Circadian Rhythm Research

The reviewed data indicate that there is a strong connection between the circadian system and psychoactive drugs. Most drugs are not only consumed at certain times of the day but affect sleep and wakefulness as well. Furthermore, there is accumulating evidence, at least for certain groups of compounds, that psychoactive drugs can alter certain circadian molecular markers and behaviors, suggesting a close mutual connection between psychoactive drugs and the circadian system. This tight interplay is expected, given that these drugs mimic the rewarding properties underlying most vital behaviors, which in turn are highly dependent on the circadian clock.

However, while human studies on drug use and sleep are increasingly available, drug-regulated circadian markers are understudied. In order to make better predictions on how to reduce drug use in a wider population or during rehabilitation, more data on circadian biomarkers and behavioral measures such as activity and wakefulness need to be obtained. In many studies, causality is missing as circadian changes and drug use are mutually influencing one another. Furthermore, group sizes are often small. Larger longitudinal studies should be promoted to allow better predictions from circadian behaviors and molecular markers to drug use and vice versa. These should include circadian interventions such as the ones mentioned above, to establish a causal link between circadian rhythms and drug use beyond measures of sleep quality or duration.

Importantly, we are also missing animal studies, that provide in-depth information on how chronic and acute drug administration affect circadian signaling on molecular and circuit levels. Specifically, the data available so far are highly biased toward cocaine, amphetamine, and alcohol, whereas almost no animal studies have been done on psychedelic drugs.

Psychoactive drugs have the advantage of being single-molecule compounds. Thus, compared to other environmental interventions that are used to study circadian changes, including a high-fat diet, light changes, or stress, drug effects can be pinpointed to a small group of well-known receptors, whose function and contribution in the body have been studied in depth. Psychoactive drugs target those brain areas, which are naturally linked to circadian behaviors.

Therefore, psychoactive drugs should be used as molecular tools: They can be applied to study acute and chronic effects on circadian systems in different phases of the circadian clock or under various circadian conditions such as shifts in circadian time (jetlag) or constant darkness. In combination with chemogenetic stimulation or inhibition of certain neural circuits, psychoactive drugs may help pinpoint the role of cell types and brain regions in circadian rhythms. The molecular changes induced by psychoactive drugs in these regions may provide us with a better understanding of how circadian rhythms are linked to reinforcing or addicting processes. In turn, thus identified molecules can be manipulated in the brain (e.g., by viral-mediated gene transfer) to alter drug-seeking behavior or relapse in rodents. Taken together, through the use of such a psychoactive drug toolbox, scientists will gain a more profound understanding of circadian rhythms in the brain. As a result, they are likely to find better preventive strategies as well as treatments for circadian/sleep problems, comorbid disorders, or drug abuse.

Acknowledgments I thank Dr. Lucile Marion-Poll (Institut du Fer à Moulin, INSERM u-839, France) and Kim Lyon-Fernandez (Friedrich-Schiller-Universität Jena, Germany) for helpful comments on this manuscript as well as the IZKF (#973685, Interdisciplinary Center for Clinical Research, University Clinics Jena, Germany) and Prof. Christian A. Hübner (Institute of Human Genetics, University Hospital Jena, Germany) for financial support and Prof. Martin Walter (Department of Psychiatry, University Hospital Jena, Germany) for outstanding mentorship.

References

Abarca C, Albrecht U, Spanagel R (2002) Cocaine sensitization and reward are under the influence of circadian genes and rhythm. Proc Natl Acad Sci USA 99:9026–9030

Abel E (1943) Marihuana, the first twelve thousand year. Springer, New York

Adan A (1994) Chronotype and personality factors in the daily consumption of alcohol and psychostimulants. Addiction 89:455–462

Agapito M, Mian N, Boyadjieva NI, Sarkar DK (2010) Period 2 gene deletion abolishes β-endorphin neuronal response to ethanol. Alcohol Clin Exp Res 34:1613–1618

Agostino PV, Do Nascimento M, Bussi IL et al (2011) Circadian modulation of interval timing in mice. Brain Res 1370:154–163

Ahnaou A, Raeymaekers L, Steckler T, Drinkenbrug WHIM (2015) Relevance of the metabotropic glutamate receptor (mGluR5) in the regulation of NREM-REM sleep cycle and homeostasis: evidence from mGluR5 (-/-) mice. Behav Brain Res 282:218–226

Akhisaroglu M, Ahmed R, Kurtuncu M et al (2004) Diurnal rhythms in cocaine sensitization and in Period1 levels are common across rodent species. Pharmacol Biochem Behav 79:37–42

Allen RP, McCann UD, Ricaurte GA (1993) Persistent effects of (±)3,4-methylenedioxymethamphetamine (MDMA, 'ecstasy') on human sleep. Sleep 16 (6):560–564

Ammon S, Mayer P, Riechert U et al (2003) Microarray analysis of genes expressed in the frontal cortex of rats chronically treated with morphine and after naloxone precipitated withdrawal. Mol Brain Res 112:113–125

Ammon-Treiber S, Höllt V (2005) Morphine-induced changes of gene expression in the brain. Addict Biol 10:81–89

Anderson T, Petranker R, Rosenbaum D et al (2019) Microdosing psychedelics: personality, mental health, and creativity differences in microdosers. Psychopharmacology (Berl) 731–740

Angarita GA, Emadi N, Hodges S, Morgan PT (2016) Sleep abnormalities associated with alcohol, cannabis, cocaine, and opiate use: a comprehensive review. Addict Sci Clin Pract 11:9

Arnedt JT, Conroy DA, Brower KJ (2007) Treatment options for sleep disturbances during alcohol recovery. J Addict Dis 26:41–54

Aronson H, Silverstein AB, Klee GD (1959) Influence of lysergic acid diethylamide (LSD-25) on subjective time. AMA Arch Gen Psychiatry 1:469–472

Asai M, Vindrola O, Zubieta M et al (1988) Diurnal variations of IR-Met-enkephalin in the brain of pentylenetetrazol-kindled rats. Brain Res 442:81–85

Asai M, Zubieta M, Matamoros-Trejo G et al (1998) Diurnal variations of opioid peptides and synenkephalin in vitro release in the amygdala of kindled rats. Neuropeptides 32:293–299

Babson KA, Sottile J, Morabito D (2017) Cannabis, cannabinoids, and sleep: a review of the literature. Curr Psychiatry Rep 19:23

Baird TJ, Gauvin DV (2000) Characterization of cocaine self-administration and pharmacokinetics as a function of time of day in the rat. Pharmacol Biochem Behav 65:289–299

Baird T, Briscoe R, Vallett M et al (1998) Alterations in circadian rhythms of temperature and activity in rats. Pharmacol Biochem Behav 61:303–315

Balogh B, Molnar E, Jakus R et al (2004) Effects of a single dose of 3,4-methylenedioxymethamphetamine on circadian patterns, motor activity and sleep in drug-naive rats and rats previously exposed to MDMA. Psychopharmacology (Berl) 173:296–309

Baño-Otálora B, Piggins HD (2017) Contributions of the lateral habenula to circadian timekeeping. Pharmacol Biochem Behav 162:46–54

Bernardi RE, Spanagel R (2013) The Clockδ19 mutation in mice fails to alter the primary and secondary reinforcing properties of nicotine. Drug Alcohol Depend 133:733–739

Biello SM, Dafters RI (2001) MDMA and fenfluramine alter the response of the circadian clock to a serotonin agonist in vitro. Brain Res 920:202–209

Bildt C, Michélsen H (2002) Gender differences in the effects from working conditions on mental health: a 4-year follow-up. Int Arch Occup Environ Health 75:252–258

Boardman WK, Goldstone S, Lhamon WT (1957) Effects of lysergic acid diethylamide (LSD) on the time sense of normals: a preliminary report. Arch Neurol Psychiatry 78:321–324

Brager AJ, Prosser RA, Glass JD (2011) Circadian and acamprosate modulation of elevated ethanol drinking in mPer2 clock gene mutant mice. Chronobiol Int 28:664–672

Brager AJ, Stowie AC, Prosser RA, Glass JD (2013) The mPer2 clock gene modulates cocaine actions in the mouse circadian system. Behav Brain Res 243:255–260

Brancaccio M, Patton AP, Chesham JE et al (2017) Astrocytes control circadian timekeeping in the suprachiasmatic nucleus via glutamatergic signaling. Neuron 93:1420–1435

Breslau N, Roth T, Rosenthal L, Andreski P (1996) Sleep disturbance and psychiatric disorders: a longitudinal epidemiological study of young adults. Biol Psychiatry 39:411–418

Brick J, Pohorecky LA, Faulkner W, Adams MN (1984) Circadian variations in behavioral and biological sensitivity to ethanol. Alcohol Clin Exp Res 8:204–211

Broms U, Kaprio J, Hublin C et al (2011) Evening types are more often current smokers and nicotine-dependent-a study of Finnish adult twins. Addiction 106:170–177

Brower KJ (2015) Assessment and treatment of insomnia in adult patients with alcohol use disorders. Alcohol 49:417–427

Brower KJ, Hoffmann R, Conroy DA et al (2011) Sleep homeostasis in alcohol-dependent, depressed and healthy control men. Eur Arch Psychiatry Clin Neurosci 261:559–566

Burke TM, Markwald RR, McHill AW et al (2015) Effects of caffeine on the human circadian clock in vivo and in vitro. Sci Transl Med 7:305ra146

Caillé S, Espejo EF, Koob GF, Stinus L (2002) Dorsal and median raphe serotonergic system lesion does not alter the opiate withdrawal syndrome. Pharmacol Biochem Behav 72:979–986

Cao M, Javaheri S (2018) Effects of chronic opioid use on sleep and wake. Sleep Med Clin 13:271–281

Carhart-Harris RL, Nutt DJ, Munafó M, Wilson SJ (2009) Current and former ecstasy users report different sleep to matched controls: a web-based questionnaire study. J Psychopharmacol 23:249–257

Carlini E, Cunha J (2013) Hypnotic and antiepileptic effects of cannabidiol. J Clin Pharmacol 21:417–427

Chagas MHN, Crippa JAS, Zuardi AW et al (2013) Effects of acute systemic administration of cannabidiol on sleep-wake cycle in rats. J Psychopharmacol 27:312–316

Chakravorty S, Chaudhary NS, Brower KJ (2016) Alcohol dependence and its relationship with insomnia and other sleep disorders. Alcohol Clin Exp Res 40:2271–2282

Cheatle MD, Webster LR (2015) Opioid therapy and sleep disorders: risks and mitigation strategies. Pain Med (United States) 16:22–26

Chen CP, Kuhn P, Advis JP, Sarkar DK (2004) Chronic ethanol consumption impairs the circadian rhythm of pro-opiomelanocortin and period genes mRNA expression in the hypothalamus of the male rat. J Neurochem 88:1547–1554

Cheng RK, MacDonald CJ, Meck WH (2006) Differential effects of cocaine and ketamine on time estimation: implications for neurobiological models of interval timing. Pharmacol Biochem Behav 85:114–122

Cheng RK, Ali YM, Meck WH (2007a) Ketamine "unlocks" the reduced clock-speed effects of cocaine following extended training: evidence for dopamine-glutamate interactions in timing and time perception. Neurobiol Learn Mem 88:149–159

Cheng RK, Etchegaray M, Meck WH (2007b) Impairments in timing, temporal memory, and reversal learning linked to neurotoxic regimens of methamphetamine intoxication. Brain Res 1186:255–266

Clark I, Landolt HP (2017) Coffee, caffeine, and sleep: a systematic review of epidemiological studies and randomized controlled trials. Sleep Med Rev 31:70–78

Colbron S, Jones M, Biello SM (2002) MDMA alters the response of the circadian clock to a photic and non-photic stimulus. Brain Res 956:45–52

Colrain IM, Baker FC (2011) Changes in sleep as a function of adolescent development. Neuropsychol Rev 21:5–21

Colrain IM, Nicholas CL, Baker FC (2014) Alcohol and the sleeping brain. Handb Clin Neurol 125:415–431

Comasco E, Nordquist N, Göktürk C et al (2010) The clock gene PER2 and sleep problems: association

with alcohol consumption among Swedish adolescents. Ups J Med Sci 115:41–48
Conroy DA, Hairston IS, Arnedt JT et al (2012) Dim light melatonin onset in alcohol-dependent men and women compared with healthy controls. Chronobiol Int 29:35–42
Cordes S, Gallistel CR (2008) Intact interval timing in circadian CLOCK mutants. Brain Res 1227:120–127
Coull JT, Cheng RK, Meck WH (2011) Neuroanatomical and neurochemical substrates of timing. Neuropsychopharmacology 36:3–25
Dafters RI, Biello SM (2003) The effect of 3,4-methylenedioxymethamphetamine ('Ecstasy') on serotonergic regulation of the mammalian circadian clock mechanism in rats: the role of dopamine and hyperthermia. Neurosci Lett 350:117–121
Dahchour A, De Witte P (1999) Effect of repeated ethanol withdrawal on glutamate microdialysate in the hippocampus. Alcohol Clin Exp Res 23:1698–1703
Daniels CW, Watterson E, Garcia R et al (2015) Revisiting the effect of nicotine on interval timing. Behav Brain Res 283:238–250
Davies J, Redfern P (1973) The effects of hallucinogenic drugs on maze exploration in the rat over a 24 hour period. Br J Pharmacol 49:121–127
Davila DG, Hurt RD, Offord KP et al (1994) Acute effects of transdermal nicotine on sleep architecture, snoring, and sleep-disordered breathing in nonsmokers. Am J Respir Crit Care Med 150:469–474
De Gregorio D, Comai S, Posa L, Gobbi G (2016) D-lysergic acid diethylamide (LSD) as a model of psychosis: mechanism of action and pharmacology. Int J Mol Sci 17:E1953
Depoortere H, Loew D (1971) Alterations in sleep/wakefulness cycle in rats following treatment with (+)-lysergic acid diethylamide (LSD-25). Proc Br Pharmacol Soc 1:402–403
Dimsdale JE, Norman D, DeJardin D, Wallace MS (2007) The effect of opioids on sleep architecture. J Clin Sleep Med 3:33–36
Dolsen MR, Harvey AG (2017) Life-time history of insomnia and hypersomnia symptoms as correlates of alcohol, cocaine and heroin use and relapse among adults seeking substance use treatment in the United States from 1991 to 1994. Addiction 112:1104–1111
Drummond SP, Gillin JC, Smith TL, DeModena A (1998) The sleep of abstinent pure primary alcoholic patients: natural course and relationship to relapse. Alcohol Clin Exp Res 22:1796–1802
Elliott KJ, Weber ET, Rea MA (2001) Adenosine A1 receptors regulate the response of the hamster circadian clock to light. Eur J Pharmacol 414:45–53
Facchinetti F, Volpe A, Farci G et al (1985) Hypothalamus-pituitary-adrenal axis of heroin addicts. Drug Alcohol Depend 15:361–366
Fakier N, Wild LG (2011) Associations among sleep problems, learning difficulties and substance use in adolescence. J Adolesc 34:717–726
Falcon E, Ozburn A, Mukherjee S et al (2013) Differential regulation of the period genes in striatal regions following cocaine exposure. PLoS One 8:e66438
Ferguson SA, Kennaway DJ, Moyer RW (1999) Nicotine phase shifts the 6-sulphatoxymelatonin rhythm and induces c- Fos in the SCN of rats. Brain Res Bull 48:527–538
Ferre S, von Euler G, Johansson B et al (1991) Stimulation of high-affinity adenosine A2 receptors decreases the affinity of dopamine D2 receptors in rat striatal membranes. Proc Natl Acad Sci USA 88:7238–7241
Ferri S, Reina RA, Spadaro C, Braga P (1977) 6-Hydroxydopamine inhibits some effects of mescaline centrally administered to rabbits. Psychopharmacology (Berl) 55:147–149
Ferris M, España RA, Chen R et al (2014) Dopamine transporters govern diurnal variation in extracellular dopamine tone. Proc Natl Acad Sci USA 111:E2751–E2759
Fisk JE, Montgomery C (2009) Sleep impairment in ecstasy/polydrug and cannabis-only users. Am J Addict 18:430–437
Fredholm BB, Bättig K, Holmén J et al (1999) Actions of caffeine in the brain with special reference to factors that contribute to its widespread use. Pharmacol Rev 51:83–133
Gardani M, Blance RN, Biello SM (2005) MDMA alters the response of the mammalian circadian clock in hamsters: effects on re-entrainment and triazolam-induced phase shifts. Brain Res 1046:105–115
Garmabi B, Vousooghi N, Vosough M et al (2016) Effect of circadian rhythm disturbance on morphine preference and addiction in male rats: involvement of period genes and dopamine D1 receptor. Neuroscience 322:104–114
Gillin JC, Lardon M, Ruiz C et al (1994) Dose-dependent effects of transdermal nicotine on early morning awakening an rapid eye movement sleep time in nonsmoking normal volunteers. J Clin Psychopharmacol 14:264–267
Glass JD, Grossman GH, Farnbauch L, DiNardo L (2003) Midbrain raphe modulation of nonphotic circadian clock resetting and 5-HT release in the mammalian suprachiasmatic nucleus. J Neurosci 23:7451–7460
Glass JD, Ruby CL, Brager AJ et al (2009) Chronic ethanol attenuates circadian photic phase resetting and alters nocturnal activity patterns in the hamster. Am J Physiol Integr Comp Physiol 297:R729–R737
Glass JD, Brager AJ, Stowie AC, Prosser RA (2012) Cocaine modulates pathways for photic and nonphotic entrainment of the mammalian SCN circadian clock. Am J Physiol Integr Comp Physiol 302:R740–R750
Hallworth R, Cato M, Colbert C, Rea MA (2002) Presynaptic adenosine A1 receptors regulate retinohypothalamic neurotransmission in the hamster suprachiasmatic nucleus. J Neurobiol 52:230–240
Hamidovic A, de Wit H (2009) Sleep deprivation increases cigarette smoking. Pharmacol Biochem Behav 93:263–269

Haney M (2009) Self-administration of cocaine, cannabis and heroin in the human laboratory: benefits and pitfalls. Addict Biol 14:9–21
Häring M, Marsicano G, Lutz B, Monory K (2007) Identification of the cannabinoid receptor type 1 in serotonergic cells of raphe nuclei in mice. Neuroscience 146:1212–1219
Hasler BP, Clark DB (2013) Circadian misalignment, reward-related brain function, and adolescent alcohol involvement. Alcohol Clin Exp Res 37:558–565
Hasler BP, Kirisci L, Clark DB (2016) Restless sleep and variable sleep timing during late childhood accelerate the onset of alcohol and other drug involvement. J Stud Alcohol Drugs 77:649–655
Heilbronner SR, Meck WH (2014) Dissociations between interval timing and intertemporal choice following administration of fluoxetine, cocaine, or methamphetamine. Behav Processes 101:123–134
Hobson JA, Datta S, Calvo JM, Quattrochi J (1993) Acetylcholine as a brain state modulator: triggering and long-term regulation of REM sleep. Prog Brain Res 98:389–404
Honma KI, Honma S, Hiroshige T (1986) Disorganization of the rat activity rhythm by chronic treatment with methamphetamine. Physiol Behav 38:687–695
Hood S, Cassidy P, Mathewson S et al (2011) Daily morphine injection and withdrawal disrupt 24-h wheel running and PERIOD2 expression patterns in the rat limbic forebrain. Neuroscience 186:65–75
Huang MC, Ho CW, Chen CH et al (2010) Reduced expression of circadian clock genes in male alcoholic patients. Alcohol Clin Exp Res 34:1899–1904
Hut RA, Van der Zee EA (2011) The cholinergic system, circadian rhythmicity, and time memory. Behav Brain Res 221:466–480
Hyman SE, Malenka RC, Neslter EJ (2006) Neural mechanisms of addiction: the role of reward-related learning and memory. Annu Rev Neurosci 29:565–598
Iijima M, Nikaido T, Akiyama M et al (2002) Methamphetamine-induced, suprachiasmatic nucleus-independent circadian rhythms of activity and mPer gene expression in the striatum of the mouse. Eur J Neurosci 16:921–929
Jarvik M, Killen JD, Varady A, Fortmann SP (1993) The favorite cigarette of the day. J Behav Med 16:423–422
Jha PK, Bouâouda H, Gourmelen S et al (2017) Sleep deprivation and caffeine treatment potentiate photic resetting of the master circadian clock in a diurnal rodent. J Neurosci 37:4343–4358
Johnson EO (2006) Epidemiology of DSM-IV insomnia in adolescence: lifetime prevalence, chronicity, and an emergent gender difference. Pediatrics 117:e247–e256
Johnson EO, Breslau N (2001) Sleep problems and substance use in adolescence. Drug Alcohol Depend 64:1–7
Jones EM, Knutson D, Haines D (2003) Common problems in patients recovering from chemical dependency. Am Fam Physician 68:1971–1978
Kakihana R, Moore JA (1976) Circadian rhythm of corticosterone in mice: the effect of chronic consumption of alcohol. Psychopharmacologia 46:301–305
Kawachi I, Willett WC, Colditz GA et al (1996) A prospective study of coffee drinking and suicide in women. Arch Intern Med 156:521–525
Kervran C, Fatséas M, Serre F et al (2015) Association between morningness/eveningness, addiction severity and psychiatric disorders among individuals with addictions. Psychiatry Res 229:1024–1030
Khalifeh AH (2017) The effect of chronotherapy on depressive symptoms: evidence-based practice. Saudi Med J 38:457–464
King C, Masserano JM, Codd E, Byrne WL (1981) Effects of β-endorphin and morphine on the sleep-wakefulness behavior of cats. Sleep 4:259–262
Kislev ME, Hartmann A, Galili E (2004) Archaeobotanical and archaeoentomological evidence from a well at Atlit-Yam indicates colder, more humid climate on the Israeli coast during the PPNC period. J Archaeol Sci 3:375–381
Klatsky AL, Armstrong MA, Friedman GD (1993) Coffee, tea, and mortality. Ann Epidemiol 3(4):375–381
Koch M, Dehghani F, Habazettl I et al (2006) Cannabinoids attenuate norepinephrine-induced melatonin biosynthesis in the rat pineal gland by reducing arylalkylamine N-acetyltransferase activity without involvement of cannabinoid receptors. J Neurochem 98:267–278
Koch M, Habazettl I, Dehghani F, Korf HW (2008) The rat pineal gland comprises an endocannabinoid system. J Pineal Res 45:351–360
Kovacic P, Somanathan R (2009) Novel, unifying mechanism for mescaline in the central nervous system. Oxid Med Cell Longev 2:181–190
Kovanen L, Saarikoski ST, Haukka J et al (2010) Circadian clock gene polymorphisms in alcohol use disorders and alcohol consumption. Alcohol Alcohol 45:303–311
Kühlwein E, Hauger RL, Irwin MR (2003) Abnormal nocturnal melatonin secretion and disordered sleep in abstinent alcoholics. Biol Psychiatry 54:1437–1443
Lafaye G, Desterke C, Marulaz L, Benyamina A (2018) Cannabidiol affects circadian clock core complex and its regulation in microglia cells. Addict Biol 24:921–934
Landolt HP, Gillin JC (2001) Sleep abnormalities during abstinence in alcohol-dependent patients: aetiology and management. CNS Drugs 15:413–425
Landolt HP, Dijk DJ, Gaus SE, Borbély AA (1995) Caffeine reduces low-frequency delta activity in the human sleep EEG. Neuropsychopharmacology 12:229–238
Lara DR (2010) Caffeine, mental health, and psychiatric disorders. J Alzheimers Dis 20:S239–S248
Lecca S, Meye FJ, Mameli M (2014) The lateral habenula in addiction and depression: an anatomical, synaptic and behavioral overview. Eur J Neurosci 39:1170–1178

Lechasseur A, Jubinville É, Routhier J et al (2017) Exposure to electronic cigarette vapors affects pulmonary and systemic expression of circadian molecular clock genes. Physiol Rep 5:e13440

Letter to HC Schumacher (1836) Alexander von Humboldt an Heinrich Christian Schumacher, 2. März 1836. Alexander von Humboldt Schumacher Lett 1979 52

Li SX, Liu LJ, Jiang WG, Lu L (2009) Morphine withdrawal produces circadian rhythm alterations of clock genes in mesolimbic brain areas and peripheral blood mononuclear cells in rats. J Neurochem 109:1668–1679

Lindberg D, Andres-Beck L, Jia YF et al (2018) Purinergic signaling in neuron-astrocyte interactions, circadian rhythms, and alcohol use disorder. Front Physiol 9:9

Lindskog M, Svenningsson P, Pozzi L et al (2002) Involvement of DARPP-32 phosphorylation in the stimulant action of caffeine. Nature 418:774–778

Liu C, Gillette M (2018) Cholinergic regulation of the suprachiasmatic nucleus circadian rhythm via a muscarinic mechanism at night. J Neurosci 16:744–751

Logan RW, Williams WP, McClung CA (2014) Circadian rhythms and addiction: mechanistic insights and future directions. Behav Neurosci 128:387–412

Luo AH, Aston-Jones G (2009) Circuit projection from suprachiasmatic nucleus to ventral tegmental area: a novel circadian output pathway. Eur J Neurosci 29:748–760

Lynch WJ, Girgenti MJ, Breslin FJ et al (2008) Gene profiling the response to repeated cocaine self-administration in dorsal striatum: a focus on circadian genes. Brain Res 1213:166–177

Madeira MD, Paula-Barbosa MM (1999) Effects of alcohol on the synthesis and expression of hypothalamic peptides. Brain Res Bull 48:3–22

Madeira MD, Andrade JP, Lieberman AR et al (1997) Chronic alcohol consumption and withdrawal do not induce cell death in the suprachiasmatic nucleus, but lead to irreversible depression of peptide immunoreactivity and mRNA levels. J Neurosci 17:1302–1319

Mamounas LA, Mullen CA, O'Hearn E, Molliver ME (1991) Dual serotoninergic projections to forebrain in the rat: morphologically distinct 5-HT axon terminals exhibit differential vulnerability to neurotoxic amphetamine derivatives. J Comp Neurol 314:558–586

Martin DA, Marona-Lewicka D, Nichols DE, Nichols CD (2014) Chronic LSD alters gene expression profiles in the mPFC relevant to schizophrenia. Neuropharmacology 83:1–8

Matell MS, Bateson M, Meck WH (2006) Single-trials analyses demonstrate that increases in clock speed contribute to the methamphetamine-induced horizontal shifts in peak-interval timing functions. Psychopharmacology (Berl) 188:201–212

Mazei-Robison MS, Nestler EJ (2012) Opiate-induced molecular and cellular plasticity of ventral tegmental area and locus coeruleus catecholamine neurons. Cold Spring Harb Perspect Med 2:a012070

McCann UD, Ricaurte GA (2007) Effects of (±) 3,4-methylenedioxymethamphetamine (MDMA) on sleep and circadian rhythms. ScientificWorldJournal 7:231–238

McCann UD, Eligulashvili V, Ricaurte GA (2000) (±)3,4-Methylenedioxymethamphetamine ('Ecstasy')-induced serotonin neurotoxicity: clinical studies. Neuropsychobiology 42:11–16

McCann UD, Peterson SC, Ricaurte GA (2007) The effect of catecholamine depletion by alpha-methyl-para-tyrosine on measures of cognitive performance and sleep in abstinent MDMA users. Neuropsychopharmacology 32:1695–1706

McClung CA (2007) Circadian rhythms, the mesolimbic dopaminergic circuit, and drug addiction. ScientificWorldJournal 7:194–202

McClung C, Sidiropoulou K, Vitaterna M et al (2005) Regulation of dopaminergic transmission and cocaine reward by the Clock gene. Proc Natl Acad Sci USA 102:9377–9381

McElroy B, Zakaria A, Glass JD, Prosser RA (2009) Ethanol modulates mammalian circadian clock phase resetting through extrasynaptic gaba receptor activation. Neuroscience 164:842–848

McGregor C, Srisurapanont M, Jittiwutikarn J et al (2005) The nature, time course and severity of methamphetamine withdrawal. Addiction 100:1320–1329

Melendez RI, McGinty JF, Kalivas PW, Becker HC (2012) Brain region-specific gene expression changes after chronic intermittent ethanol exposure and early withdrawal in C57BL/6J mice. Addict Biol 17:351–364

Miguel Asai MA, Lilian Mayagoitia LM, David García DG et al (2007) Rat brain opioid peptides-circadian rhythm is under control of melatonin. Neuropeptides 41:389–397. https://doi.org/10.1016/j.npep.2007.09.001

Miller J, Murakami D, Fuller C (1987) The response of suprachiasmatic neurons of the rat hypothalamus to photic and nicotinic stimuli. J Neurosci 7:978–986

Miller MB, Janssen T, Jackson KM (2017) The prospective association between sleep and initiation of substance use in young adolescents. J Adolesc Heal 60:154–160

Moldrich G, Wenger T (2000) Localization of the CB1 cannabinoid receptor in the rat brain. An immunohistochemical study. Peptides 21:1735–1742

Montgomery C, Fisk JE, Wareing M, Murphy P (2007) Self reported sleep quality and cognitive performance in ecstasy users. Hum Psychopharmacol 22:537–548

Morgan PT, Pace-Schott EF, Sahul ZH et al (2006) Sleep, sleep-dependent procedural learning and vigilance in chronic cocaine users: evidence for occult insomnia. Drug Alcohol Depend 82:238–249

Morley BJ, Garner LL (1990) Light-dark variation in response to chronic nicotine treatment and the density of hypothalamic α-bungarotoxin receptors. Pharmacol Biochem Behav 37:239–245

Naber D, Wirs-Justice A, Kajca M (1981) Diurnal rhythm in rat brain opiate receptor. Neurosci Lett 21:45–50
Nam HW, Lee MR, Zhu Y et al (2011) Type 1 equilibrative nucleoside transporter regulates ethanol drinking through accumbal N-methyl-D-aspartate receptor signaling. Biol Psychiatry 69:1043–1051
Nestler EJ (2013) Cellular basis of memory for addiction. Dialogues Clin Neurosci 15(4):431–443
Nguyen-Louie TT, Brumback T, Worley MJ et al (2018) Effects of sleep on substance use in adolescents: a longitudinal perspective. Addict Biol 23:750–760
Nichols DE, Johnson MW, Nichols CD (2017) Psychedelics as medicines: an emerging new paradigm. Clin Pharmacol Ther 101:209–219
Nicholson AN, Turner C, Stone BM, Robson PJ (2004) Effect of Δ-9-tetrahydrocannabinol and cannabidiol on nocturnal sleep and early-morning behavior in young adults. J Clin Psychopharmacol 24:305–313
NIDA (2019) Hallucinogens. Retrieved from https://www.drugabuse.gov/drugs-abuse/hallucinogens
Nikaido T, Akiyama M, Moriya T, Shibata S (2018) Sensitized increase of period gene expression in the mouse caudate/putamen caused by repeated injection of methamphetamine. Mol Pharmacol 59:894–900
Nutt DJ, King LA, Phillips LD (2010) Drug harms in the UK: a multicriteria decision analysis. Lancet 376:1558–1565
Ogeil RP, Rajaratnam SMW, Redman JR, Broadbear JH (2010) Acute MDMA administration alters the distribution and circadian rhythm of wheel running activity in the rat. Brain Res 1359:128–136
Ogeil RP, Rajaratnam SM, Broadbear JH (2012a) Ecstasy and sleep disturbance: progress towards elucidating a role for the circadian system. Sleep Biol Rhythms 10:3–12
Ogeil R, Kennaway D, Salkeld MD et al (2012b) MDMA induces Per1, Per2 and c-fos gene expression in rat suprachiasmatic nuclei. Psychopharmacology (Berl) 220:835–843
O'Hara BF, Edgar DM, Vinh HC et al (1998) Nicotine and nicotinic receptors in the circadian system. Psychoneuroendocrinology 23:161–173
Oike H, Kobori M, Suzuki T, Ishida N (2011) Caffeine lengthens circadian rhythms in mice. Biochem Biophys Res Commun 410:654–658
Oyefeso A, Sedgwick P, Ghodse H (1997) Subjective sleep-wake parameters in treatment-seeking opiate addicts. Drug Alcohol Depend 48:9–16
Ozburn AR, Larson EB, Self DW, McClung CA (2012) Cocaine self-administration behaviors in ClockΔ19 mice. Psychopharmacology (Berl) 223:169–177
Ozburn AR, Falcon E, Mukherjee S et al (2013) The role of clock in ethanol-related behaviors. Neuropsychopharmacology 38:2393–2400
Pačesová D, Volfová B, Červená K et al (2015) Acute morphine affects the rat circadian clock via rhythms of phosphorylated ERK1/2 and GSK3β kinases and Per1 expression in the rat suprachiasmatic nucleus. Br J Pharmacol 172:3638–3649
Parks KA, Kennedy CL (2004) Club drugs: reasons for and consequences of use. J Psychoactive Drugs 36:295–302
Passie T, Halpern JH, Stichtenoth DO et al (2008) The pharmacology of lysergic acid diethylamide: a review. CNS Neurosci Ther 14:295–314
Penetar D, McCann U, Thorne D et al (1993) Caffeine reversal of sleep deprivation effects on alertness and mood. Psychopharmacology (Berl) 112:359–365
Perreau-Lenz S, Sanchis-Segura C, Leonardi-Essmann F et al (2010) Development of morphine-induced tolerance and withdrawal: involvement of the clock gene mPer2. Eur Neuropsychopharmacol 20:509–517
Perron RR, Tyson RL, Sutherland GR (2001) Δ9-Tetrahydrocannabinol increases brain temperature and inverts circadian rhythms. Neuroreport 12:3791–3794
Petersen CC, Mistlberger RE (2017) Interval timing is preserved despite circadian desynchrony in rats: constant light and heavy water studies. J Biol Rhythms 295–308
Phillips B, Danner F (1995) Cigarette smoking and sleep disturbance. Arch Intern Med 155:734–737
Picciotto MR, Higley MJ, Mineur YS (2012) Acetylcholine as a neuromodulator: cholinergic signaling shapes nervous system function and behavior. Neuron 76:116–129
Pich EM, Pagliusi SR, Tessari M et al (1997) Common neural substrates for the addictive properties of nicotine and cocaine. Science 275:83–86
Pietilä K, Laakso I, Ahtee L (1995) Chronic oral nicotine administration affects the circadian rhythm of dopamine and 5-hydroxytryptamine metabolism in the striata of mice. Naunyn Schmiedebergs Arch Pharmacol 353:110–115
Prochazkova L, Lippelt DP, Colzato LS et al (2018) Exploring the effect of microdosing psychedelics on creativity in an open-label natural setting. Psychopharmacology (Berl) 235:3401–3413
Prosser RA, Glass JD (2009) The mammalian circadian clock exhibits acute tolerance to ethanol. Alcohol Clin Exp Res 33:2088–2093
Prosser RA, Heller HC, Dean RR et al (1993) Serotonin and the mammalian circadian system: I. In vitro phase shifts by serotonergic agonists and antagonists. J Biol Rhythms 8:1–16
Prosser RA, Mangrum CA, Glass JD (2008) Acute ethanol modulates glutamatergic and serotonergic phase shifts of the mouse circadian clock in vitro. Neuroscience 152:837–848
Prosser RA, Stowie A, Amicarelli M et al (2014) Cocaine modulates mammalian circadian clock timing by decreasing serotonin transport in the SCN. Neuroscience 275:184–193
Przewłocki R, Lasón W, Konecka AM et al (1983) The opioid peptide dynorphin, circadian rhythms, and starvation. Science 219:71–73
Rajakrishnan V, Subramanian P, Viswanathan P, Menon VP (1999) Effect of chronic ethanol ingestion on

biochemical circadian rhythms in Wistar rats. Alcohol 18:147–152
Rammsayer T (2008) Dopaminergic and serotoninergic influence on duration discrimination and vigilance. Pharmacopsychiatry 22(Suppl 1):39–43
Raymond R, Warren M, Morris R, Leikin J (1992) Periodicity of presentations of drugs of abuse and overdose in an emergency department. Clin Toxicol 30:467–478
Ribeiro JA, Sebastio AM (2010) Caffeine and adenosine. J Alzheimers Dis 20(Suppl 1):S3–S15
Roberts RE, Roberts CR, Duong HT (2009) Sleepless in adolescence: prospective data on sleep deprivation, health and functioning. J Adolesc 32:1045–1057
Robison AJ, Nestler EJ (2011) Transcriptional and epigenetic mechanisms of addiction. Nat Rev Neurosci 12:623–637
Rogers HL, Reilly SM (2002) A survey of the health experiences of international business travelers. Part One--Physiological aspects. AAOHN J 50:449–459
Roitman P, Mechoulam R, Cooper-Kazaz R, Shalev A (2014) Preliminary, open-label, pilot study of add-on oral Δ9- tetrahydrocannabinol in chronic post-traumatic stress disorder. Clin Drug Investig 34:587–591
Rosenwasser AM (2015) Chronobiology of ethanol: animal models. Alcohol 49:311–319
Rosenwasser AM, Fecteau ME, Logan RW (2005) Effects of ethanol intake and ethanol withdrawal on free-running circadian activity rhythms in rats. Physiol Behav 84:537–543
Rossetti ZL, Carboni S (1995) Ethanol withdrawal is associated with increased extracellular glutamate in the rat striatum. Eur J Pharmacol 283:177–183
Ruby CL, Verbanes NM, Palmer KN et al (2018) Caffeine delays light-entrained activity and potentiates circadian photic phase-resetting in mice. J Biol Rhythms 33:523–534
Rucker JJH, Jelen LA, Flynn S et al (2016) Psychedelics in the treatment of unipolar mood disorders: a systematic review. J. Psychopharmacol 30:1220–1229
Rupp TL, Acebo C, Carskadon MA (2007) Evening alcohol suppresses salivary melatonin in young adults. Chronobiol Int 24:463–470
Sabanayagam C, Shankar A (2011) The association between active smoking, smokeless tobacco, second-hand smoke exposure and insufficient sleep. Sleep Med 12:7–11
Saffroy R, Lafaye G, Desterke C et al (2019) Several clock genes polymorphisms are meaningful risk factors in the development and severity of cannabis addiction. Chronobiol Int 36:122–134
Sanchis-Segura C, Jancic D, Jimenez-Minchan M, Barco A (2009) Inhibition of cAMP responsive element binding protein in striatal neurons enhances approach and avoidance responses towards morphine- and morphine withdrawal-related cues. Front Behav Neurosci 3:30
Sanford AE, Castillo E, Gannon RL (2008) Cannabinoids and hamster circadian activity rhythms. Brain Res 1222:141–148
Santucci V, Storme JJ, Soubrié P, Le Fur G (1996) Arousal-enhancing properties of the CB1 cannabinoid receptor antagonist SR 141716A in rats as assessed by electroencephalographic spectral and sleep-waking cycle analysis. Life Sci 58:PL103–PL110
Seggio JA, Fixaris MC, Reed JD et al (2009) Chronic ethanol intake alters circadian phase shifting and free-running period in mice. J Biol Rhythms 24:304–312
Shannon S, Opila-Lehman J (2015) Cannabidiol oil for decreasing addictive use of marijuana: a case report. Integr Med (Encinitas) 14:31–35
Shibata S, Tahara Y, Hirao A (2010) The adjustment and manipulation of biological rhythms by light, nutrition, and abused drugs. Adv. Drug Deliv Rev 62:918–927
Shibley HL, Malcolm RJ, Veatch LM (2008) Adolescents with insomnia and substance abuse: consequences and comorbidities. J Psychiatr Pract 14:146–153
Shilo L, Sabbah H, Hadari R et al (2002) The effects of coffee consumption on sleep and melatonin secretion. Sleep Med 3:271–273
Shumay E, Fowler JS, Wang GJ et al (2012) Repeat variation in the human PER2 gene as a new genetic marker associated with cocaine addiction and brain dopamine D2 receptor availability. Transl Psychiatry 2:e86
Siegel JZ, Crockett MJ (2013) How serotonin shapes moral judgment and behavior. Ann NY Acad Sci 1299:42–51
Sigworth LA, Rea MA (2003) Adenosine A1 receptors regulate the response of the mouse circadian clock to light. Brain Res 960:246–251
Sleipness EP, Sorg BA, Jansen HT (2007) Diurnal differences in dopamine transporter and tyrosine hydroxylase levels in rat brain: dependence on the suprachiasmatic nucleus. Brain Res 1129:23–42
Smith AP (2009) Caffeine, cognitive failures and health in a non-working community sample. Hum Psychopharmacol 24:29–34
Snoeren EMS, Veening JG, Olivier B, Oosting RS (2014) Serotonin 1A receptors and sexual behavior in female rats: a review. Pharmacol Biochem Behav 121:43–52
Sorg BA, Stark G, Sergeeva A, Jansen HT (2011) Photoperiodic suppression of drug reinstatement. Neuroscience 176:284–295
Spanagel R, Rosenwasser AM, Schumann G, Sarkar DK (2005a) Alcohol consumption and the body's biological clock. Alcohol Clin Exp Res 29:1550–1557
Spanagel R, Pendyala G, Abarca C et al (2005b) The clock gene Per2 influences the glutamatergic system and modulates alcohol consumption. Nat Med 11:35–42
Stern W, Morgane P, Bronzino J (1972) LSD: effects on sleep patterns and spiking activity in the lateral geniculate nucleus. Brain Res 41:199–204
St. Hilaire MA, Lockley SW (2015) Caffeine does not entrain the circadian clock but improves daytime alertness in blind patients with non-24-hour rhythms. Sleep Med 16:800–804
Stinus L, Robert C, Karasinski P, Limoge A (1998) Continuous quantitative monitoring of spontaneous opiate

withdrawal: locomotor activity and sleep disorders. Pharmacol Biochem Behav 59:83–89
Stowie AC, Amicarelli MJ, Prosser RA, Glass JD (2015) Chronic cocaine causes long-term alterations in circadian period and photic entrainment in the mouse. Neuroscience 284:171–179
Suh S, Yang HC, Kim N et al (2017) Chronotype differences in health behaviors and health-related quality of life: a population-based study among aged and older adults. Behav Sleep Med 15:361–376
Syed YY, Keating GM (2013) Extended-release intramuscular naltrexone (VIVITROL®): a review of its use in the prevention of relapse to opioid dependence in detoxified patients. CNS Drugs 27:851–861
Tabakoff B, Jaffe R, Ritzmann R (1978) Corticosterone concentrations in mice during ethanol drinking and withdrawal. J Pharm Pharmacol 30:371–374
Takahashi TT, Vengeliene V, Spanagel R (2017) Melatonin reduces motivation for cocaine self-administration and prevents relapse-like behavior in rats. Psychopharmacology (Berl) 234:1741–1748
Tataroğlu Ö, Davidson AJ, Benvenuto LJ, Menaker M (2006) The methamphetamine-sensitive circadian oscillator (MASCO) in mice. J Biol Rhythms 21:185–194
Teplin D, Raz B, Daiter J et al (2006) Screening for substance use patterns among patients referred for a variety of sleep complaints. Am J Drug Alcohol Abuse 32:111–120
Thronhill JA, Hirst M, Gowdey CW (1976) Changes in diurnal temperature and feeding patterns of rats during repeated injections of heroin and withdrawal. Arch Int Pharmacodyn Ther 223:120–131
Trachsel L, Heller HC, Miller JD (1995) Nicotine phase-advances the circadian neuronal activity rhythm in rat suprachiasmatic nuclei explants. Neuroscience 65:797–803
Trautmann C, Burek D, Hübner CA, Girault JA, Engmann O (2020) A regulatory pathway linking caffeine action, mood and the diurnal clock. Neuropharmacology 172:108133
Treur JL, Gibson M, Taylor AE et al (2018) Investigating genetic correlations and causal effects between caffeine consumption and sleep behaviours. J Sleep Res 27:e12695
Trinkoff AM, Storr CL (1998) Work schedule characteristics and substance use in nurses. Am J Ind Med 34:266–271
Uz T, Akhisaroglu M, Ahmed R, Manev H (2003) The pineal gland is critical for circadian Period1 expression in the striatum and for circadian cocaine sensitization in mice. Neuropsychopharmacology 28:2117–2123
Uz T, Arslan AD, Kurtuncu M et al (2005a) The regional and cellular expression profile of the melatonin receptor MT1 in the central dopaminergic system. Mol Brain Res 136:45–53
Uz T, Ahmed R, Akhisaroglu M et al (2005b) Effect of fluoxetine and cocaine on the expression of clock genes in the mouse hippocampus and striatum. Neuroscience 16:560–564
van Diepen HC, Lucassen EA, Yasenkov R et al (2014) Caffeine increases light responsiveness of the mouse circadian pacemaker. Eur J Neurosci 40:3504–3511
Vaughn LK, Denning G, Stuhr KL et al (2010) Endocannabinoid signalling: has it got rhythm? Br J Pharmacol 160:530–543
Vengeliene V, Noori HR, Spanagel R (2015) Activation of melatonin receptors reduces relapse-like alcohol consumption. Neuropsychopharmacology 40:2897–2906
Vescovi PP, Coiro V, Volpi R, Passeri M (1992) Diurnal variations in plasma ACTH, cortisol and beta-endorphin levels in cocaine addicts. Horm Res Paediatr 37:221–224
Virus RM, Baglajewski T, Radulovacki M (1984) Circadian variation of [3H]N6-(l-phenylisopropyl)adenosine binding in rat brain. Neurosci Lett 46:219–222
Wackermann J, Wittmann M, Hasler F, Vollenweider FX (2008) Effects of varied doses of psilocybin on time interval reproduction in human subjects. Neurosci Lett 435:51–55
Wahlstrom KL, Owens JA (2017) School start time effects on adolescent learning and academic performance, emotional health and behaviour. Curr Opin Psychiatry 30:485–490
Walther S, Mahlberg R, Eichmann U, Kunz D (2006) Delta-9-tetrahydrocannabinol for nighttime agitation in severe dementia. Psychopharmacology (Berl) 185:524–528
Wang X, Mozhui K, Li Z et al (2012) A promoter polymorphism in the Per3 gene is associated with alcohol and stress response. Transl Psychiatry 2:e73
Wasielewski JA, Holloway FA (2001) Alcohol's interactions with circadian rhythms a focus on body temperature. Alcohol Res Health 25:94–100
Watanabe A, Moriya T, Nisikawa Y et al (1996) Adenosine A1-receptor agonist attenuates the light-induced phase shifts and fos expression in vivo and optic nerve stimulation-evoked field potentials in the suprachiasmatic nucleus in vitro. Brain Res 740:329–336
Webb IC, Baltazar RM, Wang X et al (2009) Diurnal variations in natural and drug reward, mesolimbic tyrosine hydroxylase, and clock gene expression in the male rat. J Biol Rhythms 24:465–476
Wells AM, Ridener E, Bourbonais CA et al (2017) Effects of chronic social defeat stress on sleep and circadian rhythms are mitigated by kappa-opioid receptor antagonism. J Neurosci 37:7656–7668
Wheaton AG, Chapman DP, Croft JB (2016) School start times, sleep, behavioral, health, and academic outcomes: a review of the literature. J Sch Health 86:363–381
Williamson L, Cheng R-K, Etchegaray M, Meck W (2010) "Speed" warps time: methamphetamines interactive roles in drug abuse, habit formation, and the biological clocks of circadian and interval timing. Curr Drug Abus Rev 1:203–212
Wittmann M, Dinich J, Merrow M, Roenneberg T (2006) Social jetlag: misalignment of biological and social time. Chronobiol Int 23:497–509

Wittmann G, Deli L, Kalló I et al (2007a) Distribution of type 1 cannabinoid receptor (CB1)-immunoreactive axons in the mouse hypothalamus. J Comp Neurol 503:270–279
Wittmann M, Carter O, Hasler F et al (2007b) Effects of psilocybin on time perception and temporal control of behaviour in humans. J Psychopharmacol 21:50–64
Wong MM, Brower KJ, Fitzgerald HE, Zucker RA (2004) Sleep problems in early childhood and early onset of alcohol and other drug use in adolescence. Alcohol Clin Exp Res 28:578–587
Wongchitrat P, Mukda S, Phansuwan-Pujito P, Govitrapong P (2013) Effect of amphetamine on the clock gene expression in rat striatum. Neurosci Lett 542:126–130
Wu J, Lee MR, Kim T et al (2011) Regulation of ethanol-sensitive EAAT2 expression through adenosine A1 receptor in astrocytes. Biochem Biophys Res Commun 406:47–52
Yanakieva S, Polychroni N, Family N et al (2018) The effects of microdose LSD on time perception: a randomised, double-blind, placebo-controlled trial. Psychopharmacology (Berl) 236:1159–1170
Yaw AM, Woodruff RW, Prosser RA, Glass JD (2018) Paternal cocaine disrupts offspring circadian clock function in a sex-dependent manner in mice. Neuroscience 379:257–268
Zghoul T, Abarca C, Sanchis-Segura C et al (2007) Ethanol self-administration and reinstatement of ethanol-seeking behavior in Per1Brdm1 mutant mice. Psychopharmacology (Berl) 190:13–19
Zou Y, Liao G, Liu Y et al (2008) Association of the 54-nucleotide repeat polymorphism of hPer3 with heroin dependence in Han Chinese population. Genes Brain Behav 7:26–30
Zuardi AW (2008) Cannabidiol: from an inactive cannabinoid to a drug with wide spectrum of action. Rev Bras Psiquiatr 30:21–80

Circadian Rhythms in Mood Disorders

9

Madeline R. Scott and Colleen A. McClung

Abstract

Altered behavioral rhythms are a fundamental diagnostic feature of mood disorders. Patients report worse subjective sleep and objective measures confirm this, implicating a role for circadian rhythm disruptions in mood disorder pathophysiology. Molecular clock gene mutations are associated with increased risk of mood disorder diagnosis and/or severity of symptoms, and mouse models of clock gene mutations have abnormal mood-related behaviors. The mechanism by which circadian rhythms contribute to mood disorders remains unknown, however, circadian rhythms regulate and are regulated by various biological systems that are abnormal in mood disorders and this interaction is theorized to be a key component of mood disorder pathophysiology. A growing body of evidence has begun defining how the interaction of circadian and neurotransmitter systems influences mood and behavior, including the role of current antidepressants and mood stabilizers. Additionally, the hypothalamus-pituitary-adrenal (HPA) axis interacts with both circadian and monoaminergic systems and may facilitate the contribution of environmental stressors to mood disorder pathophysiology. The central role of circadian rhythms in mood disorders has led to the development of chronotherapeutics, which are treatments designed specifically to target circadian rhythm regulators, such as sleep, light, and melatonin, to produce an antidepressant response.

Keywords

Circadian Rhythms · Mood Disorders · Major Depression · Bipolar Disorder · Chronotherapeutics

9.1 Introduction

In this chapter we will discuss how circadian rhythm dysfunction is implicated in the pathophysiology of mood disorders. Mood disorders are conditions in which the primary symptom is a disturbance in mood and include both depressive disorders and bipolar disorder (Fig. 9.1). Mood disorders are experienced by a large proportion of the population and can have a devastating impact on people's lives. In 2016, depressive disorders had a global prevalence of 268.2 million people and accounted for 44.2 million years lost to disability (YLD), while bipolar disorder (BD) had a global prevalence of 43.9 million and accounted for 9 million YLD (Vos et al. 2017). Major depressive disorder (MDD) was the fifth leading cause of YLD globally and

M. R. Scott (✉) · C. A. McClung
Translational Neuroscience Program, Department of Psychiatry, University of Pittsburgh School of Medicine, Pittsburgh, PA, USA
e-mail: scottmr4@upmc.edu

O. Engmann, M. Brancaccio (eds.), *Circadian Clock in Brain Health and Disease*, Advances in Experimental Medicine and Biology 1344, https://doi.org/10.1007/978-3-030-81147-1_9

the second in the USA (Vos et al. 2017). As such, effective treatments are necessary to combat this major global health burden. Current treatments are only effective in some patients, and many individuals experience high rates of relapse (Hirschfeld 2012), emphasizing the need to better understand the pathophysiology underlying these disorders. While the primary symptoms of mood disorders are associated with mood, a key diagnostic characteristic is altered sleep and behavioral rhythms (Wirz-Justice and Benedetti 2019). Altered behavioral rhythms, along with direct genetic evidence of molecular clock dysfunction, implicate circadian rhythms in the pathophysiology of mood disorders. Additionally, many of the pathways that are independently associated with mood disorders are known to interact with circadian rhythms, supporting a role for circadian rhythms as a central regulator of mood disorder pathophysiology.

9.2 Sleep and Mood Disorders

Altered behavioral rhythms are a characteristic feature of mood disorders, with abnormal sleep patterns being the most distinct. Patients with MDD and BD report worse sleep quality and have higher rates of biological rhythm disruption, increased nighttime activity, and later mean timing of light exposure (Slyepchenko et al. 2019). In both patient populations, severity of depression symptoms is significantly associated with insomnia severity (Le Bon et al. 2019; Palagini et al. 2019). Patients with depression are less active, have longer wake after sleep onset, and a higher probability of transitioning from rest to activity during the day (Slyepchenko et al. 2019; Tazawa et al. 2019). In BD subjects, sleep-wake patterns are abnormal during both mood episodes and euthymic periods (Steinan et al. 2016; Takaesu et al. 2016; Tazawa et al. 2019). During depressive episodes, individuals with BD report both hypersomnia and insomnia, while manic episodes are associated with reduced need for sleep and shorter REM latencies (Detre et al. 1972; Loudon et al. 1977; Casper 1985; Hudson et al. 1988; Cassidy et al. 1998; Serretti and Olgiati 2005; Steinan et al. 2016; Takaesu 2018). Additionally, sleep disturbances frequently precede the onset of mood episodes in BD (Pancheri et al. 2019). While insomnia appears to precede both manic and depressive episodes, decreased need for sleep is more likely to predict a manic episode and hypersomnia specifically seems to occur prior to depressive episodes in BD patients (Pancheri et al. 2019). Furthermore, patients with hypersomnia have higher rates of BD diagnosis and lifetime suicide attempts, as well as poorer response to antidepressants, including antidepressant resistance and increased switching to hypomania (Murru et al. 2019). MDD and BD are also often comorbid with genetic sleep disorders, such as familial advanced phase sleep syndrome (FASPS) and delayed sleep phase syndrome (DSPS), in which patients either fall asleep and wake up much earlier or much later, respectively, than desired (Shirayama et al. 2003; Xu et al. 2005; Hamet and Tremblay 2006; McClung 2007; Steinan et al. 2016; Takaesu et al. 2016; Takaesu 2018). In a study examining subjective sleep quality of over 5000 inpatients with mental illness, improvements in sleep quality were significantly associated with changes in depressive symptoms (Schennach et al. 2019). Additionally, patients with less sleep disturbance at admission had a higher likelihood of achieving remission for depressive symptoms (Schennach et al. 2019).

In addition to sleep disturbances and disorders, predisposition for the eveningness chronotype, in which an individual is generally most active/alert in the evening, is associated with increased risk of depression (Kivelä et al. 2018). Higher rates of depressive disorders and increased severity of depression symptoms are associated with eveningness in childhood, adolescence, and adulthood (Gau et al. 2007; Hirata et al. 2007; Prat and Adan 2013; Alvaro et al. 2014; Lester 2015; Chiu et al. 2017; Haraden et al. 2017, 2019; López-Soto et al. 2019). Additionally, when comparing morningness-eveningness diurnal preference and molecular oscillations in individuals, evening-type subjects with circadian phase-delays are 20 times more likely to be depressed, while individuals with mismatched behavioral

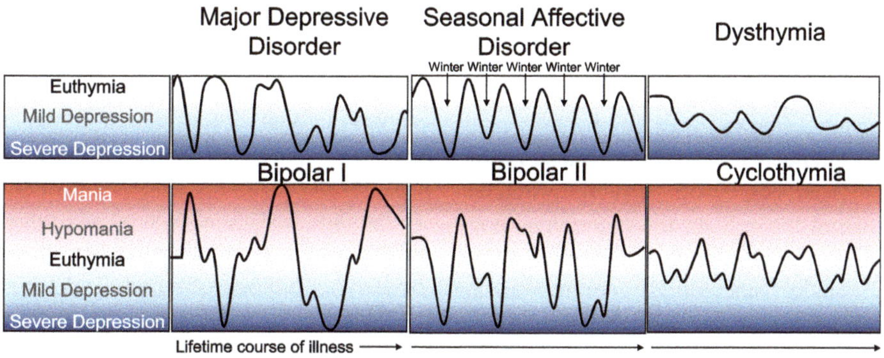

Fig. 9.1 Theoretical examples of how individuals experience mood disorders over lifetime course of illness. Depressive and bipolar disorders are characterized by combinations of periods of depression, euthymia, and mania. Major depressive disorder (MDD), seasonal affective disorder (SAD), and persistent depressive disorder (dysthymia) primarily include depressive periods, while bipolar disorder (BD) I, II and cyclothymia have periods of both severe depression and mania. MDD and SAD patients both experience severe depressive episodes, though these are specifically associated with winter in SAD. Patients with dysthymia experience chronic depression, though their symptoms do not reach the threshold of severe depression. BD I and II patients have episodes of both mania and severe depression, though BD II is characterized by less severe mania, called hypomania. Cyclothymic patients have periods of both mild depression and manic episodes that do not reach the threshold of hypomania

and molecular circadian phases are 5–8 times more likely to be depressed (Nguyen et al. 2019). Patients with SAD are also more likely to have an eveningness chronotype than comparison subjects (Natale et al. 2005; Lee et al. 2011). Studies linking chronotype to BD have had mixed results, with some demonstrating increased likelihood of an eveningness chronotype, and others showing no significant association with chronotype (Wood et al. 2009; Giglio et al. 2010; Fares et al. 2015; Melo et al. 2017; Kivelä et al. 2018). However, within bipolar patients, eveningness has been associated with more depressive episodes, increased severity of depressive symptoms, and worse functioning as assessed by the Functioning Assessment Short Test (Wood et al. 2009; Bullock et al. 2014; Melo et al. 2019).

Sleep disruptions are not only associated with mood symptoms but can also cause changes in mood. Chronic sleep deprivation exacerbates mood-related problems and individuals that have shifted work schedules are more likely to develop mood disorders (McClung 2007; Manber and Chambers 2009; Baglioni and Riemann 2012). Jet lag due to crossing multiple time zones also leads to higher reports of depression symptoms (Jauhar and Weller 1982; Young 1995; Katz et al. 2002) and, intriguingly, directionality of travel is associated with different mood symptoms, suggesting a direct connection between phase-delays, phase-advances, and mood episodes. Specifically, westward travel, causing a phase-delay, leads to depressive episodes and eastward travel, causing a phase-advance, is associated with mania (Jauhar and Weller 1982; Young 1995; Srinivasan et al. 2008).

Altogether, these studies demonstrate a clear connection between altered sleep and mood disorders, with some even suggesting a causative role for altered sleep/wake times in mood disorder pathophysiology. These findings indirectly support a role for circadian rhythms, due to their role in regulating sleep/wake patterns, and many studies have now sought to characterize molecular clock abnormalities associated with mood disorders.

9.3 The Molecular Clock in Mood Disorders

Human genetic studies have implicated genes within the molecular clock, a transcription-translation feedback loop driving circadian rhythms within the cell, in the pathophysiology of mood disorders (Mendoza and Vanotti 2019). Multiple single nucleotide polymorphisms (SNPs) in canonical molecular clock genes have been associated with increased risk of being diagnosed with MDD (NPAS2 rs11123857, CRY1 rs2287161, TIMELESS rs4630333) (Soria et al. 2010; Hua et al. 2014; Park et al. 2019), BD (CLOCK rs10462028, NR1D1 rs2314339, RORA rs782931, TIMELESS rs4630333, TIMELESS rs774045) (Kripke et al. 2009; Soria et al. 2010; Etain et al. 2014; Park et al. 2019), SAD (ARNTL rs2290035, CLOCK rs1801260, NPAS2 S471, PER2 10870, CRY2 rs10838524, CRY2 rs10838527, CRY2 rs3824872) (Partonen et al. 2007; Lavebratt et al. 2010; Kim et al. 2015), and dysthymia (CRY2 rs7121611, CRY2 rs7945565, CRY2 rs1301319, CRY2 rs10838524) (Kovanen et al. 2013). In patients with BD, SNPs in molecular clock genes have been associated with differences in symptoms, suicide attempts, daily activity, and white matter tract diffusivity (Table 9.1) (Artioli et al. 2007; Benedetti et al. 2007b; Sjöholm et al. 2010; Mccarthy et al. 2011; Pawlak et al. 2015, 2017; Bollettini et al. 2017). A variety of molecular clock SNPs have also been associated with severity of depression symptoms (ARTNL rs7107287, CLOCK rs1801260, PER3 rs57875989) (Benedetti et al. 2015; Jankowski and Dmitrzak-Weglarz 2017; Liberman et al. 2018) and likelihood of a violent suicide attempt (CLOCK rs3805148, CLOCK rs534654, TIMELESS rs2291739, TIMELESS rs11171856) (Pawlak et al. 2015). Interpretation of these results should be cautious as many of the SNPs are only identified as being associated with mood disorders in a single study. Intriguingly, when analyzed as a network rather than as individual genes, genetic differences in core clock genes are robustly associated with risk of a BD-associated illness (BD, MDD, schizophrenia, and attention deficit disorder) and with responsiveness to the mood stabilizer lithium (McCarthy et al. 2012), suggesting that dysregulation of the molecular clock as a pathway, rather than a specific gene, contributes to mood disorder pathophysiology. Altogether, these studies support a potential causative role for molecular clock dysfunction in the underlying biology of mood disorders.

Abnormalities in the circadian molecular clock go beyond gene alterations. In blood from MDD subjects, both increased mRNA expression of *CLOCK, BMAL1, PER1,* and *PER2* and an ablation of *PER1* and *CRY1* rhythmic expression have been observed (Gouin et al. 2010; Li et al. 2013b). Additionally, decreased mRNA expression of *DEC2* and *BMP* and reduced rhythmic amplitude of BMAL1, REV-ERBα, and DBP were detected in fibroblasts from BD patients (Yang et al. 2009). Twenty-four hour rhythmic gene expression is also dampened in human postmortem tissue from multiple brain regions, including the dorsolateral prefrontal cortex (DLPFC), hippocampus, nucleus accumbens (NAcc), and amygdala, from subjects with MDD (Li et al. 2013a; Ketchesin et al. 2018).

Animal studies examining the effect of molecular clock disruptions on psychiatric illness-related behavior further support a role for circadian rhythms in mood disorder pathophysiology. BMAL1 knockout (KO) in the cerebral cortex leads to increased immobility in the tail suspension test, a model of depressive-like behavior (Bering et al. 2018). Mice with the *ClockΔ19* mutation, which results in dominant-negative *Clock* function, have increased exploratory drive and impulsivity, hyperlocomotion, and increased behavioral responses to reward—consistent with a mania-like phenotype (Easton et al. 2003; McClung et al. 2005; Roybal et al. 2007; Dzirasa et al. 2010, 2011; Coque et al. 2011; Kozikowski et al. 2011; Ozburn et al. 2012, 2013; Bernardi and Spanagel 2013; van Enkhuizen et al. 2013; Bernardi and Spanagel 2014; Arey et al. 2014; Sidor et al. 2015). Individually, $Per1^{ldc-/-}$ and $Per2^{ldc-/-}$ mice do not have clear mood-related phenotypes, but when both genes are disrupted they show increased anxiety-like behavior (Bae

Table 9.1 Features associated with molecular clock SNPs within BD patients

ARNTL rs4757142	Number of depressive episodes per year (Pawlak et al. 2017)
ARNTL rs1481892	Number of depressive episodes per year (Pawlak et al. 2017)
CLOCK rs1801260	Higher evening activity levels, delayed sleep onset, reduced amount of sleep (Benedetti et al. 2007b)
CLOCK rs1801260	Increased diffusivity in white matter tracts (Bollettini et al. 2017)
PER3 rs10462021	Number of depressive episodes (Pawlak et al. 2017)
PER3 rs57875989	Increased novelty seeking (Artioli et al. 2007)
CRY1 rs8192440	Positive treatment response to lithium (Mccarthy et al. 2011)
CRY2 rs10838524	Rapid cycling (Sjöholm et al. 2010)
NR1D1 rs2071427	Positive treatment response to lithium (Mccarthy et al. 2011)
TIMELESS rs2291739	Ratio of depressive to hypo/manic episodes (Pawlak et al. 2017)
TIMELESS rs2291739	Family history of suicide attempts (Pawlak et al. 2015)
TIMELESS rs10876890	Number of depressive episodes (Pawlak et al. 2017)
TIMELESS rs10876890	Ratio of depressive to hypo/manic episodes (Pawlak et al. 2017)
TIMELESS rs1082214	Ratio of depressive to hypo/manic episodes (Pawlak et al. 2017)
TIMELESS rs2279665	Ratio of depressive to hypo/manic episodes (Pawlak et al. 2017)
TIMELESS rs11171856	Ratio of depressive to hypo/manic episodes (Pawlak et al. 2015, 2017)

et al. 2001; Spencer et al. 2013). *Rev-erbα* KO mice have a mania-like phenotype (Chung et al. 2014) and proteins that facilitate degradation of molecular clock components, like *FBXL3* and casein kinases *CK1δ* and *CK1ε*, are also associated with changes in mood behavior (Landgraf et al. 2014a).

Altogether these studies demonstrate a clear connection between molecular clock disruption and mood disorders. Mutations in molecular clock genes are associated with increased risk of mood disorder diagnosis and/or severity of symptoms, while evidence of altered circadian rhythms in expression of core clock components and molecular clock regulated transcripts has been found in both peripheral and brain tissue from human subjects. Mouse models of clock gene mutations also impact mood-related behaviors. The mechanism by which circadian rhythms contribute to mood disorders remains unknown, however, circadian rhythms regulate and are regulated by various biological systems, as is discussed in earlier chapters of this book. Many of these pathways are abnormal in mood disorders and their interaction with circadian rhythms is theorized to be a key component of mood disorder pathophysiology. In this chapter, we focus primarily on what is currently known about how the interaction of circadian rhythms with two major pathways, neurotransmission and the HPA axis, is dysregulated in mood disorders

9.4 Circadian Associated Biological Processes in Mood Disorders

9.4.1 Neurotransmission

Altered neurotransmission is a fundamental component of mood disorders, and current pharmacological treatments used in mood disorders are primarily targeted toward modifying neurotransmitter systems (Hirschfeld 2012). Various classes of antidepressants and mood stabilizing drugs target monoaminergic systems such as serotonin and norepinephrine, while the primary mechanism of action of antipsychotics is through dopamine receptors (Hirschfeld 2012). These drugs can impact rhythmic locomotor activity along with rhythmic neural activity and PER1 expression in the suprachiasmatic nucleus (SCN) of rodents. Specific serotonin reuptake inhibitors (SSRIs), a commonly used antidepressant, cause phase-advances and shortened period in locomotor activity, along with phase advanced neural activity in the SCN and shortened period of PER1 expression (Possidente et al. 1992, 1996; Prosser et al. 2006; Cuesta et al. 2008, 2009; Nomura et al. 2008). The mood stabilizers lithium and valproic acid have distinct impacts on circadian rhythms. Lithium treatment leads to a longer rhythmic period in locomotor activity, neural activity in the SCN, and PER2 expression in the SCN (Johnsson et al. 1983; LeSauter and Silver 1993; Abe et al. 2000; Johansson et al. 2011; Li et al. 2012; Noguchi et al. 2016; Yoshikawa and Honma 2016), while valproic acid causes shorter rhythmic period in locomotor activity and PER2 expression (Landgraf et al. 2016). However, both drugs increase the amplitude of PER2 rhythmic expression in the SCN (Johansson et al. 2011; Li et al. 2012; Yoshikawa and Honma 2016), suggesting that this effect may be a common mechanism underlying the treatment efficacy of mood stabilizers.

Consistent with bidirectional regulation of monoamines and circadian rhythms, expression and release of serotonin, dopamine, and norepinephrine is rhythmic, as is expression of monoamine receptors and associated enzymes (McClung 2007). Additionally, several mouse models of circadian rhythm disruption have shown that changes in mood are associated with alterations in monoamines. The manic-like phenotype of *ClockΔ19* mice specifically occurs during the day and is associated with daytime increases in dopaminergic neuronal activity, tyrosine hydroxylase (TH) expression, and dopamine synthesis in the ventral tegmental area (VTA) (Sidor et al. 2015; Logan et al. 2018). Directly manipulating *Clock* expression and/or dopaminergic neuronal activity within the VTA is sufficient to recapitulate the manic-like phenotype of the *ClockΔ19* mouse, though VTA-specific knockdown of *Clock* also results in an increase of depression-like behavior, resulting in a mixed manic/depressive model (Mukherjee et al. 2010; Sidor et al. 2011, 2015). The effect of CLOCK on dopamine appears to, at least in part, be due to its activity as a transcription factor for TH (Sidor et al. 2015; Logan et al. 2018). Supporting this, daytime-specific administration of a TH inhibitor rescues the manic-like phenotype in *ClockΔ19* mice (Sidor et al. 2015). Additionally, KO of *Rev-erbα,* another mouse model that shows a manic-like phenotype, also leads to a hyperdopaminergic state in the VTA that appears to be due to a loss of TH repression by REV-ERBα (Chung et al. 2014). Changes in the light-dark cycle, either through exposure to a short light-dark cycle (22 h) or constant darkness, can lead to depressive-like behavior in rodents (Gonzalez and Aston-Jones 2008; Hamo et al. 2016). Depression-like symptoms resulting from a short light-dark cycle are accompanied by desynchronized SCN oscillations, elevated norepinephrine and dopamine levels, and increased serotonin turnover in the prefrontal cortex, while constant darkness leads to apoptosis of noradrenergic neurons in the locus coeruleus (Gonzalez and Aston-Jones 2008; Hamo et al. 2016). Consistent with a role for norepinephrine in contributing to depression symptoms, *Bmal1* KO in the cerebral cortex, which results in depressive-like behavior in rodents, is associated with reduced norepinephrine levels (Bering et al. 2018).

In addition to monoaminergic signaling, the role of glutamatergic neurotransmission has become of interest recently due to the discovery that ketamine, a glutamate receptor antagonist, has rapid antidepressant effects in patients (Duman 2018). In NG108-15 neuronal cells, ketamine inhibits CLOCK:BMAL1-mediated transcription in a GSK3β-dependent manner (Bellet et al. 2011) and mice treated with ketamine have reduced expression of the molecular clock components *Per2, Npas4, Rorb, Dbp,* and *Ciart* (Orozco-Solis et al. 2017). Additionally, in patients with MDD, response to ketamine treatment was associated with baseline measures of slow wave sleep and actigraphy-measured activity levels (Duncan et al. 2013a, b, 2017a, b). Further research is necessary to determine the exact mechanisms by which ketamine causes changes in mood, but the association with treatment efficacy and changes in sleep/wake patterns suggests that circadian rhythms may mediate the therapeutic impact of ketamine.

9.4.2 Stress and the HPA Axis

Both MDD and BD are strongly associated with stress. High levels of childhood stress are correlated with increased risk of developing a mood disorder, and stressful life events frequently precede onset of depressive and manic episodes (Kendler et al. 1999; Landgraf et al. 2014b). Supporting a role for stress in mood disorders, BD, MDD, and SAD patients all have abnormalities in hypothalamus-pituitary-adrenal (HPA) axis function (Avery et al. 1997; Muneer 2016; Dean and Keshavan 2017). MDD specifically is associated with chronically elevated cortisol and reduced sensitivity to glucocorticoids, though atypical depression, which is characterized by lethargy, fatigue, hyperphagia, and weight gain, is associated with hypocortisolism (Landgraf et al. 2014b; Ketchesin et al. 2018). Over half of all patients with Cushing's disease, a disorder characterized by hypercortisolism, develop a comorbid mood disorder which can be treated by normalizing cortisol levels (Bratek et al. 2015), suggesting a potential causal role for HPA axis dysfunction in mood disorders. The strong connection between stress and depression is further supported by rodent models in which animals exposed to stress in a variety of ways have phenotypes associated with depression symptoms (Krishnan and Nestler 2008). For example, chronic unpredictable stress (CUS) leads to depressive-like behavior phenotypes in rats, including reductions in locomotor activity, weight gain, and sucrose consumption (Zhao and Fu 2017) and prenatal exposure to chronic maternal restraint stress results in higher anxiety- and depression-like behavior in adulthood (Darnaudéry and Maccari 2008). These models of stress-induced mood-associated behaviors are also accompanied by sleep disturbances and changes to biological rhythms in body temperature, heart rate, and locomotion (Darnaudéry and Maccari 2008; Zhao and Fu 2017), suggesting a role for circadian rhythm disruption in stress-induced mood dysregulation.

The HPA axis and circadian rhythms are known to regulate each other. Glucocorticoid expression and activity of corticotropin releasing hormone (CRH) and adrenocorticotropic hormone (ACTH) have strong circadian rhythms and various molecular clock proteins directly regulate diurnal changes in glucocorticoid receptor expression and sensitivity (Ketchesin et al. 2018). In patients with depression, diurnal variation of cortisol is blunted (Wichers et al. 2008; Doane et al. 2013), while mice with circadian disruptions have changes in both mood-related behavior and HPA axis activity (Ketchesin et al. 2018). Mutations in molecular clock proteins including *Clock, Bmal1, Cry,* and *Per* lead to altered glucocorticoid expression and rhythmicity in mice, and exposure to long photoperiods, which in mice is equivalent to a short active period, results in both depressive-like behavior and high corticosterone levels (Dulcis et al. 2013; Leliavski et al. 2014; McClung and Becker-Krail 2016; Koch et al. 2017). Conversely, the HPA axis modifies circadian rhythms. Multiple molecular clock genes have glucocorticoid response elements that allow glucocorticoids to regulate their transcription (Ketchesin et al. 2018). This is evident with *Per2* expression, for which it has been shown that glucocorticoids stimulate *Per2* rhythms and inhibiting glucocorticoid receptors alters *Per2* rhythms in the bed nucleus of the stria terminalis and the amygdala (Ketchesin et al. 2018). Furthermore, CUS alters the amplitude of *Per2* expression rhythms in a region-specific manner, with CUS leading to increased amplitude in the NAc and decreased amplitude in the SCN of mice (Logan et al. 2015). The degree of *Per2* rhythm amplitude change is directly correlated with measures of depressive-like behavior in mice that underwent CUS (Logan et al. 2015), which suggests that stress-induced behavioral changes may be working through alterations in the molecular clock. Consistent with this, when studies consider the variability of response to stress in mice they find that susceptibility to stress is associated with circadian rhythm disruptions. In one example, a chronic social defeat paradigm led to a depressive-like phenotype in some mice but not others, which is accompanied by altered biological rhythms only in the susceptible mice, but not those resilient to the stress paradigm (Krishnan and Nestler 2008). In another, stress-induced changes in restlessness, sleep disturbances, and measures of HPA axis activity were specific to mice that have high-responsiveness to stress and are not found in low responders (Touma et al. 2009).

Stress and HPA axis signaling are also interconnected with monoaminergic signaling. Acute stress increases serotonin and dopamine release and activity of serotonergic neurons (Bao and Swaab 2019) while CUS is associated with decreased monoamine levels in rats (Dean and Keshavan 2017). In human imaging studies, stress-induced cortisol release is correlated to dopamine accumulation in the ventral striatum (Bao and Swaab 2019). Both serotonin and dopamine receptors are found associated with CRH positive neurons in the paraventricular nucleus (PVN) of the hypothalamus, and serotonin and dopamine receptor ligands cause increases in circulating corticosterone levels (Bao and Swaab 2019). Monoamines may also be mediating the connection between circadian rhythm disruptions and the HPA axis. Dulcis et al. (2013) have observed that changing photoperiod length leads to interneurons in the PVN switching from producing dopamine to somatostatin, or vice versa. Long photoperiods, associated with both anxiety- and depressive-like behaviors and an elevated HPA axis response, had increased switching from dopamine to somatostatin, while short photoperiods, which produced manic-like behavior, had increased switching from somatostatin to dopamine (Dulcis et al. 2013). Supporting a fundamental role for PVN dopaminergic neurons in mood-associated behavior, when the authors ablated DA neurons in the PVN, mice displayed increased depressive- and anxiety-like behavior that were rescued by short photoperiod-induced increases in DA neurons in the PVN (Dulcis et al. 2013).

Mood disorder pathophysiology is incredibly complex and many other systems besides neurotransmission and the HPA axis are dysregulated in patients with mood disorders. Strikingly, many of the major processes implicated in mood disorders, like inflammation, mitochondrial dysfunction, and metabolic peptide abnormalities, are also known to be regulated by circadian rhythms (McClung 2013; Ketchesin et al. 2018). Along with the growing body of work that has begun specifically defining how the interaction of circadian rhythms, neurotransmitter systems, and HPA axis dysfunction influences mood and behavior, these connections emphasize the importance of considering circadian rhythm disruptions and the molecular clock when studying mood disorders and developing potential therapeutics

9.5 Chronotherapeutics

Chronotherapeutics are treatments designed specifically to target circadian rhythm regulators, such as sleep, light, and melatonin, to produce an antidepressant response. These include acute sleep deprivation (SD), which usually consists of having a patient refrain from sleep for 24 h, bright light therapy (BLT), which involves exposing patients to a blue waveform light source similar to direct sunlight, and pharmacological melatonin treatment (Wirz-Justice and Benedetti 2019). While these treatments have been shown to have antidepressant effects on their own, synergistic effects have also been observed on depression symptoms for SD when combined with either antidepressant drugs or other chronotherapeutic techniques like BLT.

SD has been repeatedly shown to induce a rapid antidepressant effect in patients with MDD (Wirz-Justice and Benedetti 2019). In MDD patients, SD that successfully results in an antidepressant effect also normalizes elevated metabolism in the medial prefrontal and anterior cingulate cortex (Wu et al. 2001). Consistent with a role for the interaction of neurotransmission and circadian rhythms in depression pathophysiology, animal models of SD show that it leads to a variety of changes in monoaminergic signaling that are comparable to treatment with antidepressants (Wirz-

Justice and Benedetti 2019). SD increases extracellular serotonin and synaptic levels of noradrenaline, along with mRNA expression of TH and the noradrenaline transporter in the locos coeruleus (Basheer et al. 1998; Lopez-Rodriguez et al. 2003; Hipólide et al. 2005). Additionally, SD increases activity of serotonergic and dopaminergic neurons and leads to an elevated behavioral response to serotonin precursors and dopamine agonists (Tufik et al. 1978; Mogilnicka 1981; Santos and Carlini 1983; Gardner et al. 1997). Early on during SD, both dopamine receptor binding sites and glutamate release increase, which is then followed by downregulation after prolonged stimulation (Wirz-Justice et al. 1981; Zwicker and Calil 1986; Dash et al. 2009). Additionally, the clinical efficacy of sleep deprivation is modulated by genotypes that regulate density of the serotonin transporter and 5HT2A receptor, expression of glutamatergic post-synaptic scaffolding proteins, and the efficiency of catechol-O-methyltransferase (COMT)-mediated clearance of noradrenaline and dopamine from the synapse (Wirz-Justice and Benedetti 2019). While the antidepressant effects of SD are rapid and striking, they are also relatively transient and, as such, don't represent a long-term strategy for dealing with chronic depression.

BLT is used to treat both SAD and non-seasonal depression (Wirz-Justice and Benedetti 2019). Therapeutic response in MDD requires longer bright light therapy than SAD, but is comparable to treatment response to antidepressants (Al-Karawi and Jubair 2016; Penders et al. 2016; Perera et al. 2016). Notably, unlike antidepressants, BLT is effective in improving depressive symptoms without increasing the risk of a manic switch in BD patients (Takaesu 2018). In SAD patients, BLT clinical efficacy is associated with its impact on serotonergic signaling (Wirz-Justice and Benedetti 2019). SAD patients have excessive seasonal increases in serotonin transporter binding, which is reduced by BLT (Tyrer et al. 2016a, b), and both rapid tryptophan/serotonin and catecholamine depletion reverse the antidepressant effects of BLT in SAD (Lam et al. 1996; Neumeister et al. 1998). The combination of SD and BLT has a significant impact on cortical activity and connectivity in MDD patients, including enhancing neural responses in the prefrontal cortex to emotional stimuli (Benedetti et al. 2007a). Additionally, the combination of these therapies improves connectivity of cortical and limbic structures in patients with MDD, for whom inefficient coupling is associated with depression symptoms (Radaelli et al. 2015; Vai et al. 2015). While the majority of chronotherapeutics target depression symptoms, dark therapy has been used to treat manic episodes in BD patients (Wirz-Justice and Benedetti 2019). Dark therapy, which involves placing patients experiencing a manic episode in a dark room during nighttime and, in one case, giving the patients blue-light blocking glasses to wear, has a striking stabilizing effect in multiple case reports (Takaesu 2018).

In addition to these behavioral interventions, the success of the drug agomelatine further emphasizes the importance of considering circadian abnormalities in treatment of mood disorders. Agomelatine is a melatonin receptor agonist and a serotonin receptor (5-HT2C) antagonist and has been shown to improve depression symptoms in patients with MDD (Komaram et al. 2015; Kennedy et al. 2016, 2018; Udristoiu et al. 2016; Che et al. 2018; Chen and Xie 2018). This response is significantly higher than placebo (Kennedy et al. 2016, 2018), and is comparable to SSRIs (Guaiana et al. 2013; Komaram et al. 2015; Udristoiu et al. 2016; Che et al. 2018; Chen and Xie 2018). Intriguingly, in several studies agomelatine had a greater effect on other symptoms associated with depression like heart-rate variability and anxiety (Yeh et al. 2016; Che et al. 2018; Chen and Xie 2018). The use of agomelatine as an antidepressant for BD patients, however, has been mixed, with some studies showing reduced depressive symptoms and others demonstrating no improvement over placebo (Fornaro et al. 2013; Ushkalova et al. 2015; Yatham et al. 2016).

9.6 Conclusion

Dysregulated sleep has long been observed as a feature of psychiatric illness. Studies have now demonstrated a clear connection between altered sleep and mood disorders, and the impact of jet lag, shift work, and chronic sleep deprivation even suggest a causative role for altered sleep/wake times in mood disorder pathophysiology. These findings implicate a role for circadian rhythms, which is directly supported by human genetic studies. Increased risk of mood disorder diagnosis and/or increased symptom severity is associated with clock gene mutations, and animal models have demonstrated altered mood-associated behaviors after molecular clock disruptions. These animal models suggest circadian rhythm disruptions contribute to mood disorders, in part, through their interaction with monoaminergic signaling. Furthermore, current pharmacological therapeutics for mood disorders, including antidepressants and mood destabilizers, have been shown to cause changes in both behavioral and biological circadian rhythms, implicating a fundamental role for the bidirectional relationship between circadian rhythms and monoaminergic neurotransmission. The HPA axis also interacts with both circadian and monoaminergic systems and may facilitate the contribution of environmental stressors to mood disorder pathophysiology. Animal models of circadian disruption have abnormal HPA axis function, and stress-induced depressive-like behavior is accompanied by changes in behavioral and biological rhythms. Further work will be necessary to fully understand the complex interactions between these various processes and circadian rhythms; however, the fundamental role of circadian rhythms in mood disorders has already led to the development of treatments that target circadian rhythm regulators like sleep, light, and melatonin. Both behavioral interventions, like light therapy, dark therapy, and sleep deprivation, and the drug agomelatine show therapeutic benefit for patients with mood disorders. These promising results further emphasize the vital need to consider the role of circadian rhythms and the molecular clock as the field continues to both investigate the complex underlying pathophysiology of mood disorders and develop effective therapies to treat this major global health burden.

References

Abe M, Herzog ED, Block GD (2000) Lithium lengthens the circadian period of individual suprachiasmatic nucleus neurons. Neuroreport. https://doi.org/10.1097/00001756-200009280-00042

Al-Karawi D, Jubair L (2016) Bright light therapy for nonseasonal depression: meta-analysis of clinical trials. J Affect Disord 198:64–71

Alvaro PK, Roberts RM, Harris JK (2014) The independent relationships between insomnia, depression, subtypes of anxiety, and chronotype during adolescence. Sleep Med. https://doi.org/10.1016/j.sleep.2014.03.019

Arey RN, Enwright JF, Spencer SM et al (2014) An important role for Cholecystokinin, a CLOCK target gene, in the development and treatment of manic-like behaviors. Mol Psychiatry. https://doi.org/10.1038/mp.2013.12

Artioli P, Lorenzi C, Pirovano A et al (2007) How do genes exert their role? Period 3 gene variants and possible influences on mood disorder phenotypes. Eur Neuropsychopharmacol. https://doi.org/10.1016/j.euroneuro.2007.03.004

Avery DH, Dahl K, Savage MV et al (1997) Circadian temperature and cortisol rhythms during a constant routine are phase-delayed in hypersomnic winter depression. Biol Psychiatry. https://doi.org/10.1016/S0006-3223(96)00210-7

Bae K, Jin X, Maywood ES et al (2001) Differential functions of mPer1, mPer2, and mPer3 in the SCN circadian clock. Neuron 30:525–536

Baglioni C, Riemann D (2012) Is chronic insomnia a precursor to major depression? Epidemiological and biological findings. Curr Psychiatry Rep. https://doi.org/10.1007/s11920-012-0308-5

Bao AM, Swaab DF (2019) The human hypothalamus in mood disorders: the HPA axis in the center. IBRO Reports

Basheer R, Magner M, McCarley RW, Shiromani PJ (1998) REM sleep deprivation increases the levels of tyrosine hydroxylase and norepinephrine transporter mRNA in the locus coeruleus. Mol Brain Res. https://doi.org/10.1016/S0169-328X(98)00088-6

Bellet MM, Vawter MP, Bunney BG et al (2011) Ketamine influences CLOCK:BMAL1 function leading to altered circadian gene expression. PLoS One. https://doi.org/10.1371/journal.pone.0023982

Benedetti F, Bernasconi A, Blasi V et al (2007a) Neural and genetic correlates of antidepressant response to sleep deprivation: a functional magnetic resonance imaging study of moral valence decision in bipolar depression. Arch Gen Psychiatry. https://doi.org/10.1001/archpsyc.64.2.179

Benedetti F, Dallaspezia S, Fulgosi MC et al (2007b) Actimetric evidence that CLOCK 3111 T/C SNP influences sleep and activity patterns in patients affected by bipolar depression. Am J Med Genet Part B Neuropsychiatr Genet. https://doi.org/10.1002/ajmg.b.30475

Benedetti F, Riccaboni R, Dallaspezia S et al (2015) Effects of CLOCK gene variants and early stress on hopelessness and suicide in bipolar depression. Chronobiol Int. https://doi.org/10.3109/07420528.2015.1060603

Bering T, Carstensen MB, Wörtwein G et al (2018) The circadian oscillator of the cerebral cortex: molecular, biochemical and behavioral effects of deleting the Arntl clock gene in cortical neurons. Cereb Cortex. https://doi.org/10.1093/cercor/bhw406

Bernardi RE, Spanagel R (2013) The Clockδ19 mutation in mice fails to alter the primary and secondary reinforcing properties of nicotine. Drug Alcohol Depend. https://doi.org/10.1016/j.drugalcdep.2013.08.024

Bernardi RE, Spanagel R (2014) Enhanced extinction of contextual fear conditioning in ClockΔ19 mutant mice. Behav Neurosci. https://doi.org/10.1037/a0037020

Bollettini I, Melloni EMT, Aggio V et al (2017) Clock genes associate with white matter integrity in depressed bipolar patients. Chronobiol Int. https://doi.org/10.1080/07420528.2016.1260026

Bratek A, Koźmin-Burzyńska A, Górniak E, Krysta K (2015) Psychiatric disorders associated with Cushing's syndrome. Psychiatr Danub 27 Suppl 1:S339–S343

Bullock B, Corlass-Brown J, Murray G (2014) Eveningness and seasonality are associated with the bipolar disorder vulnerability trait. J Psychopathol Behav Assess. https://doi.org/10.1007/s10862-014-9414-5

Casper RC (1985) Somatic symptoms in primary affective disorder. Arch Gen Psychiatry. https://doi.org/10.1001/archpsyc.1985.01790340082012

Cassidy F, Murry E, Forest K, Carroll BJ (1998) Signs and symptoms of mania in pure and mixed episodes. J Affect Disord. https://doi.org/10.1016/S0165-0327(98)00016-0

Che T, Teng X, Huang Q et al (2018) Agomelatine versus fluoxetine in glycemic control and treating depressive and anxiety symptoms in type 2 diabetes mellitus subjects: a single-blind randomized controlled trial. Neuropsychiatr Dis Treat. https://doi.org/10.2147/NDT.S167608

Chen JW, Xie SQ (2018) Agomelatine versus paroxetine in treating depressive and anxiety symptoms in patients with chronic kidney disease. Neuropsychiatr Dis Treat. https://doi.org/10.2147/NDT.S159636

Chiu WH, Yang HJ, Kuo PH (2017) Chronotype preference matters for depression in youth. Chronobiol Int. https://doi.org/10.1080/07420528.2017.1327441

Chung S, Lee EJ, Yun S et al (2014) Impact of circadian nuclear receptor REV-ERBα on midbrain dopamine production and mood regulation. Cell. https://doi.org/10.1016/j.cell.2014.03.039

Coque L, Mukherjee S, Cao JL et al (2011) Specific role of VTA dopamine neuronal firing rates and morphology in the reversal of anxiety-related, but not depression-related behavior in the clockδ19 mouse model of mania. Neuropsychopharmacology. https://doi.org/10.1038/npp.2011.33

Cuesta M, Mendoza J, Clesse D et al (2008) Serotonergic activation potentiates light resetting of the main circadian clock and alters clock gene expression in a diurnal rodent. Exp Neurol. https://doi.org/10.1016/j.expneurol.2007.11.026

Cuesta M, Clesse D, Pévet P, Challet E (2009) New light on the serotonergic paradox in the rat circadian system. J Neurochem. https://doi.org/10.1111/j.1471-4159.2009.06128.x

Darnaudéry M, Maccari S (2008) Epigenetic programming of the stress response in male and female rats by prenatal restraint stress. Brain Res Rev 57(2):571–585

Dash MB, Douglas CL, Vyazovskiy VV et al (2009) Long-term homeostasis of extracellular glutamate in the rat cerebral cortex across sleep and waking states. J Neurosci. https://doi.org/10.1523/JNEUROSCI.5486-08.2009

Dean J, Keshavan M (2017) The neurobiology of depression: an integrated view. Asian J Psychiatr 27:101–111

Detre T, Himmelhoch J, Swartzburg M et al (1972) Hypersomnia and manic-depressive disease. Am J Psychiatry. https://doi.org/10.1176/ajp.128.10.1303

Doane LD, Mineka S, Zinbarg RE et al (2013) Are flatter diurnal cortisol rhythms associated with major depression and anxiety disorders in late adolescence? the role of life stress and daily negative emotion. Dev Psychopathol. https://doi.org/10.1017/S0954579413000060

Dulcis D, Jamshidi P, Leutgeb S, Spitzer NC (2013) Neurotransmitter switching in the adult brain regulates behavior. Science. https://doi.org/10.1126/science.1234152

Duman RS (2018) Ketamine and rapid-acting antidepressants: a new era in the battle against depression and suicide [version 1; referees: 3 approved]. F1000Research. https://doi.org/10.12688/f1000research.14344.1

Duncan WC, Sarasso S, Ferrarelli F et al (2013a) Concomitant BDNF and sleep slow wave changes indicate ketamine-induced plasticity in major depressive disorder. Int J Neuropsychopharmacol. https://doi.org/10.1017/S1461145712000545

Duncan WC, Selter J, Brutsche N et al (2013b) Baseline delta sleep ratio predicts acute ketamine mood response in major depressive disorder. J Affect Disord. https://doi.org/10.1016/j.jad.2012.05.042

Duncan WC, Ballard ED, Zarate CA (2017a) Ketamine-induced glutamatergic mechanisms of sleep and wakefulness: insights for developing novel treatments for

disturbed sleep and mood. Handb Exp Pharmacol 253:337–358
Duncan WC, Slonena E, Hejazi NS et al (2017b) Motor-activity markers of circadian timekeeping are related to ketamine's rapid antidepressant properties. Biol Psychiatry. https://doi.org/10.1016/j.biopsych.2017.03.011
Dzirasa K, Coque L, Sidor MM et al (2010) Lithium ameliorates nucleus accumbens phase-signaling dysfunction in a genetic mouse model of mania. J Neurosci. https://doi.org/10.1523/JNEUROSCI.4289-10.2010
Dzirasa K, Mcgarity DL, Bhattacharya A et al (2011) Impaired limbic gamma oscillatory synchrony during anxiety-related behavior in a genetic mouse model of bipolar mania. J Neurosci. https://doi.org/10.1523/JNEUROSCI.6144-10.2011
Easton A, Arbuzova J, Turek FW (2003) The circadian Clock mutation increases exploratory activity and escape-seeking behavior. Genes Brain Behav. https://doi.org/10.1034/j.1601-183X.2003.00002.x
Etain B, Jamain S, Milhiet V et al (2014) Association between circadian genes, bipolar disorders and chronotypes. Chronobiol Int. https://doi.org/10.3109/07420528.2014.906445
Fares S, Hermens DF, Naismith SL et al (2015) Clinical correlates of chronotypes in young persons with mental disorders. Chronobiol Int. https://doi.org/10.3109/07420528.2015.1078346
Fornaro M, McCarthy MJ, De Berardis D et al (2013) Adjunctive agomelatine therapy in the treatment of acute bipolar II depression: a preliminary open label study. Neuropsychiatr Dis Treat. https://doi.org/10.2147/NDT.S41557
Gardner JP, Fornal CA, Jacobs BL (1997) Effects of sleep deprivation on serotonergic neuronal activity in the dorsal raphe nucleus of the freely moving cat. Neuropsychopharmacology. https://doi.org/10.1016/S0893-133X(97)00025-0
Gau SSF, Shang CY, Merikangas KR et al (2007) Association between morningness-eveningness and behavioral/emotional problems among adolescents. J Biol Rhythms 22(3):268–274
Giglio LMF, Magalhães PVS, Andersen ML et al (2010) Circadian preference in bipolar disorder. Sleep Breath. https://doi.org/10.1007/s11325-009-0301-3
Gonzalez MMC, Aston-Jones G (2008) Light deprivation damages monoamine neurons and produces a depressive behavioral phenotype in rats. Proc Natl Acad Sci U S A. https://doi.org/10.1073/pnas.0703615105
Gouin JP, Connors J, Kiecolt-Glaser JK et al (2010) Altered expression of circadian rhythm genes among individuals with a history of depression. J Affect Disord. https://doi.org/10.1016/j.jad.2010.04.002
Guaiana G, Gupta S, Chiodo D et al (2013) Agomelatine versus other antidepressive agents for major depression. Cochrane Database Syst Rev. https://doi.org/10.1002/14651858.CD008851.pub2
Hamet P, Tremblay J (2006) Genetics of the sleep-wake cycle and its disorders. Metabolism 55(10 Suppl 2):S7–S12
Hamo MB, Larson TA, Duge LS et al (2016) Circadian forced desynchrony of the master clock leads to phenotypic manifestation of depression in rats. eNeuro. https://doi.org/10.1523/ENEURO.0237-16.2016
Haraden DA, Mullin BC, Hankin BL (2017) The relationship between depression and chronotype: a longitudinal assessment during childhood and adolescence. Depress Anxiety. https://doi.org/10.1002/da.22682
Haraden DA, Mullin BC, Hankin BL (2019) Internalizing symptoms and chronotype in youth: a longitudinal assessment of anxiety, depression and tripartite model. Psychiatry Res. https://doi.org/10.1016/j.psychres.2018.12.117
Hipólide DC, Moreira KM, Barlow KBL et al (2005) Distinct effects of sleep deprivation on binding to norepinephrine and serotonin transporters in rat brain. Prog Neuro-Psychopharmacol Biol Psychiatry. https://doi.org/10.1016/j.pnpbp.2004.11.015
Hirata FC, Lima MCO, De Bruin VMS et al (2007) Depression in medical school: the influence of morningness-eveningness. Chronobiol Int. https://doi.org/10.1080/07420520701657730
Hirschfeld RMA (2012) The epidemiology of depression and the evolution of treatment. J Clin Psychiatry 73 Suppl 1:5–9
Hua P, Liu W, Chen D et al (2014) Cry1 and Tef gene polymorphisms are associated with major depressive disorder in the Chinese population. J Affect Disord. https://doi.org/10.1016/j.jad.2013.11.019
Hudson JI, Lipinski JF, Frankenburg FR et al (1988) Electroencephalographic sleep in mania. Arch Gen Psychiatry. https://doi.org/10.1001/archpsyc.1988.01800270085010
Jankowski KS, Dmitrzak-Weglarz M (2017) ARNTL, CLOCK and PER3 polymorphisms—links with chronotype and affective dimensions. Chronobiol Int. https://doi.org/10.1080/07420528.2017.1343341
Jauhar P, Weller MPI (1982) Psychiatric morbidity and time zone changes: a study of patients from Heathrow Airport. Br J Psychiatry. https://doi.org/10.1192/bjp.140.3.231
Johansson AS, Brask J, Owe-Larsson B et al (2011) Valproic acid phase shifts the rhythmic expression of PERIOD2::LUCIFERASE. J Biol Rhythms. https://doi.org/10.1177/0748730411419775
Johnsson A, Engelmann W, Pflug B, Klemke W (1983) Period lengthening of human circadian rhythms by lithium carbonate, a prophylactic for depressive disorders. Int J Chronobiol 8(3):129–147
Katz G, Knobler HY, Laibel Z et al (2002) Time zone change and major psychiatric morbidity: the results of a 6-year study in Jerusalem. Compr Psychiatry. https://doi.org/10.1053/comp.2002.29849
Kendler KS, Karkowski LM, Prescott CA (1999) Causal relationship between stressful life events and the onset

of major depression. Am J Psychiatry. https://doi.org/10.1176/ajp.156.6.837

Kennedy SH, Avedisova A, Belaïdi C et al (2016) Sustained efficacy of agomelatine 10 mg, 25 mg, and 25-50 mg on depressive symptoms and functional outcomes in patients with major depressive disorder. A placebo-controlled study over 6 months. Eur Neuropsychopharmacol. https://doi.org/10.1016/j.euroneuro.2015.09.006

Kennedy SH, Heun R, Avedisova A et al (2018) Effect of agomelatine 25–50 mg on functional outcomes in patients with major depressive disorder. J Affect Disord. https://doi.org/10.1016/j.jad.2018.05.060

Ketchesin KD, Becker-Krail D, McClung CA (2018) Mood-related central and peripheral clocks. Eur J Neurosci. https://doi.org/10.1111/ejn.14253

Kim HI, Lee HJ, Cho CH et al (2015) Association of CLOCK, ARNTL, and NPAS2 gene polymorphisms and seasonal variations in mood and behavior. Chronobiol Int. https://doi.org/10.3109/07420528.2015.1049613

Kivelä L, Papadopoulos MR, Antypa N (2018) Chronotype and psychiatric disorders. Curr Sleep Med Rep. https://doi.org/10.1007/s40675-018-0113-8

Koch CE, Leinweber B, Drengberg BC et al (2017) Interaction between circadian rhythms and stress. Neurobiol Stress 6:57–67

Komaram RB, Srikrishna N, Palla J et al (2015) A comparative study of efficacy and safety of agomelatine and escitalopram in major depressive disorder. J Clin Diagnostic Res. https://doi.org/10.7860/JCDR/2015/12371.6092

Kovanen L, Kaunisto M, Donner K et al (2013) CRY2 genetic variants associate with dysthymia. PLoS One. https://doi.org/10.1371/journal.pone.0071450

Kozikowski AP, Gunosewoyo H, Guo S et al (2011) Identification of a glycogen synthase kinase-3β inhibitor that attenuates hyperactivity in CLOCK mutant mice. ChemMedChem. https://doi.org/10.1002/cmdc.201100188

Kripke DF, Nievergelt CM, Joo EJ et al (2009) Circadian polymorphisms associated with affective disorders. J Circadian Rhythms. https://doi.org/10.1186/1740-3391-7-2

Krishnan V, Nestler EJ (2008) The molecular neurobiology of depression. Nature 455(7215):894–902

Lam RW, Zis AP, Grewal A et al (1996) Effects of rapid tryptophan depletion in patients with seasonal affective disorder in remission after light therapy. Arch Gen Psychiatry. https://doi.org/10.1001/archpsyc.1996.01830010043007

Landgraf D, McCarthy MJ, Welsh DK (2014a) The role of the circadian clock in animal models of mood disorders. Behav Neurosci. https://doi.org/10.1037/a0036029

Landgraf D, McCarthy MJ, Welsh DK (2014b) Circadian clock and stress interactions in the molecular biology of psychiatric disorders. Curr Psychiatry Rep 16 (10):483

Landgraf D, Joiner WJ, McCarthy MJ et al (2016) The mood stabilizer valproic acid opposes the effects of dopamine on circadian rhythms. Neuropharmacology. https://doi.org/10.1016/j.neuropharm.2016.03.047

Lavebratt C, Sjöholm LK, Soronen P et al (2010) CRY2 is associated with depression. PLoS One. https://doi.org/10.1371/journal.pone.0009407

Le Bon O, Lanquart JP, Hein M, Loas G (2019) Sleep ultradian cycling: statistical distribution and links with other sleep variables, depression, insomnia and sleepiness—a retrospective study on 2,312 polysomnograms. Psychiatry Res. https://doi.org/10.1016/j.psychres.2018.12.141

Lee HJ, Rex KM, Nievergelt CM et al (2011) Delayed sleep phase syndrome is related to seasonal affective disorder. J Affect Disord. https://doi.org/10.1016/j.jad.2011.04.046

Leliavski A, Shostak A, Husse J, Oster H (2014) Impaired glucocorticoid production and response to stress in arntl-deficient male mice. Endocrinology. https://doi.org/10.1210/en.2013-1531

LeSauter J, Silver R (1993) Lithium lengthens the period of circadian rhythms in lesioned hamsters bearing SCN grafts. Biol Psychiatry. https://doi.org/10.1016/0006-3223(93)90259-G

Lester D (2015) Morningness—eveningness, current depression, and past suicidality. Psychol Rep. https://doi.org/10.2466/16.02.PR0.116k18w5

Li J, Lu WQ, Beesley S et al (2012) Lithium impacts on the amplitude and period of the molecular circadian clockwork. PLoS One. https://doi.org/10.1371/journal.pone.0033292

Li JZ, Bunney BG, Meng F et al (2013a) Circadian patterns of gene expression in the human brain and disruption in major depressive disorder. Proc Natl Acad Sci USA. https://doi.org/10.1073/pnas.1305814110

Li SX, Liu LJ, Xu LZ et al (2013b) Diurnal alterations in circadian genes and peptides in major depressive disorder before and after escitalopram treatment. Psychoneuroendocrinology. https://doi.org/10.1016/j.psyneuen.2013.07.009

Liberman AR, Halitjaha L, Ay A, Ingram KK (2018) Modeling strengthens molecular link between circadian polymorphisms and major mood disorders. J Biol Rhythms. https://doi.org/10.1177/0748730418764540

Logan RW, Edgar N, Gillman AG et al (2015) Chronic stress induces brain region-specific alterations of molecular rhythms that correlate with depression-like behavior in mice. Biol Psychiatry. https://doi.org/10.1016/j.biopsych.2015.01.011

Logan RW, Parekh PK, Kaplan GN et al (2018) NAD+ cellular redox and SIRT1 regulate the diurnal rhythms of tyrosine hydroxylase and conditioned cocaine reward. Mol Psychiatry. https://doi.org/10.1038/s41380-018-0061-1

Lopez-Rodriguez F, Wilson CL, Maidment NT et al (2003) Total sleep deprivation increases extracellular serotonin in the rat hippocampus. Neuroscience. https://doi.org/10.1016/S0306-4522(03)00335-X

López-Soto PJ, Fabbian F, Cappadona R et al (2019) Chronotype, nursing activity, and gender: a systematic review. J Adv Nurs 75(4):734–748

Loudon JB, Ashworth CM, Blackburn IM (1977) A study of the symptomatology and course of manic illness using a new scale. Psychol Med. https://doi.org/10.1017/s0033291700006395

Manber R, Chambers AS (2009) Insomnia and depression: a multifaceted interplay. Curr Psychiatry Rep 11 (6):437–442

Mccarthy MJ, Nievergelt CM, Shekhtman T et al (2011) Functional genetic variation in the Rev-Erbα pathway and lithium response in the treatment of bipolar disorder. Genes Brain Behav. https://doi.org/10.1111/j.1601-183X.2011.00725.x

McCarthy MJ, Nievergelt CM, Kelsoe JR, Welsh DK (2012) A survey of genomic studies supports association of circadian clock genes with bipolar disorder spectrum illnesses and lithium response. PLoS One. https://doi.org/10.1371/journal.pone.0032091

McClung CA (2007) Circadian genes, rhythms and the biology of mood disorders. Pharmacol Ther 114 (2):222–232

McClung CA (2013) How might circadian rhythms control mood? Let me count the ways. Biol Psychiatry 74 (4):242–249

McClung C, Becker-Krail D (2016) Implications of circadian rhythm and stress in addiction vulnerability. F1000Research 5:59

McClung CA, Sidiropoulou K, Vitaterna M et al (2005) Regulation of dopaminergic transmission and cocaine reward by the Clock gene. Proc Natl Acad Sci U S A. https://doi.org/10.1073/pnas.0503584102

Melo MCA, Abreu RLC, Linhares Neto VB et al (2017) Chronotype and circadian rhythm in bipolar disorder: a systematic review. Sleep Med Rev 34:46–58

Melo MC, Garcia RF, de Araújo CF et al (2019) Chronotype in bipolar disorder: an 18-month prospective study. Braz J Psychiatry. https://doi.org/10.1590/1516-4446-2019-0489

Mendoza J, Vanotti G (2019) Circadian neurogenetics of mood disorders. Cell Tissue Res 377(1):81–94

Mogilnicka E (1981) REM sleep deprivation changes behavioral response to catecholaminergic and serotonergic receptor activation in rats. Pharmacol Biochem Behav. https://doi.org/10.1016/0091-3057(81)90355-5

Mukherjee S, Coque L, Cao JL et al (2010) Knockdown of clock in the ventral tegmental area through RNA interference results in a mixed state of mania and depression-like behavior. Biol Psychiatry. https://doi.org/10.1016/j.biopsych.2010.04.031

Muneer A (2016) The neurobiology of bipolar disorder: an integrated approach. Chonnam Med J. https://doi.org/10.4068/cmj.2016.52.1.18

Murru A, Guiso G, Barbuti M et al (2019) The implications of hypersomnia in the context of major depression: results from a large, international, observational study. Eur Neuropsychopharmacol. https://doi.org/10.1016/j.euroneuro.2019.02.011

Natale V, Adan A, Scapellato P (2005) Are seasonality of mood and eveningness closely associated? Psychiatry Res. https://doi.org/10.1016/j.psychres.2004.12.010

Neumeister A, Turner EH, Matthews JR et al (1998) Effects of tryptophan depletion vs catecholamine depletion in patients with seasonal affective disorder in remission with light therapy. Arch Gen Psychiatry. https://doi.org/10.1001/archpsyc.55.6.524

Nguyen C, Murray G, Anderson S et al (2019) In vivo molecular chronotyping, circadian misalignment, and high rates of depression in young adults. J Affect Disord 250:425–431. https://doi.org/10.1016/j.jad.2019.03.050

Noguchi T, Lo K, Diemer T, Welsh DK (2016) Lithium effects on circadian rhythms in fibroblasts and suprachiasmatic nucleus slices from Cry knockout mice. Neurosci Lett. https://doi.org/10.1016/j.neulet.2016.02.030

Nomura K, Castanon-Cervantes O, Davidson A, Fukuhara C (2008) Selective serotonin reuptake inhibitors and raft inhibitors shorten the period of Period1-driven circadian bioluminescence rhythms in rat-1 fibroblasts. Life Sci. https://doi.org/10.1016/j.lfs.2008.03.024

Orozco-Solis R, Montellier E, Aguilar-Arnal L et al (2017) A circadian genomic signature common to ketamine and sleep deprivation in the anterior cingulate cortex. Biol Psychiatry. https://doi.org/10.1016/j.biopsych.2017.02.1176

Ozburn AR, Larson EB, Self DW, McClung CA (2012) Cocaine self-administration behaviors in ClockΔ19 mice. Psychopharmacology (Berl). https://doi.org/10.1007/s00213-012-2704-2

Ozburn AR, Falcon E, Mukherjee S et al (2013) The role of clock in ethanol-related behaviors. Neuropsychopharmacology. https://doi.org/10.1038/npp.2013.138

Palagini L, Cipollone G, Masci I et al (2019) Insomnia symptoms predict emotional dysregulation, impulsivity and suicidality in depressive bipolar II patients with mixed features. Compr Psychiatry. https://doi.org/10.1016/j.comppsych.2018.12.009

Pancheri C, Verdolini N, Pacchiarotti I et al (2019) A systematic review on sleep alterations anticipating the onset of bipolar disorder. Eur Psychiatry 58:45–53

Park M, Kim SA, Yee J et al (2019) Significant role of gene–gene interactions of clock genes in mood disorder. J Affect Disord 257:510–517. https://doi.org/10.1016/j.jad.2019.06.056

Partonen T, Treutlein J, Alpman A et al (2007) Three circadian clock genes Per2, Arntl, and Npas2 contribute to winter depression. Ann Med. https://doi.org/10.1080/07853890701278795

Pawlak J, Dmitrzak-Weglarz M, Maciukiewicz M et al (2015) Suicidal behavior in the context of disrupted rhythmicity in bipolar disorder-Data from an association study of suicide attempts with clock genes. Psychiatry Res. https://doi.org/10.1016/j.psychres.2015.01.010

Pawlak J, Szczepankiewicz A, Kapelski P et al (2017) Suicidal behavior in the context of disrupted rhythmicity in bipolar disorder—complementary research of clock genes with suicide risks factors and course of disease. Psychiatry Res. https://doi.org/10.1016/j.psychres.2017.07.057

Penders TM, Stanciu CN, Schoemann AM et al (2016) Bright light therapy as augmentation of pharmacotherapy for treatment of depression: a systematic review and meta-analysis. Prim Care Companion CNS Disord 18(5)

Perera S, Eisen R, Bhatt M et al (2016) Light therapy for non-seasonal depression: systematic review and meta-analysis. BJPsych Open. https://doi.org/10.1192/bjpo.bp.115.001610

Possidente B, Lumia AR, McEldowney S, Rapp M (1992) Fluoxetine shortens circadian period for wheel running activity in mice. Brain Res Bull. https://doi.org/10.1016/0361-9230(92)90114-D

Possidente B, Lumia AR, McGinnis MY et al (1996) Effects of fluoxetine and olfactory bulbectomy on mouse circadian activity rhythms. Brain Res. https://doi.org/10.1016/0006-8993(95)01490-X

Prat G, Adan A (2013) Relationships among circadian typology, psychological symptoms, and sensation seeking. Chronobiol Int. https://doi.org/10.3109/07420528.2013.790044

Prosser RA, Lee HM, Wehner A (2006) Serotonergic pre-treatments block in vitro serotonergic phase shifts of the mouse suprachiasmatic nucleus circadian clock. Neuroscience. https://doi.org/10.1016/j.neuroscience.2006.06.014

Radaelli D, Sferrazza Papa G, Vai B et al (2015) Fronto-limbic disconnection in bipolar disorder. Eur Psychiatry. https://doi.org/10.1016/j.eurpsy.2014.04.001

Roybal K, Theobold D, Graham A et al (2007) Mania-like behavior induced by disruption of CLOCK. Proc Natl Acad Sci U S A. https://doi.org/10.1073/pnas.0609625104

Santos R, Carlini EA (1983) Serotonin receptor activation in rats previously deprived of REM sleep. Pharmacol Biochem Behav. https://doi.org/10.1016/0091-3057(83)90271-X

Schennach R, Feige B, Riemann D et al (2019) Pre- to post-inpatient treatment of subjective sleep quality in 5,481 patients with mental disorders: a longitudinal analysis. J Sleep Res. https://doi.org/10.1111/jsr.12842

Serretti A, Olgiati P (2005) Profiles of "manic" symptoms in bipolar I, bipolar II and major depressive disorders. J Affect Disord. https://doi.org/10.1016/j.jad.2003.09.011

Shirayama M, Shirayama Y, Iida H et al (2003) The psychological aspects of patients with delayed sleep phase syndrome (DSPS). Sleep Med. https://doi.org/10.1016/S1389-9457(03)00101-1

Sidor MM, Spencer S, Tye K, Warden M, Dzirasa K, Deisseroth K et al (2011) Optogenetic control of mesolimbic dopamine neural activity recapitulates the anxiety-related phenotype of the clock-delta19 mouse model of mania. Neuropsychopharmacology 36:S92

Sidor MM, Spencer SM, Dzirasa K et al (2015) Daytime spikes in dopaminergic activity drive rapid mood-cycling in mice. Mol Psychiatry. https://doi.org/10.1038/mp.2014.167

Sjöholm LK, Backlund L, Cheteh EH et al (2010) CRY2 is associated with rapid cycling in bipolar disorder-patients. PLoS One. https://doi.org/10.1371/journal.pone.0012632

Slyepchenko A, Allega OR, Leng X et al (2019) Association of functioning and quality of life with objective and subjective measures of sleep and biological rhythms in major depressive and bipolar disorder. Aust N Z J Psychiatry. https://doi.org/10.1177/0004867419829228

Soria V, Martínez-Amorós È, Escaramís G et al (2010) Differential association of circadian genes with mood disorders: CRY1 and NPAS2 are associated with unipolar major depression and clock and VIP with bipolar disorder. Neuropsychopharmacology. https://doi.org/10.1038/npp.2009.230

Spencer S, Falcon E, Kumar J et al (2013) Circadian genes Period 1 and Period 2 in the nucleus accumbens regulate anxiety-related behavior. Eur J Neurosci. https://doi.org/10.1111/ejn.12010

Srinivasan V, Spence DW, Pandi-Perumal SR et al (2008) Jet lag: therapeutic use of melatonin and possible application of melatonin analogs. Travel Med Infect Dis. https://doi.org/10.1016/j.tmaid.2007.12.002

Steinan MK, Morken G, Lagerberg TV et al (2016) Delayed sleep phase: an important circadian subtype of sleep disturbance in bipolar disorders. J Affect Disord. https://doi.org/10.1016/j.jad.2015.11.025

Takaesu Y (2018) Circadian rhythm in bipolar disorder: a review of the literature. Psychiatry Clin Neurosci 72(9):673–682

Takaesu Y, Inoue Y, Murakoshi A et al (2016) Prevalence of circadian rhythm sleep-wake disorders and associated factors in euthymic patients with bipolar disorder. PLoS One. https://doi.org/10.1371/journal.pone.0159578

Tazawa Y, Wada M, Mitsukura Y et al (2019) Actigraphy for evaluation of mood disorders: a systematic review and meta-analysis. J Affect Disord 253:257–269. https://doi.org/10.1016/j.jad.2019.04.087

Touma C, Fenzl T, Ruschel J et al (2009) Rhythmicity in mice selected for extremes in stress reactivity: behavioural, endocrine and sleep changes resembling endophenotypes of major depression. PLoS One. https://doi.org/10.1371/journal.pone.0004325

Tufik S, Lindsey CJ, Carlini EA (1978) Does REM sleep deprivation induce a supersensitivity of dopaminergic receptors in the rat brain. Pharmacology. https://doi.org/10.1159/000136753

Tyrer AE, Levitan RD, Houle S et al (2016a) Serotonin transporter binding is reduced in seasonal affective disorder following light therapy. Acta Psychiatr Scand. https://doi.org/10.1111/acps.12632

Tyrer AE, Levitan RD, Houle S et al (2016b) Increased seasonal variation in serotonin transporter binding in seasonal affective disorder. Neuropsychopharmacology. https://doi.org/10.1038/npp.2016.54

Udristoiu T, Dehelean P, Nuss P et al (2016) Early effect on general interest, and short-term antidepressant efficacy and safety of agomelatine (25-50 mg/day) and escitalopram (10-20 mg/day) in outpatients with Major Depressive Disorder. A 12-week randomised double-blind comparative study. J Affect Disord. https://doi.org/10.1016/j.jad.2016.03.048

Ushkalova AV, Kostyukova EG, Mosolov SN (2015) A comparative study of different types of pharmacotherapy in treatment of depressive phase of bipolar II disorder. Zhurnal Nevrol i Psihiatr Im SS Korsakova. https://doi.org/10.17116/jnevro20151151223-30

Vai B, Poletti S, Radaelli D et al (2015) Successful antidepressant chronotherapeutics enhance fronto-limbic neural responses and connectivity in bipolar depression. Psychiatry Res Neuroimaging. https://doi.org/10.1016/j.pscychresns.2015.07.015

van Enkhuizen J, Minassian A, Young JW (2013) Further evidence for Clockδ19 mice as a model for bipolar disorder mania using cross-species tests of exploration and sensorimotor gating. Behav Brain Res. https://doi.org/10.1016/j.bbr.2013.04.023

Vos T, Abajobir AA, Abbafati C et al (2017) Global, regional, and national incidence, prevalence, and years lived with disability for 328 diseases and injuries for 195 countries, 1990-2016: a systematic analysis for the Global Burden of Disease Study 2016. Lancet. https://doi.org/10.1016/S0140-6736(17)32154-2

Wichers MC, Myin-Germeys I, Jacobs N et al (2008) Susceptibility to depression expressed as alterations in cortisol day curve: a cross-twin, cross-trait study. Psychosom Med. https://doi.org/10.1097/PSY.0b013e31816b1eee

Wirz-Justice A, Benedetti F (2019) Perspectives in affective disorders: clocks and sleep. Eur J Neurosci. https://doi.org/10.1111/ejn.14362

Wirz-Justice A, Tobler I, Kafka MS et al (1981) Sleep deprivation: effects on circadian rhythms of rat brain neurotransmitter receptors. Psychiatry Res. https://doi.org/10.1016/0165-1781(81)90062-7

Wood J, Birmaher B, Axelson D et al (2009) Replicable differences in preferred circadian phase between bipolar disorder patients and control individuals. Psychiatry Res. https://doi.org/10.1016/j.psychres.2008.03.003

Wu JC, Buchsbaum M, Bunney WE (2001) Clinical neurochemical implications of sleep deprivation's effects on the anterior cingulate of depressed responders. Neuropsychopharmacology. https://doi.org/10.1016/S0893-133X(01)00336-0

Xu Y, Padiath QS, Shapiro RE et al (2005) Functional consequences of a CKIδ mutation causing familial advanced sleep phase syndrome. Nature. https://doi.org/10.1038/nature03453

Yang S, Van Dongen HPA, Wang K et al (2009) Assessment of circadian function in fibroblasts of patients with bipolar disorder. Mol Psychiatry. https://doi.org/10.1038/mp.2008.10

Yatham LN, Vieta E, Goodwin GM et al (2016) Agomelatine or placebo as adjunctive therapy to a mood stabiliser in bipolar I depression: randomised double-Blind placebo-Controlled trial. Br J Psychiatry. https://doi.org/10.1192/bjp.bp.114.147587

Yeh TC, Kao LC, Tzeng NS et al (2016) Heart rate variability in major depressive disorder and after antidepressant treatment with agomelatine and paroxetine: findings from the Taiwan Study of Depression and Anxiety (TAISDA). Prog Neuro-Psychopharmacol Biol Psychiatry. https://doi.org/10.1016/j.pnpbp.2015.07.007

Yoshikawa T, Honma S (2016) Lithium lengthens circadian period of cultured brain slices in area specific manner. Behav Brain Res. https://doi.org/10.1016/j.bbr.2016.07.045

Young DM (1995) Psychiatric morbidity in travelers to Honolulu, Hawaii. Compr Psychiatry. https://doi.org/10.1016/0010-440X(95)90086-B

Zhao Y, Fu Y (2017) Effects of chronic stress depression on the circadian rhythm of peripheral neuroendocrine hormone of rats. Zhongguo Ying Yong Sheng Li Xue Za Zhi. https://doi.org/10.12047/j.cjap.5529.2017.096

Zwicker AP, Calil HM (1986) The effects of REM sleep deprivation on striatal dopamine receptor sites. Pharmacol Biochem Behav. https://doi.org/10.1016/0091-3057(86)90415-6

The Reciprocal Interaction Between Sleep and Alzheimer's Disease

10

Samuel S. Harris, Tom Schwerd-Kleine, Byung Il Lee, and Marc Aurel Busche

Abstract

It is becoming increasingly recognized that patients with a variety of neurodegenerative diseases exhibit disordered sleep/wake patterns. While sleep impairments have typically been thought of as sequelae of underlying neurodegenerative processes in sleep-wake cycle regulating brain regions, including the brainstem, hypothalamus, and basal forebrain, emerging evidence now indicates that sleep deficits may also act as pathophysiological drivers of brain-wide disease progression. Specifically, recent work has indicated that impaired sleep can impact on neuronal activity, brain clearance mechanisms, pathological build-up of proteins, and inflammation. Altered sleep patterns may therefore be novel (potentially reversible) dynamic functional markers of proteinopathies and modifiable targets for early therapeutic intervention using non-invasive stimulation and behavioral techniques. Here we highlight research describing a potentially reciprocal interaction between impaired sleep and circadian patterns and the accumulation of pathological signs and features in Alzheimer's disease, the most prevalent neurodegenerative disease in the elderly.

Keywords

Sleep impairment · Sleep-wake cycle · Slow-wave sleep · Learning and memory · Alzheimer's disease · Amyloid-beta · Tau · Clinical · Translational

10.1 Background

Sleep is a highly complex and regulated brain state which, although incompletely understood, has been implicated in cellular and network restitution, learning and synaptic plasticity, removal of neurotoxic waste products, and modulation of endocrine and immune functions, among other functions (Diekelmann and Born 2010; Klinzing et al. 2019; Tononi and Cirelli 2014; Zada et al. 2019; Xie et al. 2013; Besedovsky et al. 2019). Sleep deprivation (SD), and even sleep fragmentation, can lead to the emergence of poor memory, impaired cognition and epileptic seizures, and even frank psychosis with disorganized thinking and a loss of ability to perceive reality (Krause et al. 2017; Joiner 2019). Sleep is therefore not a passive and quiescent, steady state, but rather is associated with specific and alternating neuronal rhythms that may be broadly classified as rapid-eye movement (REM) or non-REM phases. REM sleep is associated with oscillations in the theta

S. S. Harris · T. Schwerd-Kleine · B. I. Lee · M. A. Busche (✉)
UK Dementia Research Institute at UCL, London, UK
e-mail: samuel.harris@ucl.ac.uk; tom.schwerd-kleine@uni-tuebingen.de; byungil.lee@ucl.ac.uk; m.busche@ucl.ac.uk

O. Engmann, M. Brancaccio (eds.), *Circadian Clock in Brain Health and Disease*, Advances in Experimental Medicine and Biology 1344, https://doi.org/10.1007/978-3-030-81147-1_10

and gamma frequency bands, whereas cortical slow-wave oscillations (<1 Hz), delta waves (1–4.5 Hz), thalamocortical spindles, and hippocampal sharp-wave ripples occur during non-REM (NREM) sleep. The different sleep stages and oscillations are believed to be related to aspects of sleep physiology and linked to a variety of physiological parameters such as changes in hormone release, cardiovascular control, regulation of breathing, convulsive thresholds, and gastrointestinal physiology (Taheri et al. 2002). In this regard, it is important to note that our knowledge of the function and role of REM sleep, beyond its inferred contribution to implicit procedural and emotional memory consolidation (Diekelmann and Born 2010; Tononi and Cirelli 2014; Klinzing et al. 2019) and forgetting (Izawa et al. 2019), significantly lags behind our emerging understanding of NREM sleep, and that of its deepest phase, slow-wave sleep (SWS), which is the primary focus of the current chapter.

Memory consolidation is thought to occur during SWS and requires effective hippocampal-cortical communication involving the fine-scale temporal coupling between hippocampal ripples, neocortical slow-wave oscillations, and thalamocortical spindles (Diekelmann and Born 2010; Sirota et al. 2003; Latchoumane et al. 2017; Maingret et al. 2016; Helfrich et al. 2019). Distinct patterns of firing activity in specific hippocampal neuronal ensembles during awake behavior are reactivated during subsequent SWS and manifest as hippocampal sharp-wave ripples (Wilson and McNaughton 1994; Rasch et al. 2007). Hippocampal memories are then posited to be transferred to neocortical regions for long-term storage, facilitated by a priming effect of thalamocortical spindles, and gated by the induction of up and down states (membrane depolarization and hyperpolarization, respectively) by neocortical slow oscillations (Huber et al. 2004; Sirota and Buzsáki 2005; Sirota et al. 2003; Klinzing et al. 2019; but see Yonelinas et al. 2019). Interestingly, whereas slow-wave activity has been associated with memory consolidation, delta waves, on the other hand, have recently been implicated in forgetting and related to the differences in temporal coupling of spindles to both forms of activity (Kim et al. 2019). In turn, the synaptic homeostasis hypothesis theorizes that slow-wave activity during sleep enables the brain to renormalize synaptic strength, that is potentiated during wakefulness, through relational synaptic downscaling that enables neural circuits to operate energy- and space-efficiently, and which promotes learning and memory (Tononi and Cirelli 2014; Tononi and Cirelli 2006). It is important to note that the memory consolidation and homeostatic functions of NREM sleep, often tacitly perceived as independent, may not in fact be mutually exclusive, and may act differentially on neuronal populations according to their intrinsic firing rate properties (Levenstein et al. 2017).

Sleep is also associated with a dramatic increase in interstitial space due to shrinkage of glial cells, which allows for influx of cerebrospinal fluid (CSF) into the brain parenchyma along paravascular spaces surrounding penetrating vasculature, driven by slow-wave and, subsequently, hemodynamic oscillations, and enabling convective exchange between CSF and interstitial fluid (ISF) leading to "glymphatic" clearance of waste and metabolites (Xie et al. 2013; Iliff et al. 2013; Fultz et al. 2019; Kiviniemi et al. 2016). Finally, it has also been suggested that sleep and immunity are bi-directionally related, with SWS associated with the promotion of inflammatory homeostasis and cytokine production, and sleep loss leading to low-grade systemic inflammation and immunodeficiency (Besedovsky et al. 2012, 2019), as well as higher expression of genes characteristic of aged-microglia and microglial activation (Kaneshwaran et al. 2019).

10.2 Sleep Disruption in Alzheimer's Disease

Ageing alters sleep architecture, most evidently as a reduction of non-REM-associated SWS (Ohayon et al. 2004), but also in a variety of other manners, including an advance in circadian phase and reduced circadian amplitude, decreased REM sleep, reduced sleep efficiency, increased arousals, enhanced sleep fragmentation, and sleep-disordered breathing (e.g., obstructive sleep apnea, OSA). Sleep disturbances are exacerbated in AD relative to normal ageing (Prinz et al. 1982; Bombois et al. 2010) and become more severe with disease progression and are associated with increased cerebrospinal fluid (CSF) levels of orexin (also known as hypocretin), a hypothalamic neuropeptide that regulates sleep and arousal states but also appetite, with increased levels being linked to wakefulness, and predicted by CSF levels of Aβ and tau, the two hallmark proteins in AD (Liguori et al. 2014; Gabelle et al. 2017). Clinical sleep phenotypes in AD dementia include insomnia, nighttime wakefulness and wandering, excessive daytime sleepiness, and sundowning (a tendency to become confused and agitated towards the evening) (Cipriani et al. 2015) and have been associated with differential Aβ deposition patterns, e.g. Aβ deposition in brainstem and precuneus was reportedly linked to daytime sleepiness and nocturnal wakefulness, respectively (You et al. 2019). Sleep phenotypes may in turn be exacerbated by cognitive and behavioral symptoms and other factors such as medication, lack of exercise, and nighttime lighting. Sleep appears to be both a global phenomenon, involving neuronal networks in brain stem, hypothalamus and basal forebrain, and pathological sleep phenotypes may thus arise due to disruption or degeneration of specific sleep-related circuits, for example those involving wake-promoting neurons in locus coeruleus, orexin-producing neurons in the lateral hypothalamic area (LHA), and histaminergic neurons in the tuberomammillary nucleus (TMN) (Swaab et al. 1985; Clark and Warren 2013; Oh et al. 2019a, b), although local processes within specific brain areas, particularly the cortex, may also play a critical, albeit overlooked, role in sleep modulation (Krueger et al. 2019).

Sleep disturbances are thus prevalent in AD and have, historically, been considered an epiphenomenon of the associated neurodegenerative process in the disorder. However, it is now being increasingly realized that sleep perturbations are also manifest in early AD prior to the emergence of widespread neurodegeneration and cognitive symptoms (Musiek et al. 2015; Sprecher et al. 2017) and even in those at risk of developing the disorder (e.g., the presence of the apolipoprotein E epsilon4 (ApoE4) genotype, a risk factor for AD, is also associated with increased risk of sleep-disordered breathing, Kadotani et al. 2001). In addition, recent data indicate that sleep impairments, via direct or indirect processes, are, in of themselves, able to instigate the abnormal release and build-up of the pathogenic proteins characteristic of AD, Aβ, and tau, thus intensifying the risk of developing the disorder and accelerating disease progression. This reciprocal interaction between sleep and AD is exemplified by the observation that sleep disturbances increase the risk of AD, whereas enhanced sleep hygiene has the antagonistic effect (Osorio et al. 2011; Yaffe et al. 2011; Lim et al. 2013a, b). In the following sections, we highlight the groundswell of research which supports the mediating and reciprocal role of sleep and circadian dysregulation in the development of Aβ and tau pathology.

10.3 Sleep Impairment Promotes the Emergence of AD-Related Pathological Features

10.3.1 Clinical Evidence

Cerebrospinal fluid (CSF) Aβ and tau levels in healthy humans display inherent diurnal variations (decreases during sleep and increases during wakefulness) (Kang et al. 2009; Holth et al. 2019). Interestingly, similar oscillatory behavior in CSF Aβ levels was also observed in human subjects harboring a PS1 mutation (a major cause of familial AD), but which was lost in subjects also exhibiting Aβ deposition by positron emission tomography (PET) imaging (Roh et al. 2012) and supported by a previous report of diurnal CSF Aβ oscillations that becomes attenuated with Aβ deposition (Huang et al. 2012). These reports indicate that CSF Aβ and tau levels exhibit a sleep-wake cycle which likely reflects associated broad state changes in neuronal activity (note that while overall firing rates and metabolism are decreased during sleep, sleep sub-states differentially affect neurons according to their firing properties with markedly heterogenous effects, see Watson et al. (2016)) and activity-dependent variations in production and/or extracellular release of Aβ and tau (Cirrito et al. 2005; Yamada et al. 2014; Pooler et al. 2013) as well as fluctuations in their clearance into the CSF, for example via the glymphatic/paravascular system (Xie et al. 2013) (Fig. 10.1). As described below, impaired sleep and prolonged wakefulness in cognitively normal individuals markedly disrupt these behaviors and may initiate a deleterious cascade leading to AD-like pathophysiology and cognitive deficits.

Self-reported measures of excessive daytime sleepiness, which may arise due to sleep-disordered breathing (e.g., OSA) or fragmented/insufficient sleep, but also the loss of wake-promoting neurons (Oh et al. 2019a), in cognitively normal adults (>60 years), was associated with greater than double the odds of PET Aβ deposition at follow-up ~15 years later (Spira et al. 2018), with increased sleep latency and sleep fragmentation also linked to cortical PET Aβ load in cognitively normal individuals (Ettore et al. 2019). One night of total SD, monitored using polysomnography, in healthy middle-aged men, was associated with an increase in Aβ42 cerebrospinal fluid (CSF) levels (Ooms et al. 2014) and, similarly, overnight levels of Aβ38, Aβ40, and Aβ42 in CSF markedly increased (~30%) in sleep-deprived cognitively normal adults (30–60 years) relative to controls (Lucey et al. 2018). Natural overnight decreases in Aβ42 plasma levels were also attenuated by sleep fragmentation in psychiatrists following 90 days of being on-call (Grimmer et al. 2020). Short-term (24 h) SD in healthy adults (mean 27.3 years) was also associated with significantly increased morning plasma Aβ40 and serum malondialdehyde levels (a marker for oxidative stress), significantly decreased Aβ42/Aβ40 ratio, and serum superoxide dismutase levels (a marker of antioxidant activity), and significantly decreased plasma lipoprotein receptor-related protein 1 (LRP-1) and receptor for advanced glycation end products (RAGE) concentrations, that were correlated to Aβ42 and Aβ40 levels, suggesting increased oxidative stress and impaired peripheral clearance of Aβ (Wei et al. 2017). Self-reported diminished sleep quality in cognitively normal adults was also found to be associated with increased PET Aβ load (Choe et al. 2019), with the number of nocturnal awakenings in healthy older adults also inversely correlated to insular gray matter volume, a region notably activated during sleep spindles (Branger et al. 2016). PET Aβ burden in healthy controls following one night of SD increased in the right hippocampus and thalamus, with baseline burden in subcortical areas and precuneus negatively correlated to reported sleep time during rested sleep, but which was not related to genetic risk for AD (APOE genotype) (Shokri-Kojori et al. 2018). Genetic variations in the water-channel protein aquaporin-4 (AQP4), expressed in astrocytic end-feet, in cognitively normal older adults, however, were associated with disrupted self-reported sleep quality (Pittsburgh Sleep Quality Index, PSQI) and found to moderate the coupling between sleep latency/duration and PET Aβ burden (Rainey-Smith et al. 2018; Brown et al. 2016). Given the implication of AQP4 in glymphatic transport (Mestre et al. 2018), as well as the modulatory effect of AQP4-haplotype on NREM slow-wave energy (Larsen et al. 2019), these results support the notion of paravascular clearance of Aβ and suggest that genetic variations in AQP4 modulate the efficacy of this process.

SD was also observed to increase CSF tau by over half in a human cohort (Holth et al. 2019) with changes in phosphorylation being highly site-specific (Barthelemy et al. 2020). Acute sleep loss was similarly associated with increases in blood total tau in healthy young men (Benedict et al. 2020), and self-reports of poor sleep and daytime sleepiness in cognitively normal subjects (mean age 63 years) was associated with lower CSF Aβ42/Aβ40 and higher total-tau/Aβ42 (a ratio highly concordant with Aβ PET measures and predictive of cognitive decline, Hansson et al. 2018; Fagan et al. 2007), and ratio of chitinase-3-like protein 1 (YKL-40, a glial marker of

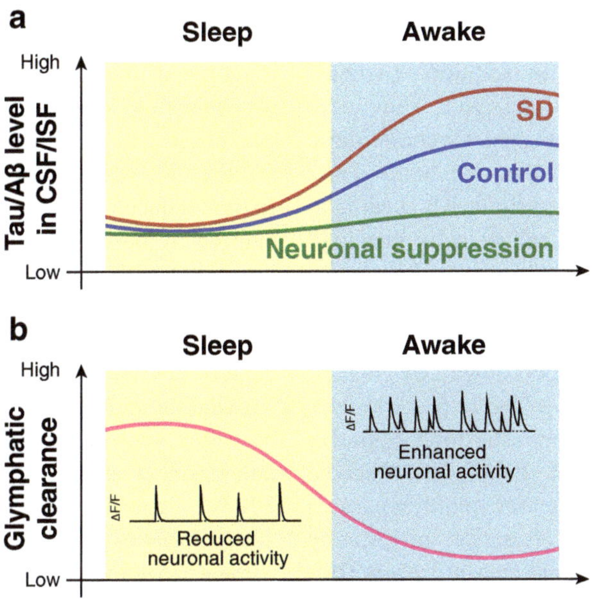

Fig. 10.1 Schematic of sleep-wake cycle fluctuations in interstitial and cerebrospinal fluid protein levels, neuronal activity, and glymphatic clearance. (**a**) Levels of Aβ and tau in interstitial and cerebrospinal fluid (ISF/CSF) undergo sleep-wake cycle variations under normal conditions (control, blue trace), with an increase during wakefulness and a decrease during sleep. Sleep deprivation (SD) exacerbates (red trace), while suppression of neuronal activity attenuates (green trace), increases in Aβ and tau ISF/CSF levels during wakefulness. (**b**) Overall levels of neuronal activation are enhanced during wakefulness (right inset) but diminished during sleep (left inset), the latter period also associated with an increase in glymphatic clearance (blue trace)

neuroinflammation) to Aβ42 (Sprecher et al. 2017). Abnormal nighttime behavior, as assessed by the Neuropsychiatric Inventory Sleep (NPI-sleep) inventory, was associated with increased PET measured accumulation of Aβ in precuneus, posterior cingulate and medial orbitofrontal, and tau in entorhinal cortex, in clinically normal elderly subjects (Shokouhi 2019).

OSA, in which the airway becomes transiently blocked, is associated with recurrent arousals from sleep and hypoxemia, increases the risk of developing dementia, and is associated with a hastened age of cognitive decline (Osorio et al. 2015), increased neuronal activity, as well as reduced slow-wave activity and sleep spindle density (Ju et al. 2016; Ondze et al. 2003). OSA treatment, using positive airway pressure (PAP), may delay progression of cognitive decline (Osorio et al. 2015) and was notably linked to an enhancement in slow-wave activity that was correlated to reduced post-treatment CSF Aβ levels (Ju et al. 2019). Sleep disordered breathing in cognitively normal older adults was associated with increased Aβ PET load and neuronal activity (as measured by functional magnetic resonance imaging, fMRI) most notably in the posterior cingulate cortex and precuneus (André et al. 2020), whereas witnessed apneas in healthy elderly individuals were also linked to elevated tau-PET signals in entorhinal cortex and inferior temporal lobe (Carvalho et al. 2020). Patients with subjective cognitive impairment (SCI) and OSA exhibited lower CSF Aβ42 levels, higher lactate and total-tau/Aβ42 levels, and reduced sleep quality, in comparison with SCI controls and those with OSA and concurrent CPAP treatment (Liguori et al. 2017). Reduced CSF Aβ levels in OSA patients have been ascribed to the breathing disorder inducing internal high-pressure fluctuations disrupting paravascular/glymphatic flow during sleep between the interstitial fluid (ISF) and CSF (Xie et al. 2013; Iliff et al. 2013; Kiviniemi et al. 2016), and leading to increased ISF and reduced CSF Aβ levels, respectively (Ju et al. 2016). In addition, disrupted NREM sleep and impaired slow-wave oscillations would be expected to compromise CSF influx and efflux within the brain (Fultz et al. 2019) leading to interstitial Aβ accumulation that could be further enhanced by increased neuronal activity during OSA-related arousals. Interestingly, diminished reductions in circadian blood pressure during sleep were also associated with disrupted cerebral blood flow (CBF) regulation and increased PET Aβ load in posterior cingulate of patients with amnestic mild cognitive impairment (Tarumi et al. 2015). The observed impairment of CBF dynamics could suggest a mechanism by which clearance of Aβ by glymphatic/paravascular or other processes (Iliff et al. 2013; Xie et al. 2013; Kiviniemi et al. 2016) is disrupted by disordered sleep, particularly in light of the recent observation of a link between slow activity, hemodynamic oscillations, and flow of CSF within the brain (Fultz et al. 2019). Nevertheless, it is difficult to exclude the possibility that sleep disturbances, such as breathing disorders, which impair such clearance (Xie et al. 2013), and have been associated with the absence of nocturnal BP reductions (Wolf et al. 2010), underpin these results.

One night of total SD in healthy young men also resulted in a significant increase in morning serum levels of neuron-specific enolase (NSE) and S100 calcium binding protein B (S-100B) (Benedict et al. 2014). Since these factors are typically localized to neuronal and glial cytoplasm, these findings could reflect neuronal damage and/or disruption to the blood brain barrier (BBB) during sleep loss (He et al. 2014). Partial SD (maximum of 4 h sleep) in healthy adults (20–40 years), with preserved SWS (monitored using polysomnography and actigraphy), was also associated with increased CSF orexin-A (an isoform of orexin) concentrations (Olsson et al. 2018). Furthermore, CSF orexin-A was found to be upregulated in cognitively normal elder individuals and correlated to CSF Aβ42, phosphorylated-tau, and total-tau levels (Osorio et al. 2016).

Disruption of NREM slow-wave activity has been associated with age-related memory impairment (Mander et al. 2013) and even mild disruption and suppression of slow-wave activity can negatively impact memory performance in healthy individuals (Van Der Werf et al. 2009). In turn, reduced and fragmented slow-wave sleep, evinced by polysomnography, was shown to be associated with increases in CSF Aβ42 levels in cognitively normal elderly individuals at low risk of AD (Varga et al. 2016). Importantly, the degree of impairment of slow-wave activity correlates with PET Aβ burden in medial prefrontal cortex and predicts overnight memory retention (Mander et al. 2015; and see Winer et al. 2019), suggesting that cortical Aβ pathology affects memory by disturbing hippocampal-cortical communication (as confirmed and extended in our translational work, described below). An inverse relationship between NREM slow-wave activity measured using single channel electroencephalography (EEG) and tau-PET levels was also found in several brain areas of predominantly cognitively normal participants (>60 years), including entorhinal, parahippocampal, orbital frontal, precuneus, inferior parietal and temporal regions, and also with CSF tau/Aβ42 levels (Lucey et al. 2019). The degree of impairment in slow oscillation-sleep spindle coupling also predicted tau burden in medial temporal lobe, but not cortical Aβ load, such that weaker coupling was associated with increased tau (Winer et al. 2019). Targeted suppression of slow-wave sleep in healthy older adults (35–65 years, confirmed using polysomnography) increased CSF Aβ40, with worse home sleep quality (measured by actigraphy) also associated with increased CSF tau (Ju et al. 2017). Indeed, sleep spindle density during NREM sleep in clinically normal elderly subjects undergoing polysomnography was significantly and inversely correlated with CSF total tau levels and suggested to be a mechanism by which tau may disrupt memory consolidation (Kam et al. 2019a). Adults with amnestic mild cognitive impairment displayed a significant relationship between impaired SWS and increased Aβ42 plasma levels, whereas, interestingly, reduced REM sleep in the same population was correlated to thinning of the posterior cingulate and precuneus (Sanchez-Espinosa et al. 2014), the functional hubs of the default mode network (DMN). Since slow-wave activity is also associated with reduced activity of the DMN, a distributed brain network in which Aβ accumulation is initiated (Samann et al. 2011; Palmqvist et al. 2017), it is possible that a loss in slow-wave activity leads to heightened activity of the DMN and subsequently more Aβ accumulation.

10.3.2 On the Interaction Between Sleep, Memory, and Epilepsy in AD

Epileptiform activity is a prevalent phenomenon in AD patients, the incidence of which exceeds that observed in the general population, and is associated with an earlier onset of cognitive deficits, exacerbated neurodegeneration, and an enhanced risk of mortality (Forstl et al. 1992; Scarmeas et al. 2009; Vossel et al. 2013, 2016). Notably, recent work has suggested that the prevalence of epileptiform activity in AD is likely underestimated, as many may be sub-clinical in nature, localized to deep brain regions undetectable to surface recordings, and to markedly preponderate during sleep (Lam et al. 2017; Vossel et al. 2016). Sleep and epilepsy have long been associated as familiar bedfellows (Derry and Duncan 2013), with reports of increased epileptiform activity during sleep in focal epilepsy (Malow et al. 1998) and an association between nocturnal seizures, respiratory disorders (such as OSA), and sudden death in epilepsy (SUDEP) (Ryvlin et al. 2013; Lamberts et al. 2012). The presence of pathological epileptiform activity during sleep, a period critical for memory consolidation (Diekelmann and Born 2010), may therefore underpin the hastened cognitive decline seen in AD patients with epilepsy. Indeed, emerging research has now posited that physiological sleep circuits are "hijacked" by epileptic activity (Beenhakker and Huguenard 2009). More specifically, recent work has indicated that post-ictal sleep is associated with the reactivation of interictal discharges (IEDs) and seizure-related neuronal activity patterns, mimicking natural processes involved in memory consolidation during slow-wave sleep following a behavioral experience (Bower et al. 2015, 2017; Wilson and McNaughton 1994; Diekelmann and Born 2010; Klinzing et al. 2019). In this context, it is interesting to note that hippocampal IEDs in a rodent model of temporal lobe epilepsy have been shown to co-opt hippocampal-cortical communication during NREM sleep, essential for memory consolidation (Colgin 2011), by replacing hippocampal ripples and autonomously driving thalamocortical spindles in prefrontal cortex, the extent to which being correlated to memory impairment (Gelinas et al. 2016). Importantly, IEDs were also found to induce cortical spindles with high temporal reliability in humans with focal epilepsy, suggesting that the mechanisms subserving physiological sleep-related memory consolidation are usurped by epileptic processes and underscoring the potential of therapeutically targeting aberrant oscillatory network activity in AD (Gelinas et al. 2016).

10.3.3 Mechanistic Lessons from Translational Models

Prior to Aβ aggregation into plaques, young APP/PS1 mice (a transgenic mouse model of AD which overproduces Aβ) display diurnal variations in hippocampal ISF Aβ levels (decreases during sleep and increases during wakefulness) that are correlated to ISF lactate levels, but which become markedly attenuated with the emergence of plaque pathology, though can be "rescued" by Aβ immunization (Roh et al. 2012; Kang et al. 2009). Acute SD and orexin infusion (to promote wakefulness) were found to enhance ISF Aβ levels in mice, while chronic (21 day) sleep restriction increased plaque deposition and was counteracted by treatment with a dual orexin receptor antagonist (Kang et al. 2009). Accordingly, APP/PS1 mice harboring an orexin gene knockout exhibited reduced Aβ pathology and increased sleep time, which was reversed by SD and rescue of orexigenic neurons,

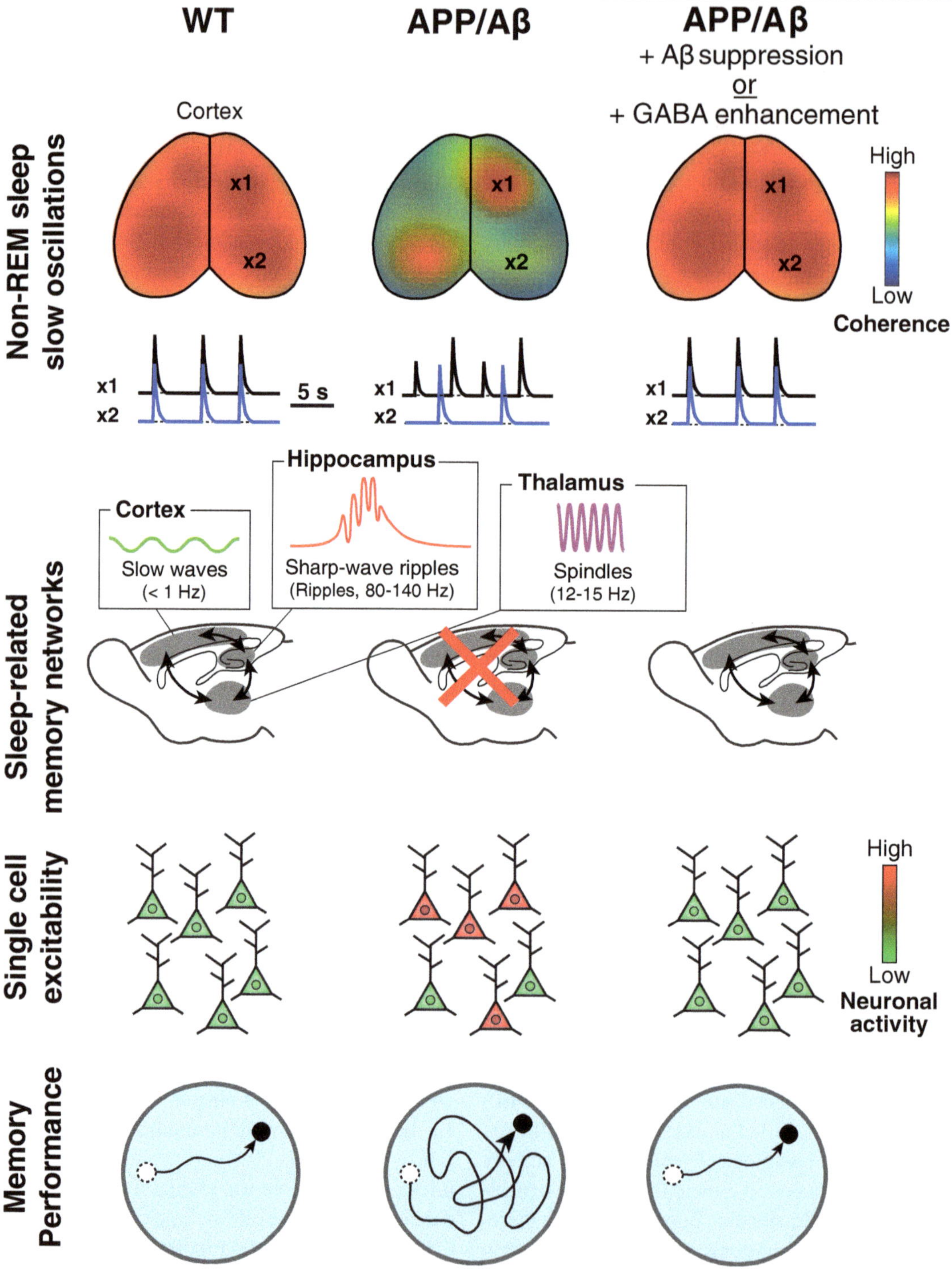

Fig. 10.2 The effects of Aβ on non-REM sleep slow-wave activity, sleep-related neuronal networks, neuronal hyperexcitability, and memory performance. Top panel: Slow-wave oscillations, and their long-range coherence, observed in wild-type (WT) mice (left), are impaired in mice which overexpress the amyloid precursor protein (APP) and overproduce Aβ (middle), but can be rescued by suppression of Aβ or enhancement of GABAergic inhibition in the same animals (right). Second panel from top: The impairment of slow waves results in a breakdown of long-range coupling of activity between cortex, hippocampus, and thalamus (denoted by cortical slow waves,

although hippocampal overexpression of orexin did not recapitulate these effects (Roh et al. 2014). As well as increased Aβ deposition being correlated to induced sleep fragmentation in APP/PS1 mice (Minakawa et al. 2017), chronic SD in wild-type rats and mice was associated with increased expression of beta-site amyloid precursor protein cleaving enzyme 1 (BACE1) (Chen et al. 2017; Zhao et al. 2017), but decreased levels of plasma Aβ levels and plasma soluble LRP-1 (Zhao et al. 2019). Notably, LRP-1 has been implicated in modulating amyloid precursor protein (APP) processing (Ulery et al. 2000), mediating Aβ transport across the blood brain barrier (Storck et al. 2016), and recently shown to regulate endocytosis of tau and subsequent inter-neuronal propagation (Rauch et al. 2020). Intracerebroventricular injection of Aβ oligomers into 8–10 week wild-type mice disrupted sleep patterns, and a 1 month chronic sleep restriction protocol was associated with significant reduction in synaptophysin and postsynaptic density protein 95 (PSD-95) (markers of pre and post synaptic integrity, respectively) in hippocampus, but not frontal cortex (Kincheski et al. 2017). Reduced and fragmented sleep was also observed in a *Drosophila* model of AD which overexpresses Aβ pan-neuronally, with SD enhancing intrinsic neuronal hyperexcitability and increasing Aβ burden that was rescued through suppression of neuronal excitability (Tabuchi et al. 2015).

SD exacerbates ISF tau sleep-wake cycle fluctuations, and chronic SD, interestingly, was found to promote the spread of tau pathology from hippocampus to locus coeruleus (LC) in a P301S mouse model of tauopathy (Holth et al. 2019). Notably, since tetrodotoxin (TTX) abolished the SD-induced elevation in ISF tau levels, and in light of other reports of enhanced tau propagation with increased neuronal activity (Wu et al. 2016) and activity-dependent tau release (Pooler et al. 2013; Yamada et al. 2014), this suggests a putative process by which tau increases with wakefulness and SD (Holth et al. 2019). SD in APP/PS1 mice was associated with phosphorylation of endogenous tau (alongside increased plaque deposition) and mitochondrial dysfunction (Qiu et al. 2016), the latter also observed in frontal cortex in WT mice subjected to sleep restriction (Zhao et al. 2016). SD in young 3xTg mice, a transgenic model of AD with both plaques and tangles, induced a decline in learning and memory alongside a significant increase in total insoluble tau and MC-1 immunoreactivity, indicating an effect on tau solubility and conformation, respectively (Di Meco et al. 2014). Interestingly, no effects on Aβ were found, although the authors did report a significant increase in glial fibrillary acidic protein (GFAP) expression, a marker for astrocytosis, as well as a significant reduction in PSD-95 in sleep-deprived mice similarly to Kincheski et al. (2017) (Di Meco et al. 2014). Notably, chronic short sleep in P301S mice was also seen to prompt an early increase in AT8 and MC-1, indicative of increased tau phosphorylation and pathological conformational changes, in the brainstem locus coeruleus (a putative site for early tau pathology), as well as increased microglial (Iba-1) and astrocytic activation (GFAP) in hippocampus (Zhu et al. 2018). It is interesting to note that tau itself may play a role in the regulation of the sleep-wake cycle, with tau knockout mice exhibiting increased wakefulness and decreased NREM sleep time (Cantero et al. 2010). Moreover, tau deficient *Drosophila* exhibit dysregulation of

Fig. 10.2 (continued) hippocampal sharp-wave ripples, and thalamocortical spindles) in mice which overexpress APP/overproduce Aβ (middle) compared to WT animals (left), but can again be restored following suppression of Aβ or enhancement of GABAergic inhibition (right). Bottom panels: Neuronal hyperexcitability and impaired memory performance is also observed in APP/Aβ mice compared to controls and can be similarly rescued (labeling convention follows that in upper panels). The model is supported by mediation analysis (see supplementary figure S9 in Keskin et al. 2017)

circadian and sleep patterns alongside disruption of circadian pacemaker neurons (Arnes et al. 2019), and expression of the 0N4R isoform of tau in the Drosophila clock network was reported to result in elevated locomotor activity and loss of nighttime sleep, as well as increased diurnal and nocturnal spiking in large lateral ventral clock neurons (Buhl et al. 2019).

Our recent work has provided evidence of A-β-dependent neuronal hyperactivity (Busche et al. 2008, 2012; Busche and Hyman 2020; Harris et al. 2020) as well as impairment in slow-wave oscillations in APP mouse models of AD, which correlates with deficits in learning and memory and can be rescued by BACE inhibition (i.e., Aβ suppression) or enhancement of $GABA_A$ergic inhibition (Busche et al. 2015; Keskin et al. 2017) (Fig. 10.2). In particular, these slow-wave oscillations, which manifest as propagating travelling waves similarly to that observed in humans (Massimini et al. 2004; Muller et al. 2018), were disrupted by both endogenous and exogenous Aβ, resulting in impaired long-range coherence of cortical slow-wave activity, as well as large-scale functional decoupling of coherent activity between cortex and hippocampus, and cortex and thalamus (Busche et al. 2015; Keskin et al. 2017) (Fig. 10.2). These findings are consistent with another report that optogenetic activation of parvalbumin-positive interneurons in sleep deprived mice rescued contextual fear memory consolidation (Ognjanovski et al. 2018). Our results are also in line with another report of disrupted slow-wave connectivity between hippocampal CA1 and medial frontal cortex during NREM sleep in APP/PS1 mice, and decreased coupling between cortical spindles and hippocampal ripples relative to WT animals (Zhurakovskaya et al. 2019). Impairment of slow-wave oscillations, characterized by prolonged down states and reduced neuronal firing, has also been reported in the Tg4510 mouse model of tauopathy (Menkes-Caspi et al. 2015), and the coupling between sleep spindles and cortical slow oscillations was observed to be markedly reduced in the PS19 mouse model of tauopathy (Kam et al. 2019b). These experimental reports therefore support the notion that disrupted NREM slow-wave activity is a feature of several AD mouse models and recapitulate clinical findings. While less is known on the relationship between AD pathologies and REM sleep in animal models, it is interesting to note that intra-hippocampal injection of Aβ was observed to induce a pronounced decrease in theta-band frequency power during REM sleep in rats (Maleysson et al. 2019). Optogenetic inhibition of hippocampal theta oscillations during REM sleep also impaired object recognition and contextual fear memory (Boyce et al. 2016).

10.4 A Putative Multi-level Feedback Model of Pathophysiology

Taken together, the above clinical and translational literature indicates that sleep impairment, per se, is associated with aberrant production/release and, ultimately, deposition of Aβ and tau pathology. These effects have multiple probable etiologies, including an increase in neurometabolic rate due to disrupted sleep and increased wakefulness (Scalise et al. 2006; Buchsbaum et al. 1989), leading to enhanced activity-dependent release of Aβ and tau (Cirrito et al. 2005; Yamada et al. 2014; Pooler et al. 2013), as well as oxidative stress promoting further production of Aβ (Frederikse et al. 1996; Gabuzda et al. 1994) and tau phosphorylation (Melov et al. 2007). In turn, increased levels of Aβ in the brain induce neuronal hyperexcitability (Busche et al. 2008, 2012) providing positive feedback to the aforementioned effects, and impairing slow-wave activity and long-range networks, thus affecting memory processing (Busche et al. 2015; Keskin et al. 2017), as well as glymphatic clearance mechanisms that rely on slow-wave oscillations to drive brain metabolites into the periphery (Fultz et al. 2019; Iliff et al. 2013; Kiviniemi et al. 2016; Xie et al. 2013; van Veluw et al. 2020). In a toxic pas de deux, disrupted sleep, in of itself, exacerbates deficits in clearance of pathogenic proteins by the glymphatic system and/or via the BBB (the permeability of which is vulnerable to inflammation, disordered sleep and subject to the sleep-wake cycle) (He et al. 2014; Cuddapah et al. 2019; Haruwaka et al. 2019), and thwarts restorative cellular processes, including nuclear maintenance, that might ameliorate the above pathological processes (Everson et al. 2014; Zada et al. 2019), and affects inflammatory homeostasis, further contributing to a vicious cycle leading to the accumulation of Aβ and tau pathology. Finally, several of these processes, including disordered sleep, impairment of slow-wave activity, Aβ-dependent neuronal hyperactivity, perturbed DNA repair, and reduced BBB integrity, conspire to promote the emergence of epileptiform activity, that appropriates and recurrently amplifies these pathological mechanisms and sets in motion a feedback system leading to accelerated accumulation of AD-related peptides and, ultimately, neurodegeneration and dementia (Fig. 10.3).

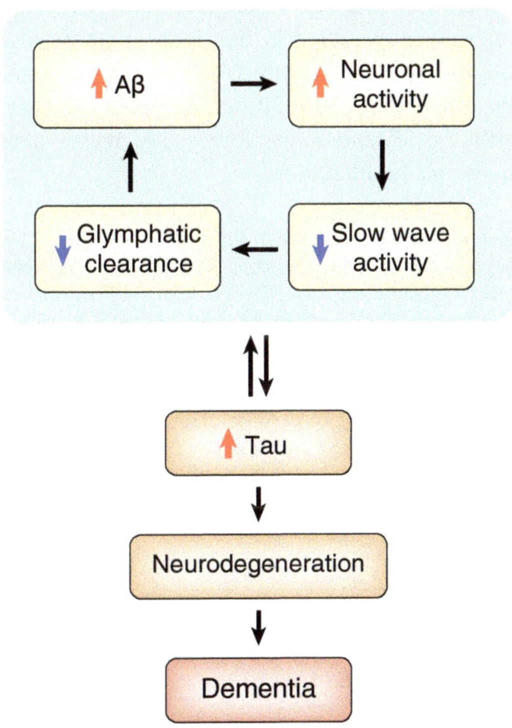

Fig. 10.3 A putative multi-level feedback model describing possible interactions between several pathological mechanisms that are self-amplifying and exacerbated by sleep impairments, leading to the development of AD features and accelerated AD progression

10.5 Methodological Considerations

Polysomnography remains the gold standard for quantitative sleep assessment, but it is manually scored by sleep-specialists and based only on short intervals of data with only summary statistics provided. This results in large amounts of potentially informative data being discarded. New methods in deep learning may help to access the full richness of the data in the future, but currently the technique cannot be performed in a home environment, and this limitation will introduce confounds and alterations in sleeping behaviors within the clinical setting. Actigraphy, in turn, currently only provides crude measures of rest and activity, and it is thus paramount that more advanced home monitoring systems are developed which allow continuous recording of sleep variables in the patient's natural environment. In addition, sleep assessments based on clinical history and patient self-report scales are inherently unreliable, perhaps more so in the elderly and in patients with neurodegenerative disorders, and there is a lack of standardized instruments that allow comprehensive screening for a spectrum of sleep problems in the context of dementia. Quantitative unbiased biomarkers, including those for sleep debt, hypoxia, and circadian phase, are therefore urgently needed. However, in this regard, methodological approaches remain imperfect, with the recent report of erroneous CSF AD biomarkers resulting from repeated lumbar punctures (Olsson et al. 2019), and the limitations of PET imaging approaches to quantify Aβ and tau deposition at the earliest stages, being two cases in point. In addition, further research is needed to clarify what forms of AD-related proteins are most critical to disease progression, with recent work by us indicating that soluble forms of Aβ and tau, as opposed to Aβ plaques and neurofibrillary tangles, are key drivers of neuronal dysfunction (Busche et al. 2012, 2019; Keskin et al. 2017; Busche and Hyman 2020; Harris et al. 2020). Moreover, in the case of translational experiments, transparent efforts must be made to disambiguate the effects of induced sleep disruption from those arising epiphenomenologically as a result of other factors, such as stress (Kang et al. 2007).

10.6 Conclusions and Future Directions

It is evident that sleep and its physiological mechanisms are disrupted in AD, but it is becoming increasingly recognized that sleep impairment and its sequelae can manifest before widespread neurodegeneration and cognitive symptoms emerge. Growing evidence now suggests that sleep perturbations, either directly or indirectly, modulate the release and accumulation of pathogenic proteins in AD, through a myriad of recurrent processes, and thereby increase the risk of development of AD and accelerate disease progression. Importantly, these data, concerning the relationship between tau pathology and sleep disruption in particular, also highlight the value in examining the role of pathological sleep phenotypes in other tauopathies such as frontotemporal dementia, in which evidence of marked sleep impairment has recently emerged and become the focus of intense research (Warren and Clark 2017). Despite the complexity of the interaction between sleep and AD, and while it is currently not possible to point to the initiating mechanism which drives this pathological coupling, sleep has emerged as a potentially modifiable target for early therapy in AD, for which no disease-modifying treatment has yet been found. For example, non-invasive circuit-based interventions that enhance slow-wave sleep such as transcranial magnetic and direct current stimulation have been shown to improve memory and cognitive performance in AD patients and healthy older individuals (Marshall et al. 2006) (Nguyen et al. 2017; Westerberg et al. 2015; Ladenbauer et al. 2017; Diep et al. 2020). Other approaches to enhance cognition have included the use of pharmacological agents (e.g., GABA modulators or orexin receptor antagonists, see Herring et al. 2020) to increase sleep spindle density (Mednick et al. 2013) or the use of closed-loop auditory stimulation of slow-wave activity and sleep spindles (Ngo et al. 2013), albeit there is some debate whether the latter technique is able to reliably improve memory performance (Henin et al. 2019). In addition, sleep impairment is also, importantly, amenable to cognitive behavioral therapies (Geiger-Brown et al. 2015; Morin and Benca 2012) as well as environmental adjustments and interventions (Herberger et al. 2019). In turn, a greater understanding of sleep in AD may render sleep readouts as valuable diagnostic tools to identify risk of developing the disease and/or disease stage, particularly when integrated with blood-based, CSF, or neuroimaging biomarkers. It remains to be seen whether sleep disruption can explain, in part, the marked clinical heterogeneity in AD, such as age of disease onset and disease course, and whether it is truly causal for disease progression. For that level of understanding it will be necessary to better elucidate the mechanisms and functions of sleep, and the role of different sleep stages, about which we still know surprisingly little. This, unfortunately, parallels our lack of knowledge on how variant AD affects the awake brain and its physiology, and further technical advances will be essential to addressing these open questions in the future.

Acknowledgements SSH, TSK, BIL, and MAB are supported by the UK Dementia Research Institute which receives its funding from DRI Ltd, funded by the Medical Research Council, Alzheimer's Society and Alzheimer Research UK. We acknowledge the donors of Alzheimer's Disease Research, a program of BrightFocus Foundation, for support of this research (Grant Number: A2019112S). SSH is further supported by a UK DRI Pilot Studies programme award. MAB is further supported by a UKRI Future Leaders Fellowship (Grant Number: MR/S017003/1).

References

André C, Rehel S, Kuhn E, Landeau B, Moulinet I, Touron E, Ourry V, Le Du G, Mézenge F, Tomadesso C, de Flores R, Bejanin A, Sherif S, Delcroix N, Manrique A, Abbas A, Marchant NL, Lutz A, Klimecki OM, Collette F, Arenaza-Urquijo EM, Poisnel G, Vivien D, Bertran F, de la Sayette V, Chételat G, Rauchs G, for the Medit-Ageing Research Group (2020) Association of sleep-disordered breathing with Alzheimer disease biomarkers in community-dwelling older adults: a secondary analysis of a randomized clinical trial. JAMA Neurol. https://doi.org/10.1001/jamaneurol.2020.0311

Arnes M, Alaniz ME, Karam CS, Cho JD, Lopez G, Javitch JA, Santa-Maria I (2019) Role of Tau protein in remodeling of circadian neuronal circuits and sleep. Front Aging Neurosci 11(320):320. https://doi.org/10.3389/fnagi.2019.00320

Barthelemy NR, Liu H, Lu W, Kotzbauer PT, Bateman RJ, Lucey BP (2020) Sleep deprivation affects tau phosphorylation in human cerebrospinal fluid. Ann Neurol 87(5):700–709. https://doi.org/10.1002/ana.25702

Beenhakker MP, Huguenard JR (2009) Neurons that fire together also conspire together: is normal sleep circuitry hijacked to generate epilepsy? Neuron 62 (5):612–632. https://doi.org/10.1016/j.neuron.2009.05.015

Benedict C, Cedernaes J, Giedraitis V, Nilsson EK, Hogenkamp PS, Vagesjo E, Massena S, Pettersson U, Christoffersson G, Phillipson M, Broman JE, Lannfelt L, Zetterberg H, Schioth HB (2014) Acute sleep deprivation increases serum levels of neuron-specific enolase (NSE) and S100 calcium binding protein B (S-100B) in healthy young men. Sleep 37 (1):195–198. https://doi.org/10.5665/sleep.3336

Benedict C, Blennow K, Zetterberg H, Cedernaes J (2020) Effects of acute sleep loss on diurnal plasma dynamics of CNS health biomarkers in young men. Neurology 94(11):e1181–e1189. https://doi.org/10.1212/WNL.0000000000008866

Besedovsky L, Lange T, Born J (2012) Sleep and immune function. Pflugers Arch 463(1):121–137. https://doi.org/10.1007/s00424-011-1044-0

Besedovsky L, Lange T, Haack M (2019) The sleep-immune crosstalk in health and disease. Physiol Rev 99(3):1325–1380. https://doi.org/10.1152/physrev.00010.2018

Bombois S, Derambure P, Pasquier F, Monaca C (2010) Sleep disorders in aging and dementia. J Nutr Health Aging 14(3):212–217. https://doi.org/10.1007/s12603-010-0052-7

Bower MR, Stead M, Bower RS, Kucewicz MT, Sulc V, Cimbalnik J, Brinkmann BH, Vasoli VM, St Louis EK, Meyer FB, Marsh WR, Worrell GA (2015) Evidence for consolidation of neuronal assemblies after seizures in humans. J Neurosci 35(3):999–1010. https://doi.org/10.1523/JNEUROSCI.3019-14.2015

Bower MR, Kucewicz MT, St Louis EK, Meyer FB, Marsh WR, Stead M, Worrell GA (2017) Reactivation of seizure-related changes to interictal spike shape and synchrony during postseizure sleep in patients. Epilepsia 58(1):94–104. https://doi.org/10.1111/epi.13614

Boyce R, Glasgow SD, Williams S, Adamantidis A (2016) Causal evidence for the role of REM sleep theta rhythm in contextual memory consolidation. Science (New York, NY) 352(6287):812. https://doi.org/10.1126/science.aad5252

Branger P, Arenaza-Urquijo EM, Tomadesso C, Mezenge F, Andre C, de Flores R, Mutlu J, de La Sayette V, Eustache F, Chetelat G, Rauchs G (2016) Relationships between sleep quality and brain volume, metabolism, and amyloid deposition in late adulthood. Neurobiol Aging 41:107–114. https://doi.org/10.1016/j.neurobiolaging.2016.02.009

Brown BM, Rainey-Smith SR, Villemagne VL, Weinborn M, Bucks RS, Sohrabi HR, Laws SM, Taddei K, Macaulay SL, Ames D, Fowler C, Maruff P, Masters CL, Rowe CC, Martins RN, AIBL Research Group (2016) The relationship between sleep quality and brain amyloid burden. Sleep 39 (5):1063–1068. https://doi.org/10.5665/sleep.5756

Buchsbaum MS, Gillin JC, Wu J, Hazlett E, Sicotte N, Dupont RM, Bunney WE Jr (1989) Regional cerebral glucose metabolic rate in human sleep assessed by positron emission tomography. Life Sci 45 (15):1349–1356. https://doi.org/10.1016/0024-3205(89)90021-0

Buhl E, Higham JP, Hodge JJL (2019) Alzheimer's disease-associated tau alters Drosophila circadian activity, sleep and clock neuron electrophysiology. Neurobiol Dis 130:104507. https://doi.org/10.1016/j.nbd.2019.104507

Busche MA, Hyman BT (2020) Synergy between amyloid-β and tau in Alzheimer's disease. Nat Neurosci 23 (10):1183–1193. https://doi.org/10.1038/s41593-020-0687-6. Epub 2020 Aug 10

Busche MA, Eichhoff G, Adelsberger H, Abramowski D, Wiederhold KH, Haass C, Staufenbiel M, Konnerth A, Garaschuk O (2008) Clusters of hyperactive neurons near amyloid plaques in a mouse model of Alzheimer's disease. Science (New York, NY) 321 (5896):1686–1689. https://doi.org/10.1126/science.1162844

Busche MA, Chen X, Henning HA, Reichwald J, Staufenbiel M, Sakmann B, Konnerth A (2012) Critical role of soluble amyloid-beta for early hippocampal hyperactivity in a mouse model of Alzheimer's disease. Proc Natl Acad Sci U S A 109(22):8740–8745. https://doi.org/10.1073/pnas.1206171109

Busche MA, Kekus M, Adelsberger H, Noda T, Forstl H, Nelken I, Konnerth A (2015) Rescue of long-range circuit dysfunction in Alzheimer's disease models. Nat Neurosci 18(11):1623–1630. https://doi.org/10.1038/nn.4137

Busche MA, Wegmann S, Dujardin S, Commins C, Schiantarelli J, Klickstein N, Kamath TV, Carlson GA, Nelken I, Hyman BT (2019) Tau impairs neural circuits, dominating amyloid-β effects, in Alzheimer models in vivo. Nat Neurosci 22(1):57–64. https://doi.org/10.1038/s41593-018-0289-8

Cantero JL, Hita-Yanez E, Moreno-Lopez B, Portillo F, Rubio A, Avila J (2010) Tau protein role in sleep-wake cycle. J Alzheimers Dis 21(2):411–421. https://doi.org/10.3233/JAD-2010-100285

Carvalho DZ, St Louis EK, Schwarz CG, Lowe VJ, Boeve BF, Przybelski SA, Reddy A, Mielke MM, Knopman

DS, Petersen RC, Jack CR Jr, Vemuri P (2020) Witnessed apneas are associated with elevated tau-PET levels in cognitively unimpaired elderly. Neurology 94(17):e1793–e1802. https://doi.org/10.1212/WNL.0000000000009315
Chen L, Huang J, Yang L, Zeng XA, Zhang Y, Wang X, Chen M, Li X, Zhang Y, Zhang M (2017) Sleep deprivation accelerates the progression of Alzheimer's disease by influencing Abeta-related metabolism. Neurosci Lett 650:146–152. https://doi.org/10.1016/j.neulet.2017.04.047
Choe YM, Byun MS, Yi D, Lee JH, Jeon SY, Sohn BK, Kim YK, Shin SA, Sohn C-H, Lee YJ, Lee DY, for the KRG (2019) Sleep experiences during different lifetime periods and in vivo Alzheimer pathologies. Alzheimers Res Ther 11(1):79. https://doi.org/10.1186/s13195-019-0536-6
Cipriani G, Lucetti C, Danti S, Nuti A (2015) Sleep disturbances and dementia. Psychogeriatrics 15 (1):65–74. https://doi.org/10.1111/psyg.12069. Epub 2014 Dec 16
Cirrito JR, Yamada KA, Finn MB, Sloviter RS, Bales KR, May PC, Schoepp DD, Paul SM, Mennerick S, Holtzman DM (2005) Synaptic activity regulates interstitial fluid amyloid-beta levels in vivo. Neuron 48 (6):913–922. https://doi.org/10.1016/j.neuron.2005.10.028
Clark CN, Warren JD (2013) A hypnic hypothesis of Alzheimer's disease. Neurodegener Dis 12 (4):165–176. https://doi.org/10.1159/000350060
Colgin LL (2011) Oscillations and hippocampal-prefrontal synchrony. Curr Opin Neurobiol 21(3):467–474. https://doi.org/10.1016/j.conb.2011.04.006
Cuddapah VA, Zhang SL, Sehgal A (2019) Regulation of the blood-brain barrier by circadian rhythms and sleep. Trends Neurosci 42(7):500–510. https://doi.org/10.1016/j.tins.2019.05.001
Derry CP, Duncan S (2013) Sleep and epilepsy. Epilepsy Behav 26(3):394–404. https://doi.org/10.1016/j.yebeh.2012.10.033
Diekelmann S, Born J (2010) The memory function of sleep. Nat Rev Neurosci 11(2):114–126. https://doi.org/10.1038/nrn2762
Diep C, Ftouni S, Manousakis JE, Nicholas CL, Drummond SPA, Anderson C (2020) Acoustic slow wave sleep enhancement via a novel, automated device improves executive function in middle-aged men. Sleep 43(1). https://doi.org/10.1093/sleep/zsz197
Di Meco A, Joshi YB, Pratico D (2014) Sleep deprivation impairs memory, tau metabolism, and synaptic integrity of a mouse model of Alzheimer's disease with plaques and tangles. Neurobiol Aging 35 (8):1813–1820. https://doi.org/10.1016/j.neurobiolaging.2014.02.011
Ettore E, Bakardjian H, Solé M, Levy Nogueira M, Habert M-O, Gabelle A, Dubois B, Robert P, David R (2019) Relationships between objectives sleep parameters and brain amyloid load in subjects at risk for Alzheimer's disease: the INSIGHT-preAD Study. Sleep 42(9). https://doi.org/10.1093/sleep/zsz137
Everson CA, Henchen CJ, Szabo A, Hogg N (2014) Cell injury and repair resulting from sleep loss and sleep recovery in laboratory rats. Sleep 37(12):1929–1940. https://doi.org/10.5665/sleep.4244
Fagan AM, Roe CM, Xiong C, Mintun MA, Morris JC, Holtzman DM (2007) Cerebrospinal fluid tau/β-amyloid42 ratio as a prediction of cognitive decline in nondemented older adults. Arch Neurol 64 (3):343–349
Forstl H, Burns A, Levy R, Cairns N, Luthert P, Lantos P (1992) Neurologic signs in Alzheimer's disease. Results of a prospective clinical and neuropathologic study. Arch Neurol 49(10):1038–1042. https://doi.org/10.1001/archneur.1992.00530340054018
Frederikse PH, Garland D, Zigler JJ, Piatigorsky J (1996) Oxidative stress increases production of beta-amyloid precursor protein and beta-amyloid (Abeta) in mammalian lenses, and Abeta has toxic effects on lens epithelial cells. J Biol Chem 271(17):10169–10174
Fultz NE, Bonmassar G, Setsompop K, Stickgold RA, Rosen BR, Polimeni JR, Lewis LD (2019) Coupled electrophysiological, hemodynamic, and cerebrospinal fluid oscillations in human sleep. Science (New York, NY) 366(6465):628–631. https://doi.org/10.1126/science.aax5440
Gabelle A, Jaussent I, Hirtz C, Vialaret J, Navucet S, Grasselli C, Robert P, Lehmann S, Dauvilliers Y (2017) Cerebrospinal fluid levels of orexin-A and histamine, and sleep profile within the Alzheimer process. Neurobiol Aging 53:59–66. https://doi.org/10.1016/j.neurobiolaging.2017.01.011
Gabuzda D, Busciglio J, Chen LB, Matsudaira P, Yankner BA (1994) Inhibition of energy metabolism alters the processing of amyloid precursor protein and induces a potentially amyloidogenic derivative. J Biol Chem 269 (18):13623–13628
Geiger-Brown JM, Rogers VE, Liu W, Ludeman EM, Downton KD, Diaz-Abad M (2015) Cognitive behavioral therapy in persons with comorbid insomnia: a meta-analysis. Sleep Med Rev 23:54–67. https://doi.org/10.1016/j.smrv.2014.11.007
Gelinas JN, Khodagholy D, Thesen T, Devinsky O, Buzsaki G (2016) Interictal epileptiform discharges induce hippocampal-cortical coupling in temporal lobe epilepsy. Nat Med 22(6):641–648. https://doi.org/10.1038/nm.4084
Grimmer T, Laub T, Hapfelmeier A, Eisele T, Fatke B, Hölzle P, Lüscher S, Parchmann A-M, Rentrop M, Schwerthöffer D, Müller-Sarnowski F, Ortner M, Goldhardt O, Kurz A, Förstl H, Alexopoulos P (2020) The overnight reduction of amyloid β 1-42 plasma levels is diminished by the extent of sleep fragmentation, sAPP-β, and APOE ε4 in psychiatrists on call. Alzheimers Dement. https://doi.org/10.1002/alz.12072
Hansson O, Seibyl J, Stomrud E, Zetterberg H, Trojanowski JQ, Bittner T, Lifke V, Corradini V, Eichenlaub U, Batrla R, Buck K, Zink K, Rabe C, Blennow K, Shaw LM (2018) CSF biomarkers of Alzheimer's disease concord with amyloid-β PET and

predict clinical progression: a study of fully automated immunoassays in BioFINDER and ADNI cohorts. Alzheimers Dement 14(11):1470–1481. https://doi.org/10.1016/j.jalz.2018.01.010

Harris SS, Wolf F, De Strooper B, Busche MA (2020) Tipping the scales: peptide-dependent dysregulation of neural circuit dynamics in Alzheimer's disease. Neuron 107(3):417–435. https://doi.org/10.1016/j.neuron.2020.06.005. Epub 2020 Jun 23

Haruwaka K, Ikegami A, Tachibana Y, Ohno N, Konishi H, Hashimoto A, Matsumoto M, Kato D, Ono R, Kiyama H, Moorhouse AJ, Nabekura J, Wake H (2019) Dual microglia effects on blood brain barrier permeability induced by systemic inflammation. Nat Commun 10(1):5816. https://doi.org/10.1038/s41467-019-13812-z

He J, Hsuchou H, He Y, Kastin AJ, Wang Y, Pan W (2014) Sleep restriction impairs blood-brain barrier function. J Neurosci 34(44):14697–14706. https://doi.org/10.1523/JNEUROSCI.2111-14.2014

Helfrich RF, Lendner JD, Mander BA, Guillen H, Paff M, Mnatsakanyan L, Vadera S, Walker MP, Lin JJ, Knight RT (2019) Bidirectional prefrontal-hippocampal dynamics organize information transfer during sleep in humans. Nat Commun 10(1):3572. https://doi.org/10.1038/s41467-019-11444-x

Henin S, Borges H, Shankar A, Sarac C, Melloni L, Friedman D, Flinker A, Parra LC, Buzsaki G, Devinsky O, Liu A (2019) Closed-loop acoustic stimulation enhances sleep oscillations but not memory performance. eneuro 6 (6). https://doi.org/10.1523/ENEURO.0306-19.2019

Herberger S, Kräuchi K, Glos M, Lederer K, Assmus L, Hein J, Penzel T, Fietze I (2019) Effects of sleep on a high heat capacity mattress on sleep stages, EEG power spectra, cardiac interbeat intervals and body temperatures in healthy middle-aged men. Sleep. https://doi.org/10.1093/sleep/zsz271

Herring WJ, Ceesay P, Snyder E, Bliwise D, Budd K, Hutzelmann J, Stevens J, Lines C, Michelson D (2020) Polysomnographic assessment of suvorexant in patients with probable Alzheimer's disease dementia and insomnia: a randomized trial. Alzheimers Dement 16(3):541–551. https://doi.org/10.1002/alz.12035

Holth JK, Fritschi SK, Wang C, Pedersen NP, Cirrito JR, Mahan TE, Finn MB, Manis M, Geerling JC, Fuller PM, Lucey BP, Holtzman DM (2019) The sleep-wake cycle regulates brain interstitial fluid tau in mice and CSF tau in humans. Science (New York, NY) 363 (6429):880–884. https://doi.org/10.1126/science.aav2546

Huang Y, Potter R, Sigurdson W, Santacruz A, Shih S, Ju YE, Kasten T, Morris JC, Mintun M, Duntley S, Bateman RJ (2012) Effects of age and amyloid deposition on Abeta dynamics in the human central nervous system. Arch Neurol 69(1):51–58. https://doi.org/10.1001/archneurol.2011.235

Huber R, Ghilardi MF, Massimini M, Tononi GJN (2004) Local sleep and learning. Nature 430 (6995):78–81

Iliff JJ, Wang M, Zeppenfeld DM, Venkataraman A, Plog BA, Liao Y, Deane R, Nedergaard M (2013) Cerebral arterial pulsation drives paravascular CSF-interstitial fluid exchange in the murine brain. J Neurosci 33 (46):18190–18199. https://doi.org/10.1523/JNEUROSCI.1592-13.2013

Izawa S, Chowdhury S, Miyazaki T, Mukai Y, Ono D, Inoue R, Ohmura Y, Mizoguchi H, Kimura K, Yoshioka M, Terao A, Kilduff TS, Yamanaka A (2019) REM sleep–active MCH neurons are involved in forgetting hippocampus-dependent memories. Science (New York, NY) 365 (6459):1308. https://doi.org/10.1126/science.aax9238

Joiner WJ (2019) Neuroscience: sleep fragmentation impairs memory formation. Curr Biol 29(22):R1181–R1184. https://doi.org/10.1016/j.cub.2019.09.060

Ju YE, Finn MB, Sutphen CL, Herries EM, Jerome GM, Ladenson JH, Crimmins DL, Fagan AM, Holtzman DM (2016) Obstructive sleep apnea decreases central nervous system-derived proteins in the cerebrospinal fluid. Ann Neurol 80(1):154–159. https://doi.org/10.1002/ana.24672

Ju YS, Ooms SJ, Sutphen C, Macauley SL, Zangrilli MA, Jerome G, Fagan AM, Mignot E, Zempel JM, Claassen J, Holtzman DM (2017) Slow wave sleep disruption increases cerebrospinal fluid amyloid-beta levels. Brain 140(8):2104–2111. https://doi.org/10.1093/brain/awx148

Ju YS, Zangrilli MA, Finn MB, Fagan AM, Holtzman DM (2019) Obstructive sleep apnea treatment, slow wave activity, and amyloid-β. Ann Neurol 85(2):291–295. https://doi.org/10.1002/ana.25408. Epub 2019 Jan 17

Kadotani H, Kadotani T, Young T, Peppard PE, Finn L, Colrain IM, Murphy GM Jr, Mignot E (2001) Association between apolipoprotein E epsilon4 and sleep-disordered breathing in adults. Jama 285 (22):2888–2890. https://doi.org/10.1001/jama.285.22.2888

Kam K, Parekh A, Sharma RA, Andrade A, Lewin M, Castillo B, Bubu OM, Chua NJ, Miller MD, Mullins AE, Glodzik L, Mosconi L, Gosselin N, Prathamesh K, Chen Z, Blennow K, Zetterberg H, Bagchi N, Cavedoni B, Rapoport DM, Ayappa I, de Leon MJ, Petkova E, Varga AW, Osorio RS (2019a) Sleep oscillation-specific associations with Alzheimer's disease CSF biomarkers: novel roles for sleep spindles and tau. Mol Neurodegener 14(1):10. https://doi.org/10.1186/s13024-019-0309-5

Kam K, Pettibone W, Shim K, Varga A (2019b) 0296 human tau-related impairment of sleep spindle-slow oscillation coupling in the PS19 mouse model of tauopathy. Sleep 42(Suppl_1):A121–A121. https://doi.org/10.1093/sleep/zsz067.295

Kaneshwaran K, Olah M, Tasaki S, Yu L, Bradshaw EM, Schneider JA, Buchman AS, Bennett DA, De Jager PL, Lim ASP (2019) Sleep fragmentation, microglial aging, and cognitive impairment in adults with and without Alzheimer's dementia. Sci Adv 5(12): eaax7331. https://doi.org/10.1126/sciadv.aax7331

Kang JE, Cirrito JR, Dong H, Csernansky JG, Holtzman DM (2007) Acute stress increases interstitial fluid amyloid-beta via corticotropin-releasing factor and neuronal activity. Proc Natl Acad Sci U S A 104 (25):10673–10678. https://doi.org/10.1073/pnas.0700148104

Kang JE, Lim MM, Bateman RJ, Lee JJ, Smyth LP, Cirrito JR, Fujiki N, Nishino S, Holtzman DM (2009) Amyloid-beta dynamics are regulated by orexin and the sleep-wake cycle. Science (New York, NY) 326 (5955):1005–1007. https://doi.org/10.1126/science.1180962

Keskin AD, Kekus M, Adelsberger H, Neumann U, Shimshek DR, Song B, Zott B, Peng T, Forstl H, Staufenbiel M, Nelken I, Sakmann B, Konnerth A, Busche MA (2017) BACE inhibition-dependent repair of Alzheimer's pathophysiology. Proc Natl Acad Sci U S A 114(32):8631–8636. https://doi.org/10.1073/pnas.1708106114

Kim J, Gulati T, Ganguly K (2019) Competing roles of slow oscillations and delta waves in memory consolidation versus forgetting. Cell 179(2):514–526.e513. https://doi.org/10.1016/j.cell.2019.08.040

Kincheski GC, Valentim IS, Clarke JR, Cozachenco D, Castelo-Branco MTL, Ramos-Lobo AM, Rumjanek V, Donato J Jr, De Felice FG, Ferreira ST (2017) Chronic sleep restriction promotes brain inflammation and synapse loss, and potentiates memory impairment induced by amyloid-beta oligomers in mice. Brain Behav Immun 64:140–151. https://doi.org/10.1016/j.bbi.2017.04.007

Kiviniemi V, Wang X, Korhonen V, Keinanen T, Tuovinen T, Autio J, LeVan P, Keilholz S, Zang YF, Hennig J, Nedergaard M (2016) Ultra-fast magnetic resonance encephalography of physiological brain activity—glymphatic pulsation mechanisms? J Cereb Blood Flow Metab 36(6):1033–1045. https://doi.org/10.1177/0271678X15622047

Klinzing JG, Niethard N, Born J (2019) Mechanisms of systems memory consolidation during sleep. Nat Neurosci 22(10):1598–1610. https://doi.org/10.1038/s41593-019-0467-3

Krause AJ, Simon EB, Mander BA, Greer SM, Saletin JM, Goldstein-Piekarski AN, Walker MP (2017) The sleep-deprived human brain. Nat Rev Neurosci 18 (7):404–418. https://doi.org/10.1038/nrn.2017.55

Krueger JM, Nguyen JT, Dykstra-Aiello CJ, Taishi P (2019) Local sleep. Sleep Med Rev 43:14–21. https://doi.org/10.1016/j.smrv.2018.10.001

Ladenbauer J, Ladenbauer J, Kulzow N, de Boor R, Avramova E, Grittner U, Floel A (2017) Promoting sleep oscillations and their functional coupling by transcranial stimulation enhances memory consolidation in mild cognitive impairment. J Neurosci 37 (30):7111–7124. https://doi.org/10.1523/JNEUROSCI.0260-17.2017

Lam AD, Deck G, Goldman A, Eskandar EN, Noebels J, Cole AJ (2017) Silent hippocampal seizures and spikes identified by foramen ovale electrodes in Alzheimer's disease. Nat Med 23(6):678–680. https://doi.org/10.1038/nm.4330

Lamberts RJ, Thijs RD, Laffan A, Langan Y, Sander JW (2012) Sudden unexpected death in epilepsy: people with nocturnal seizures may be at highest risk. Epilepsia 53(2):253–257. https://doi.org/10.1111/j.1528-1167.2011.03360.x

Larsen SMU, Landolt H-P, Berger W, Nedergaard M, Knudsen GM, Holst SC (2019) Haplotype of the astrocytic water channel AQP4 modulates slow wave energy in human NREM Sleep. bioRxiv https://doi.org/10.1101/2019.12.23.886952

Latchoumane CV, Ngo HV, Born J, Shin HS (2017) Thalamic spindles promote memory formation during sleep through triple phase-locking of cortical, thalamic, and hippocampal rhythms. Neuron 95(2):424–435. e426. https://doi.org/10.1016/j.neuron.2017.06.025

Levenstein D, Watson BO, Rinzel J, Buzsáki G (2017) Sleep regulation of the distribution of cortical firing rates. Curr Opin Neurobiol 44:34–42. https://doi.org/10.1016/j.conb.2017.02.013

Liguori C, Romigi A, Nuccetelli M, Zannino S, Sancesario G, Martorana A, Albanese M, Mercuri NB, Izzi F, Bernardini S, Nitti A, Sancesario GM, Sica F, Marciani MG, Placidi F (2014) Orexinergic system dysregulation, sleep impairment, and cognitive decline in Alzheimer disease. JAMA Neurol 71 (12):1498–1505. https://doi.org/10.1001/jamaneurol.2014.2510

Liguori C, Mercuri NB, Izzi F, Romigi A, Cordella A, Sancesario G, Placidi F (2017) Obstructive sleep apnea is associated with early but possibly modifiable Alzheimer's disease biomarkers changes. Sleep 40(5). https://doi.org/10.1093/sleep/zsx011

Lim AS, Kowgier M, Yu L, Buchman AS, Bennett DA (2013a) Sleep fragmentation and the risk of incident Alzheimer's disease and cognitive decline in older persons. Sleep 36(7):1027–1032. https://doi.org/10.5665/sleep.2802

Lim AS, Yu L, Kowgier M, Schneider JA, Buchman AS, Bennett DA (2013b) Modification of the relationship of the apolipoprotein E epsilon4 allele to the risk of Alzheimer disease and neurofibrillary tangle density by sleep. JAMA Neurol 70(12):1544–1551. https://doi.org/10.1001/jamaneurol.2013.4215

Lucey BP, Hicks TJ, McLeland JS, Toedebusch CD, Boyd J, Elbert DL, Patterson BW, Baty J, Morris JC, Ovod V, Mawuenyega KG, Bateman RJ (2018) Effect of sleep on overnight cerebrospinal fluid amyloid beta kinetics. Ann Neurol 83(1):197–204. https://doi.org/10.1002/ana.25117

Lucey BP, McCullough A, Landsness EC, Toedebusch CD, McLeland JS, Zaza AM, Fagan AM, McCue L, Xiong C, Morris JC, Benzinger TLS, Holtzman DM (2019) Reduced non-rapid eye movement sleep is associated with tau pathology in early Alzheimer's disease. Sci Transl Med 11(474):eaau6550. https://doi.org/10.1126/scitranslmed.aau6550

Maingret N, Girardeau G, Todorova R, Goutierre M, Zugaro M (2016) Hippocampo-cortical coupling mediates memory consolidation during sleep. Nat Neurosci 19(7):959–964. https://doi.org/10.1038/nn.4304

Maleysson V, Page G, Janet T, Klein RL, Haida O, Maurin A, Richard S, Champeroux P, Fauconneau B (2019) Relevance of electroencephalogram assessment in amyloid and tau pathology in rat. Behav Brain Res 359:127–134. https://doi.org/10.1016/j.bbr.2018.10.026

Malow BA, Lin X, Kushwaha R, Aldrich MS (1998) Interictal spiking increases with sleep depth in temporal lobe epilepsy. Epilepsia 39(12):1309–1316. https://doi.org/10.1111/j.1528-1157.1998.tb01329.x

Mander BA, Rao V, Lu B, Saletin JM, Lindquist JR, Ancoli-Israel S, Jagust W, Walker MP (2013) Prefrontal atrophy, disrupted NREM slow waves and impaired hippocampal-dependent memory in aging. Nat Neurosci 16(3):357–364. https://doi.org/10.1038/nn.3324

Mander BA, Marks SM, Vogel JW, Rao V, Lu B, Saletin JM, Ancoli-Israel S, Jagust WJ, Walker MP (2015) beta-amyloid disrupts human NREM slow waves and related hippocampus-dependent memory consolidation. Nat Neurosci 18(7):1051–1057. https://doi.org/10.1038/nn.4035

Marshall L, Helgadottir H, Molle M, Born J (2006) Boosting slow oscillations during sleep potentiates memory. Nature 444(7119):610–613. https://doi.org/10.1038/nature05278

Massimini M, Huber R, Ferrarelli F, Hill S, Tononi G (2004) The sleep slow oscillation as a traveling wave. J Neurosci 24(31):6862–6870. https://doi.org/10.1523/JNEUROSCI.1318-04.2004

Mednick SC, McDevitt EA, Walsh JK, Wamsley E, Paulus M, Kanady JC, Drummond SP (2013) The critical role of sleep spindles in hippocampal-dependent memory: a pharmacology study. J Neurosci 33(10):4494–4504. https://doi.org/10.1523/jneurosci.3127-12.2013

Melov S, Adlard PA, Morten K, Johnson F, Golden TR, Hinerfeld D, Schilling B, Mavros C, Masters CL, Volitakis I, Li Q-X, Laughton K, Hubbard A, Cherny RA, Gibson B, Bush AI (2007) Mitochondrial oxidative stress causes hyperphosphorylation of tau. PLoS One 2(6):e536. https://doi.org/10.1371/journal.pone.0000536

Menkes-Caspi N, Yamin HG, Kellner V, Spires-Jones TL, Cohen D, Stern EA (2015) Pathological tau disrupts ongoing network activity. Neuron 85(5):959–966. https://doi.org/10.1016/j.neuron.2015.01.025

Mestre H, Hablitz LM, Xavier AL, Feng W, Zou W, Pu T, Monai H, Murlidharan G, Castellanos Rivera RM, Simon MJ, Pike MM, Pla V, Du T, Kress BT, Wang X, Plog BA, Thrane AS, Lundgaard I, Abe Y, Yasui M, Thomas JH, Xiao M, Hirase H, Asokan A, Iliff JJ, Nedergaard M (2018) Aquaporin-4-dependent glymphatic solute transport in the rodent brain. Elife 7: e40070. https://doi.org/10.7554/eLife.40070

Minakawa EN, Miyazaki K, Maruo K, Yagihara H, Fujita H, Wada K, Nagai Y (2017) Chronic sleep fragmentation exacerbates amyloid beta deposition in Alzheimer's disease model mice. Neurosci Lett 653:362–369. https://doi.org/10.1016/j.neulet.2017.05.054

Morin CM, Benca R (2012) Chronic insomnia. Lancet 379 (9821):1129–1141. https://doi.org/10.1016/S0140-6736(11)60750-2

Muller L, Chavane F, Reynolds J, Sejnowski TJ (2018) Cortical travelling waves: mechanisms and computational principles. Nat Rev Neurosci 19(5):255–268. https://doi.org/10.1038/nrn.2018.20

Musiek ES, Xiong DD, Holtzman DM (2015) Sleep, circadian rhythms, and the pathogenesis of Alzheimer disease. Exp Mol Med 47(3):e148. https://doi.org/10.1038/emm.2014.121

Ngo HV, Martinetz T, Born J, Mölle M (2013) Auditory closed-loop stimulation of the sleep slow oscillation enhances memory. Neuron 78(3):545–553. https://doi.org/10.1016/j.neuron.2013.03.006

Nguyen JP, Suarez A, Kemoun G, Meignier M, Le Saout E, Damier P, Nizard J, Lefaucheur JP (2017) Repetitive transcranial magnetic stimulation combined with cognitive training for the treatment of Alzheimer's disease. Neurophysiologie clinique (Clin Neurophysiol) 47(1):47–53. https://doi.org/10.1016/j.neucli.2017.01.001

Ognjanovski N, Broussard C, Zochowski M, Aton SJ (2018) Hippocampal network oscillations rescue memory consolidation deficits caused by sleep loss. Cereb Cortex (New York, NY) 28(10):3711–3723. https://doi.org/10.1093/cercor/bhy174

Oh J, Eser RA, Ehrenberg AJ, Morales D, Petersen C, Kudlacek J, Dunlop SR, Theofilas P, Resende E, Cosme C, Alho EJL, Spina S, Walsh CM, Miller BL, Seeley WW, Bittencourt JC, Neylan TC, Heinsen H, Grinberg LT (2019a) Profound degeneration of wake-promoting neurons in Alzheimer's disease. Alzheimers Dement 15(10):1253–1263. https://doi.org/10.1016/j.jalz.2019.06.3916

Oh J, Petersen C, Walsh CM, Bittencourt JC, Neylan TC, Grinberg LT (2019b) The role of co-neurotransmitters in sleep and wake regulation. Mol Psychiatry 24 (9):1284–1295. https://doi.org/10.1038/s41380-018-0291-2

Ohayon MM, Carskadon MA, Guilleminault C, Vitiello MV (2004) Meta-analysis of quantitative sleep parameters from childhood to old age in healthy individuals: developing normative sleep values across the human lifespan. Sleep 27(7):1255–1273. https://doi.org/10.1093/sleep/27.7.1255

Olsson M, Arlig J, Hedner J, Blennow K, Zetterberg H (2018) Sleep deprivation and cerebrospinal fluid biomarkers for Alzheimer's disease. Sleep 41(5). https://doi.org/10.1093/sleep/zsy025

Olsson M, Arlig J, Hedner J, Blennow K, Zetterberg H (2019) Repeated lumbar punctures within 3 days may affect CSF biomarker levels. Fluids Barriers CNS 16 (1):37. https://doi.org/10.1186/s12987-019-0157-2

Ondze B, Espa F, Dauvilliers Y, Billiard M, Besset A (2003) Sleep architecture, slow wave activity and sleep spindles in mild sleep disordered breathing. Clin Neurophysiol 114(5):867–874. https://doi.org/10.1016/s1388-2457(02)00389-9

Ooms S, Overeem S, Besse K, Rikkert MO, Verbeek M, Claassen JA (2014) Effect of 1 night of total sleep deprivation on cerebrospinal fluid beta-amyloid 42 in healthy middle-aged men: a randomized clinical trial. JAMA Neurol 71(8):971–977. https://doi.org/10.1001/jamaneurol.2014.1173

Osorio RS, Pirraglia E, Agüera-Ortiz LF, During EH, Sacks H, Ayappa I, Walsleben J, Mooney A, Hussain A, Glodzik L, Frangione B, Martínez-Martín P, de Leon MJ (2011) Greater risk of Alzheimer's disease in older adults with insomnia. J Am Geriatr Soc 59(3):559–562. https://doi.org/10.1111/j.1532-5415.2010.03288.x

Osorio RS, Gumb T, Pirraglia E, Varga AW, Lu SE, Lim J, Wohlleber ME, Ducca EL, Koushyk V, Glodzik L, Mosconi L, Ayappa I, Rapoport DM, de Leon MJ (2015) Alzheimer's disease neuroimaging initiative. Sleep-disordered breathing advances cognitive decline in the elderly. Neurology 84(19):1964–1971. https://doi.org/10.1212/WNL.0000000000001566. Epub 2015 Apr 15

Osorio RS, Ducca EL, Wohlleber ME, Tanzi EB, Gumb T, Twumasi A, Tweardy S, Lewis C, Fischer E, Koushyk V, Cuartero-Toledo M, Sheikh MO, Pirraglia E, Zetterberg H, Blennow K, Lu SE, Mosconi L, Glodzik L, Schuetz S, Varga AW, Ayappa I, Rapoport DM, de Leon MJ (2016) Orexin-A is associated with increases in cerebrospinal fluid phosphorylated-tau in cognitively normal elderly subjects. Sleep 39(6):1253–1260. https://doi.org/10.5665/sleep.5846

Palmqvist S, Schöll M, Strandberg O, Mattsson N, Stomrud E, Zetterberg H, Blennow K, Landau S, Jagust W, Hansson O (2017) Earliest accumulation of β-amyloid occurs within the default-mode network and concurrently affects brain connectivity. Nat Commun 8 (1):1214. https://doi.org/10.1038/s41467-017-01150-x

Pooler AM, Phillips EC, Lau DHW, Noble W, Hanger DP (2013) Physiological release of endogenous tau is stimulated by neuronal activity. EMBO Rep 14 (4):389–394. https://doi.org/10.1038/embor.2013.15

Prinz PN, Vitaliano PP, Vitiello MV, Bokan J, Raskind M, Peskind E, Gerber C (1982) Sleep, EEG and mental function changes in senile dementia of the Alzheimer's type. Neurobiol Aging 3(4):361–370. https://doi.org/10.1016/0197-4580(82)90024-0

Qiu H, Zhong R, Liu H, Zhang F, Li S, Le W (2016) Chronic sleep deprivation exacerbates learning-memory disability and Alzheimer's disease-like pathologies in AbetaPP(swe)/PS1(DeltaE9) mice. J Alzheimers Dis 50(3):669–685. https://doi.org/10.3233/JAD-150774

Rainey-Smith SR, Mazzucchelli GN, Villemagne VL, Brown BM, Porter T, Weinborn M, Bucks RS, Milicic L, Sohrabi HR, Taddei K, Ames D, Maruff P, Masters CL, Rowe CC, Salvado O, Martins RN, Laws SM, AIBL Research Group (2018) Genetic variation in Aquaporin-4 moderates the relationship between sleep and brain Abeta-amyloid burden. Transl Psychiatry 8 (1):47. https://doi.org/10.1038/s41398-018-0094-x

Rasch B, Büchel C, Gais S, Born JJS (2007) Odor cues during slow-wave sleep prompt declarative memory consolidation. Science 315 (5817):1426–1429

Rauch JN, Luna G, Guzman E, Audouard M, Challis C, Sibih YE, Leshuk C, Hernandez I, Wegmann S, Hyman BT, Gradinaru V, Kampmann M, Kosik KS (2020) LRP1 is a master regulator of tau uptake and spread. Nature 580(7803):381–385. https://doi.org/10.1038/s41586-020-2156-5

Roh JH, Huang Y, Bero AW, Kasten T, Stewart FR, Bateman RJ, Holtzman DM (2012) Disruption of the sleep-wake cycle and diurnal fluctuation of beta-amyloid in mice with Alzheimer's disease pathology. Sci Transl Med 4 (150):150ra122. https://doi.org/10.1126/scitranslmed.3004291

Roh JH, Jiang H, Finn MB, Stewart FR, Mahan TE, Cirrito JR, Heda A, Snider BJ, Li M, Yanagisawa M, de Lecea L, Holtzman DM (2014) Potential role of orexin and sleep modulation in the pathogenesis of Alzheimer's disease. J Exp Med 211(13):2487–2496. https://doi.org/10.1084/jem.20141788

Ryvlin P, Nashef L, Lhatoo SD, Bateman LM, Bird J, Bleasel A, Boon P, Crespel A, Dworetzky BA, Hogenhaven H, Lerche H, Maillard L, Malter MP, Marchal C, Murthy JM, Nitsche M, Pataraia E, Rabben T, Rheims S, Sadzot B, Schulze-Bonhage A, Seyal M, So EL, Spitz M, Szucs A, Tan M, Tao JX, Tomson T (2013) Incidence and mechanisms of cardiorespiratory arrests in epilepsy monitoring units (MORTEMUS): a retrospective study. Lancet Neurol 12(10):966–977. https://doi.org/10.1016/S1474-4422(13)70214-X

Samann PG, Wehrle R, Hoehn D, Spoormaker VI, Peters H, Tully C, Holsboer F, Czisch M (2011) Development of the brain's default mode network from wakefulness to slow wave sleep. Cereb Cortex (New York, NY) 21(9):2082–2093. https://doi.org/10.1093/cercor/bhq295

Sanchez-Espinosa MP, Atienza M, Cantero JL (2014) Sleep deficits in mild cognitive impairment are related to increased levels of plasma amyloid-beta and cortical thinning. NeuroImage 98:395–404. https://doi.org/10.1016/j.neuroimage.2014.05.027

Scalise A, Desiato MT, Gigli GL, Romigi A, Tombini M, Marciani MG, Izzi F, Placidi F (2006) Increasing cortical excitability: a possible explanation for the proconvulsant role of sleep deprivation. Sleep 29 (12):1595–1598. https://doi.org/10.1093/sleep/29.12.1595

Scarmeas N, Honig LS, Choi H, Cantero J, Brandt J, Blacker D, Albert M, Amatniek JC, Marder K, Bell K, Hauser WA, Stern Y (2009) Seizures in Alzheimer disease: who, when, and how common? Arch Neurol 66(8):992–997. https://doi.org/10.1001/archneurol.2009.130

Shokouhi S (2019) Associations of informant-based sleep reports with Alzheimer's disease pathologies. Clin Interv Aging 14:1631–1642. https://doi.org/10.2147/CIA.S215208

Shokri-Kojori E, Wang GJ, Wiers CE, Demiral SB, Guo M, Kim SW, Lindgren E, Ramirez V, Zehra A, Freeman C, Miller G, Manza P, Srivastava T, De Santi S, Tomasi D, Benveniste H, Volkow ND (2018) beta-Amyloid accumulation in the human brain after one night of sleep deprivation. Proc Natl Acad Sci U S A 115(17):4483–4488. https://doi.org/10.1073/pnas.1721694115

Sirota A, Buzsáki G (2005) Interaction between neocortical and hippocampal networks via slow oscillations. Thalamus Relat Syst 3(4):245–259

Sirota A, Csicsvari J, Buhl D, Buzsáki G (2003) Communication between neocortex and hippocampus during sleep in rodents. Proc Natl Acad Sci U S A 100 (4):2065–2069

Spira AP, An Y, Wu MN, Owusu JT, Simonsick EM, Bilgel M, Ferrucci L, Wong DF, Resnick SM (2018) Excessive daytime sleepiness and napping in cognitively normal adults: associations with subsequent amyloid deposition measured by PiB PET. Sleep 41 (10). https://doi.org/10.1093/sleep/zsy152

Sprecher KE, Koscik RL, Carlsson CM, Zetterberg H, Blennow K, Okonkwo OC, Sager MA, Asthana S, Johnson SC, Benca RM, Bendlin BB (2017) Poor sleep is associated with CSF biomarkers of amyloid pathology in cognitively normal adults. Neurology 89 (5):445–453. https://doi.org/10.1212/WNL.0000000000004171

Storck SE, Meister S, Nahrath J, Meissner JN, Schubert N, Di Spiezio A, Baches S, Vandenbroucke RE, Bouter Y, Prikulis I, Korth C, Weggen S, Heimann A, Schwaninger M, Bayer TA, Pietrzik CU (2016) Endothelial LRP1 transports amyloid-beta(1-42) across the blood-brain barrier. J Clin Invest 126(1):123–136. https://doi.org/10.1172/JCI81108

Swaab DF, Fliers E, Partiman TS (1985) The suprachiasmatic nucleus of the human brain in relation to sex, age and senile dementia. Brain Res 342 (1):37–44. https://doi.org/10.1016/0006-8993(85)91350-2

Tabuchi M, Lone SR, Liu S, Liu Q, Zhang J, Spira AP, Wu MN (2015) Sleep interacts with abeta to modulate intrinsic neuronal excitability. Curr Biol 25 (6):702–712. https://doi.org/10.1016/j.cub.2015.01.016

Taheri S, Zeitzer JM, Mignot E (2002) The role of hypocretins (orexins) in sleep regulation and narcolepsy. Annu Rev Neurosci 25(1):283–313. https://doi.org/10.1146/annurev.neuro.25.112701.142826

Tarumi T, Harris TS, Hill C, German Z, Riley J, Turner M, Womack KB, Kerwin DR, Monson NL, Stowe AM, Mathews D, Cullum CM, Zhang R (2015) Amyloid burden and sleep blood pressure in amnestic mild cognitive impairment. Neurology 85(22):1922–1929. https://doi.org/10.1212/WNL.0000000000002167

Tononi G, Cirelli C (2006) Sleep function and synaptic homeostasis. Sleep Med Rev 10(1):49–62. https://doi.org/10.1016/j.smrv.2005.05.002

Tononi G, Cirelli C (2014) Sleep and the price of plasticity: from synaptic and cellular homeostasis to memory consolidation and integration. Neuron 81(1):12–34. https://doi.org/10.1016/j.neuron.2013.12.025

Ulery PG, Beers J, Mikhailenko I, Tanzi RE, Rebeck GW, Hyman BT, Strickland DK (2000) Modulation of β-amyloid precursor protein processing by the low density lipoprotein receptor-related protein (LRP). Evidence that LRP contributes to the pathogenesis of Alzheimer's disease. J Biol Chem 275(10):7410–7415

Van Der Werf YD, Altena E, Schoonheim MM, Sanz-Arigita EJ, Vis JC, De Rijke W, Van Someren EJ (2009) Sleep benefits subsequent hippocampal functioning. Nat Neurosci 12(2):122–123. https://doi.org/10.1038/nn.2253

van Veluw SJ, Hou SS, Calvo-Rodriguez M, Arbel-Ornath M, Snyder AC, Frosch MP, Greenberg SM, Bacskai BJ (2020) Vasomotion as a driving force for paravascular clearance in the awake mouse brain. Neuron 105(3):549–561.e545. https://doi.org/10.1016/j.neuron.2019.10.033

Varga AW, Wohlleber ME, Gimenez S, Romero S, Alonso JF, Ducca EL, Kam K, Lewis C, Tanzi EB, Tweardy S, Kishi A, Parekh A, Fischer E, Gumb T, Alcolea D, Fortea J, Lleo A, Blennow K, Zetterberg H, Mosconi L, Glodzik L, Pirraglia E, Burschtin OE, de Leon MJ, Rapoport DM, Lu SE, Ayappa I, Osorio RS (2016) Reduced slow-wave sleep is associated with high cerebrospinal fluid Abeta42 levels in cognitively normal elderly. Sleep 39(11):2041–2048. https://doi.org/10.5665/sleep.6240

Vossel KA, Beagle AJ, Rabinovici GD, Shu H, Lee SE, Naasan G, Hegde M, Cornes SB, Henry ML, Nelson AB, Seeley WW, Geschwind MD, Gorno-Tempini ML, Shih T, Kirsch HE, Garcia PA, Miller BL, Mucke L (2013) Seizures and epileptiform activity in the early stages of Alzheimer disease. JAMA Neurol 70(9):1158–1166. https://doi.org/10.1001/jamaneurol.2013.136

Vossel KA, Ranasinghe KG, Beagle AJ, Mizuiri D, Honma SM, Dowling AF, Darwish SM, Van Berlo V, Barnes DE, Mantle M, Karydas AM, Coppola G, Roberson ED, Miller BL, Garcia PA, Kirsch HE, Mucke L, Nagarajan SS (2016) Incidence and impact of subclinical epileptiform activity in Alzheimer's disease. Ann Neurol 80(6):858–870. https://doi.org/10.1002/ana.24794

Warren JD, Clark CN (2017) A new hypnic paradigm of neurodegenerative proteinopathies. Sleep Med

32:282–283. https://doi.org/10.1016/j.sleep.2016.12.006
Watson BO, Levenstein D, Greene JP, Gelinas JN, Buzsaki G (2016) Network homeostasis and state dynamics of neocortical sleep. Neuron 90 (4):839–852. https://doi.org/10.1016/j.neuron.2016.03.036
Wei M, Zhao B, Huo K, Deng Y, Shang S, Liu J, Li Y, Ma L, Jiang Y, Dang L, Chen C, Wei S, Zhang J, Yang H, Gao F, Qu Q (2017) Sleep deprivation induced plasma amyloid-beta transport disturbance in healthy young adults. J Alzheimers Dis 57 (3):899–906. https://doi.org/10.3233/JAD-161213
Westerberg CE, Florczak SM, Weintraub S, Mesulam MM, Marshall L, Zee PC, Paller KA (2015) Memory improvement via slow-oscillatory stimulation during sleep in older adults. Neurobiol Aging 36 (9):2577–2586. https://doi.org/10.1016/j.neurobiolaging.2015.05.014
Wilson MA, McNaughton BL (1994) Reactivation of hippocampal ensemble memories during sleep. Science (New York, NY) 265(5172):676–679. https://doi.org/10.1126/science.8036517
Winer JR, Mander BA, Helfrich RF, Maass A, Harrison TM, Baker SL, Knight RT, Jagust WJ, Walker MP (2019) Sleep as a potential biomarker of tau and beta-amyloid burden in the human brain. J Neurosci 39 (32):6315–6324. https://doi.org/10.1523/JNEUROSCI.0503-19.2019
Wolf J, Hering D, Narkiewicz K (2010) Non-dipping pattern of hypertension and obstructive sleep apnea syndrome. Hypertens Res 33(9):867–871. https://doi.org/10.1038/hr.2010.153
Wu JW, Hussaini SA, Bastille IM, Rodriguez GA, Mrejeru A, Rilett K, Sanders DW, Cook C, Fu H, Boonen RA, Herman M, Nahmani E, Emrani S, Figueroa YH, Diamond MI, Clelland CL, Wray S, Duff KE (2016) Neuronal activity enhances tau propagation and tau pathology in vivo. Nat Neurosci 19 (8):1085–1092. https://doi.org/10.1038/nn.4328
Xie L, Kang H, Xu Q, Chen MJ, Liao Y, Thiyagarajan M, O'Donnell J, Christensen DJ, Nicholson C, Iliff JJ, Takano T, Deane R, Nedergaard M (2013) Sleep drives metabolite clearance from the adult brain. Science (New York, NY) 342(6156):373–377. https://doi.org/10.1126/science.1241224
Yaffe K, Laffan AM, Harrison SL, Redline S, Spira AP, Ensrud KE, Ancoli-Israel S, Stone KL (2011) Sleep-disordered breathing, hypoxia, and risk of mild cognitive impairment and dementia in older women. JAMA 306(6):613–619. https://doi.org/10.1001/jama.2011.1115
Yamada K, Holth JK, Liao F, Stewart FR, Mahan TE, Jiang H, Cirrito JR, Patel TK, Hochgrafe K, Mandelkow EM, Holtzman DM (2014) Neuronal activity regulates extracellular tau in vivo. J Exp Med 211(3):387–393. https://doi.org/10.1084/jem.20131685
Yonelinas AP, Ranganath C, Ekstrom AD, Wiltgen BJ (2019) A contextual binding theory of episodic memory: systems consolidation reconsidered. Nat Rev Neurosci 20(6):364–375. https://doi.org/10.1038/s41583-019-0150-4
You JC, Jones E, Cross DE, Lyon AC, Kang H, Newberg AB, Lippa CF (2019) Association of β-amyloid burden with sleep dysfunction and cognitive impairment in elderly individuals with cognitive disorders. JAMA Netw Open 2(10):e1913383. https://doi.org/10.1001/jamanetworkopen.2019.13383
Zada D, Bronshtein I, Lerer-Goldshtein T, Garini Y, Appelbaum L (2019) Sleep increases chromosome dynamics to enable reduction of accumulating DNA damage in single neurons. Nat Commun 10(1):895. https://doi.org/10.1038/s41467-019-08806-w
Zhao H, Wu H, He J, Zhuang J, Liu Z, Yang Y, Huang L, Zhao Z (2016) Frontal cortical mitochondrial dysfunction and mitochondria-related beta-amyloid accumulation by chronic sleep restriction in mice. Neuroreport 27(12):916–922. https://doi.org/10.1097/WNR.0000000000000631
Zhao HY, Wu HJ, He JL, Zhuang JH, Liu ZY, Huang LQ, Zhao ZX (2017) Chronic sleep restriction induces cognitive deficits and cortical beta-amyloid deposition in mice via BACE1-antisense activation. CNS Neurosci Ther 23(3):233–240. https://doi.org/10.1111/cns.12667
Zhao B, Liu P, Wei M, Li Y, Liu J, Ma L, Shang S, Jiang Y, Huo K, Wang J, Qu Q (2019) Chronic sleep restriction induces abeta accumulation by disrupting the balance of abeta production and clearance in rats. Neurochem Res 44(4):859–873. https://doi.org/10.1007/s11064-019-02719-2
Zhu Y, Zhan G, Fenik P, Brandes M, Bell P, Francois N, Shulman K, Veasey S (2018) Chronic sleep disruption advances the temporal progression of tauopathy in P301S mutant mice. J Neurosci 38(48):10255–10270. https://doi.org/10.1523/JNEUROSCI.0275-18.2018
Zhurakovskaya E, Ishchenko I, Gureviciene I, Aliev R, Grohn O, Tanila H (2019) Impaired hippocampal-cortical coupling but preserved local synchrony during sleep in APP/PS1 mice modeling Alzheimer's disease. Sci Rep 9(1):5380. https://doi.org/10.1038/s41598-019-41851-5

GPSR Compliance
The European Union's (EU) General Product Safety Regulation (GPSR) is a set of rules that requires consumer products to be safe and our obligations to ensure this.

If you have any concerns about our products, you can contact us on

ProductSafety@springernature.com

In case Publisher is established outside the EU, the EU authorized representative is:

Springer Nature Customer Service Center GmbH
Europaplatz 3
69115 Heidelberg, Germany

www.ingramcontent.com/pod-product-compliance
Ingram Content Group UK Ltd.
Pitfield, Milton Keynes, MK11 3LW, UK
UKHW052005051026
15624UKWH00003B/568
* 9 7 8 3 0 3 0 8 1 1 4 6 4 *